METHODS IN MOLECULAR BIOLOGY

Series Editor
John M. Walker
School of Life and Medical Sciences
University of Hertfordshire
Hatfield, Hertfordshire, AL10 9AB, UK

For further volumes:
http://www.springer.com/series/7651

Carbohydrate-Based Vaccines

Methods and Protocols

Edited by

Bernd Lepenies

University of Veterinary Medicine, Research Center for Emerging Infections and Zoonoses,
Hannover, Germany

 Humana Press

Editor
Bernd Lepenies
University of Veterinary Medicine
Research Center for Emerging Infections and Zoonoses
Hannover, Germany

ISSN 1064-3745   ISSN 1940-6029   (electronic)
Methods in Molecular Biology
ISBN 978-1-4939-4919-9   ISBN 978-1-4939-2874-3   (eBook)
DOI 10.1007/978-1-4939-2874-3

Springer New York Heidelberg Dordrecht London
© Springer Science+Business Media New York 2015
Softcover reprint of the hardcover 1st edition 2015

Printed on acid-free paper

Humana Press is a brand of Springer
Springer Science+Business Media LLC New York is part of Springer Science+Business Media (www.springer.com)

Preface

Carbohydrates are the most abundant and structurally diverse molecules in nature. They are displayed on all cells in our body and form the so-called glycocalyx. However, carbohydrates are also present on pathogens such as viruses, bacteria, parasites, or fungi. These unique carbohydrate structures on the pathogen surface serve as "pathogen signatures" that are recognized as foreign by the host immune system and may finally induce a protective immune response. Since numerous glycan epitopes are highly pathogen-specific, they are promising candidates for carbohydrate-based vaccines.

The history of carbohydrate-based vaccines dates back to 1923 when Avery and Heidelberger identified the carbohydrate nature of the pneumococcal capsule derived from *Streptococcus pneumoniae*. Since then, tremendous progress has been made in the development of carbohydrate-based vaccines against infectious diseases as well as cancer. Some vaccines such as the *Haemophilus influenzae* type B, *Neisseria meningitidis*, or *S. pneumoniae* vaccine have already entered the clinic, while others are in the preclinical stage. This book aims to summarize the current status in this exciting field and details cutting-edge methods related to carbohydrate-based vaccines—from the identification of a suitable carbohydrate antigen via the preparation of glycoconjugate vaccines to the characterization of vaccine candidates for their use in preclinical and clinical studies.

In Chapter 1, *Hütter* and *Lepenies* give a historical overview on the development and success story of carbohydrate-based vaccines. Chapter 2 by *Zimmermann* and *Lepenies* discusses immunological aspects of polysaccharides and glycoconjugate vaccines and also highlights recent advances in the design of carbohydrate-based adjuvants. The central part of this book, the Methods section, starts with Chapter 3. A prerequisite for the preparation of carbohydrate-based vaccines is the identification of relevant pathogen-related glycans as described in Chapters 3 and 4. In Chapter 3, *Xia* and *Gildersleeve* present the glycan array platform to identify carbohydrate antigens including glycan microarray fabrication, microarray binding assays, and the analysis of microarray data. In Chapter 4, *Ramsland*, *Yuriev*, and colleagues describe a protocol for the computational analysis of carbohydrate–protein interactions using the AutoMap software which might be a helpful tool for a rational selection of carbohydrate antigens. In Chapter 5, *Anish*, *Seeberger*, and colleagues provide a protocol for the generation of anti-carbohydrate monoclonal antibodies of high specificity, selectivity, and affinity that can be used for diagnostic and therapeutic purposes. The protocol given in Chapter 6 by *Segura* and colleagues details the opsonophagocytic assay as a correlate for protection to measure the functional capacities of vaccine candidate-raised antibodies.

The determination of pathogen-specific glycosylation patterns by suitable analytical tools and techniques is essential for carbohydrate antigen selection. Exemplary protocols are given in Chapters 7 and 8. In Chapter 7, *Crispin* and colleagues describe the glycan analysis of viral glycoproteins by ion mobility mass spectrometry, whereas in Chapter 8 *Rapp* and colleagues focus on the multiplexed capillary gel electrophoresis with laser-induced fluorescence detection technology (xCGE-LIF) for high-throughput glycan analysis.

The preparation of carbohydrate-based vaccines is the focus of Chapters 9 and 10. In Chapter 9, *Lipinski* and *Bundle* provide a strategy for the synthesis of glycoconjugate vaccines. In Chapter 10, *Chiodo* and *Marradi* present the preparation of gold nanoparticles as carriers for carbohydrate-based vaccines. In addition to the vaccine antigen, adjuvants are often crucial and impact vaccine efficacy. The protocol by *Johannssen* and *Lepenies* in Chapter 11 details the identification and characterization of carbohydrate-based adjuvants.

Chapters 12 and 13 deal with the characterization of carbohydrate-based vaccines. In Chapter 12, *Berti* and *Ravenscroft* focus on the characterization of carbohydrate vaccines by NMR spectroscopy whereas in Chapter 13 *Harding* and colleagues review the characterization of capsular polysaccharides and glycoconjugate vaccines by hydrodynamic methods. The final Chapter 14 by *Jones* reviews regulatory aspects of carbohydrate-based vaccines—a valuable and highly relevant addition to the book.

Although the present book is not an all-encompassing compendium of all methods related to carbohydrate-based vaccines and adjuvants, it contains a broad selection of relevant protocols. Thus, I expect this volume to be a valuable manual that will facilitate research in the field of carbohydrate-based vaccines.

Hannover, Germany   *Bernd Lepenies*

Contents

Preface.. *v*

Contributors.. *ix*

1   Carbohydrate-Based Vaccines: An Overview 1
    Julia Hütter and Bernd Lepenies

2   Glycans as Vaccine Antigens and Adjuvants:
    Immunological Considerations 11
    Stephanie Zimmermann and Bernd Lepenies

3   The Glycan Array Platform as a Tool to Identify
    Carbohydrate Antigens 27
    Li Xia and Jeffrey C. Gildersleeve

4   Antibody-Carbohydrate Recognition from Docked Ensembles
    Using the AutoMap Procedure 41
    Tamir Dingjan, Mark Agostino, Paul A. Ramsland,
    and Elizabeth Yuriev

5   Generation of Monoclonal Antibodies Against Defined
    Oligosaccharide Antigens 57
    Felix Broecker, Chakkumkal Anish, and Peter H. Seeberger

6   Murine Whole-Blood Opsonophagocytosis Assay to Evaluate Protection
    by Antibodies Raised Against Encapsulated Extracellular Bacteria 81
    Guillaume Goyette-Desjardins, René Roy, and Mariela Segura

7   Determination of *N*-linked Glycosylation in Viral Glycoproteins
    by Negative Ion Mass Spectrometry and Ion Mobility................. 93
    David Bitto, David J. Harvey, Steinar Halldorsson, Katie J. Doores,
    Laura K. Pritchard, Juha T. Huiskonen, Thomas A. Bowden,
    and Max Crispin

8   *N*-Glycosylation Fingerprinting of Viral Glycoproteins by xCGE-LIF 123
    René Hennig, Erdmann Rapp, Robert Kottler, Samanta Cajic,
    Matthias Borowiak, and Udo Reichl

9   Temporary Conversion of Protein Amino Groups to Azides:
    A Synthetic Strategy for Glycoconjugate Vaccines 145
    Tomasz Lipinski and David R. Bundle

10   Gold Nanoparticles as Carriers for Synthetic Glycoconjugate Vaccines....... 159
    Fabrizio Chiodo and Marco Marradi

11   Identification and Characterization of Carbohydrate-Based Adjuvants 173
    Timo Johannssen and Bernd Lepenies

12   Characterization of Carbohydrate Vaccines by NMR Spectroscopy 189
    Francesco Berti and Neil Ravenscroft

13   Characterization of Capsular Polysaccharides and Their
     Glycoconjugates by Hydrodynamic Methods.   211
     Stephen E. Harding, Ali Saber Abdelhameed, Richard B. Gillis,
     Gordon A. Morris, and Gary G. Adams

14   Glycoconjugate Vaccines: The Regulatory Framework.   229
     Christopher Jones

Index.   253

Contributors

ALI SABER ABDELHAMEED • *Department of Pharmaceutical Chemistry, College of Pharmacy, King Saud University, Riyadh, Kingdom of Saudi Arabia*

GARY G. ADAMS • *National Centre for Macromolecular Hydrodynamics, University of Nottingham, Sutton Bonington, UK; School of Health Sciences, Faculty of Medicine, University of Nottingham, Nottingham, UK*

MARK AGOSTINO • *CHIRI Biosciences and Curtin Institute for Computation, School of Biomedical Sciences, Curtin University, Perth, WA, Australia; Centre for Biomedical Research, Burnet Institute, Melbourne, VIC, Australia; BSC-IRB Joint Program in Computational Biology, Barcelona, Spain*

CHAKKUMKAL ANISH • *Department of Biomolecular Systems, Max Planck Institute of Colloids and Interfaces, Potsdam, Germany; Bacterial Vaccines Discovery and Early Development, Janssen Pharmaceuticals (Johnson & Johnson), Leiden, The Netherlands*

FRANCESCO BERTI • *Research, GSK Vaccines, Siena, Italy*

DAVID BITTO • *Division of Structural Biology, Wellcome Trust Centre for Human Genetics, University of Oxford, Oxford, UK*

MATTHIAS BOROWIAK • *glyXera GmbH, Magdeburg, Germany*

THOMAS A. BOWDEN • *Division of Structural Biology, University of Oxford, Oxford, UK*

FELIX BROECKER • *Department of Biomolecular Systems, Max Planck Institute of Colloids and Interfaces, Potsdam, Germany; Institute of Chemistry and Biochemistry, Freie Universität Berlin, Berlin, Germany*

DAVID R. BUNDLE • *Department of Chemistry, Alberta Glycomics Centre, University of Alberta, Edmonton, AB, Canada*

SAMANTA CAJIC • *Max Planck Institute for Dynamics of Complex Technical Systems, Magdeburg, Germany; glyXera GmbH, Magdeburg, Germany*

FABRIZIO CHIODO • *Biofunctional Nanomaterials Unit, Laboratory of GlycoNanotechnology, CIC biomaGUNE, San Sebastian, Spain; Department of Parasitology, Leiden University Medical Center, Leiden, The Netherlands*

MAX CRISPIN • *Oxford Glycobiology Institute, Department of Biochemistry, University of Oxford, Oxford, UK*

TAMIR DINGJAN • *Medicinal Chemistry, Monash Institute of Pharmaceutical Sciences, Monash University, Parkville, VIC, Australia*

KATIE J. DOORES • *School of Medicine at Guy's, King's and St Thomas' Hospitals, Guy's Hospital, King's College London, Great Maze Pond, London, UK*

JEFFREY C. GILDERSLEEVE • *Chemical Biology Laboratory, National Cancer Institute, Frederick, MD, USA*

RICHARD B. GILLIS • *National Centre for Macromolecular Hydrodynamics, University of Nottingham, Sutton Bonington, UK*

GUILLAUME GOYETTE-DESJARDINS • *Laboratory of Immunology, Faculty of Veterinary Medicine, University of Montreal, Saint-Hyacinthe, QC, Canada*

STEINAR HALLDORSSON • *Division of Structural Biology, University of Oxford, Oxford, UK*

STEPHEN E. HARDING • *National Centre for Macromolecular Hydrodynamics, University of Nottingham, Sutton Bonington, UK*

DAVID J. HARVEY • *Department of Biochemistry, Oxford Glycobiology Institute, University of Oxford, Oxford, UK*

RENÉ HENNIG • *Max Planck Institute for Dynamics of Complex Technical Systems, Magdeburg, Germany; glyXera GmbH, Magdeburg, Germany*

JULIA HÜTTER • *Department of Biomolecular Systems, Max Planck Institute of Colloids and Interfaces, Potsdam, Germany; Department of Biology, Chemistry and Pharmacy, Institute of Chemistry and Biochemistry, Freie Universität Berlin, Berlin, Germany; Section for Virology, National Veterinary Institute, Technical University of Denmark, Frederiksberg, Denmark*

JUHA T. HUISKONEN • *Division of Structural Biology, University of Oxford, Oxford, UK*

TIMO JOHANNSSEN • *Department of Biomolecular Systems, Max Planck Institute of Colloids and Interfaces, Potsdam, Germany; Department of Biology, Chemistry and Pharmacy, Institute of Chemistry and Biochemistry, Freie Universität Berlin, Berlin, Germany*

CHRISTOPHER JONES • *Laboratory for Molecular Structure, National Institute for Biological Standards and Control, South Mimms, Hertfordshire, UK*

ROBERT KOTTLER • *Max Planck Institute for Dynamics of Complex Technical Systems, Magdeburg, Germany*

BERND LEPENIES • *Department of Biomolecular Systems, Max Planck Institute of Colloids and Interfaces, Potsdam, Germany; Department of Biology, Chemistry and Pharmacy, Institute of Chemistry and Biochemistry, Freie Universität Berlin, Berlin, Germany; University of Veterinary Medicine, Research Center for Emerging Infections and Zoonoses, Hannover, Germany*

TOMASZ LIPINSKI • *Department of Chemistry, Alberta Glycomics Centre, University of Alberta, Edmonton, AB, Canada; Institute of Immunology and Experimental Therapy, Polish Academy of Sciences, Wroclaw, Poland*

MARCO MARRADI • *Biomaterials Unit, Materials Division, IK-4-CIDETEC, San Sebastian, Spain, Biofunctional Nanomaterials Unit, Laboratory of GlycoNanotechnology, CIC biomaGUNE, San Sebastian, Spain; Radiochemistry Department and Biosurfaces Unit, CIC biomaGUNE, San Sebastian, Spain*

GORDON A. MORRIS • *Department of Chemical Sciences, School of Applied Science, University of Huddersfield, Huddersfield, UK*

LAURA K. PRITCHARD • *Department of Biochemistry, Oxford Glycobiology Institute, University of Oxford, Oxford, UK*

PAUL A. RAMSLAND • *Centre for Biomedical Research, Burnet Institute, Melbourne, VIC, Australia; Department of Surgery Austin Health, University of Melbourne, Heidelberg, VIC, Australia; Department of Immunology, Alfred Medical Research and Education Precinct, Monash University, Melbourne, VIC, Australia*

ERDMANN RAPP • *Max Planck Institute for Dynamics of Complex Technical Systems, Magdeburg, Germany; glyXera GmbH, Magdeburg, Germany*

NEIL RAVENSCROFT • *Department of Chemistry, University of Cape Town, Rondebosch, South Africa*

UDO REICHL • *Max Planck Institute for Dynamics of Complex Technical Systems, Magdeburg, Germany; Otto-von-Guericke University, Magdeburg, Germany*

RENÉ ROY • *Department of Chemistry, Université du Québec à Montréal, Montreal, QC, Canada*

Peter H. Seeberger • *Department of Biomolecular Systems, Max Planck Institute of Colloids and Interfaces, Potsdam, Germany, Institute of Chemistry and Biochemistry, Freie Universität Berlin, Berlin, Germany*

Mariela Segura • *Laboratory of Immunology, Faculty of Veterinary Medicine, University of Montreal, Saint-Hyacinthe, QC, Canada*

Li Xia • *Chemical Biology Laboratory, National Cancer Institute, Frederick, MD, USA*

Elizabeth Yuriev • *Medicinal Chemistry, Monash Institute of Pharmaceutical Sciences, Monash University, Parkville, VIC, Australia*

Stephanie Zimmermann • *Department of Biomolecular Systems, Max Planck Institute of Colloids and Interfaces, Potsdam, Germany; Institute of Chemistry and Biochemistry, Freie Universität Berlin, Berlin, Germany*

Chapter 1

Carbohydrate-Based Vaccines: An Overview

Julia Hütter and Bernd Lepenies

Abstract

Vaccination is one of the key developments in the fight against infectious diseases. It is based on the principle that immunization with pathogen-derived antigens provides protection from the respective infection by inducing an antigen-specific immune response. The discovery by Avery and Heidelberger in the 1920s that capsular polysaccharides (CPS) from *Streptococcus pneumoniae* are immunoreactive was the starting point of the development of carbohydrate-based vaccines. CPS-specific neutralizing antibodies were found to mediate protection against S. *pneumoniae* infection. Since the majority of bacterial pathogens carry a dense array of polysaccharides on their surface, the carbohydrate-based vaccine approach was applied to a variety of bacterial strains. The first CPS-based vaccines against S. *pneumoniae* were licensed in the 1940s. The increasing emergence of antibiotic-resistant bacterial strains since the 1960s boosted the development of carbohydrate-based vaccines and led to the approval of CPS-based vaccines against *Neisseria meningitidis*, *Haemophilus influenzae* type b (Hib), and *Salmonella typhi*. Meanwhile, it was observed that CPS generally do not elicit protective antibody responses in children below the age of 2 years who are at the greatest risk of infection. As a consequence, studies refocused on the conjugation of oligosaccharides to proteins in order to increase vaccine immunogenicity which led to the introduction of the first glycoconjugate vaccine against Hib in 1987. Due to the success of the first glycoconjugate vaccines, higher valent formulations were developed against numerous bacterial infections to achieve broad serotype coverage. Current research also focuses on the development of carbohydrate-based vaccines against other pathogens such as viruses, fungi, protozoan parasites, or helminths.

Key words Glycans, Carbohydrates, Vaccines, Polysaccharides, Glycoconjugates

1 Carbohydrate-Based Vaccines Until the 1930s

The development of modern vaccines was a major triumph in the fight against a variety of infectious agents [1]. Vaccination with pathogen-derived antigens provides protection against the respective infectious disease by inducing an antigen-specific primary immune response including the production of long-lasting antibodies and the formation of memory B and T cells [2, 3]. As a result, the immune system is rapidly activated upon re-encounter with the same pathogen. The history of vaccines started with Edward Jenner's discovery in 1796 that people are protected

Bernd Lepenies (ed.), *Carbohydrate-Based Vaccines: Methods and Protocols*, Methods in Molecular Biology, vol. 1331, DOI 10.1007/978-1-4939-2874-3_1, © Springer Science+Business Media New York 2015

against smallpox by prior injection of the similar but less virulent cowpox virus [4]. This finding manifested the potential of vaccines for prophylactic protection of the human population against infectious diseases. Initially, inactivated preparations of whole pathogens were used to elicit potent immune responses and to induce protection, but later also less virulent, so-called attenuated, strains were introduced that mimic the live pathogen and induce strong protective immunity [3, 5].

The story of carbohydrate-based vaccines started with the discovery of Avery and Heidelberger in the 1920s that capsular polysaccharides (CPS) are immunoreactive components of *Streptococcus pneumoniae* [6, 7]. This Gram-positive bacterium causes otitis media, pneumonia, and meningitis and is responsible for high morbidity and mortality rates [8]. These early findings were supported by studies of Dubos and Avery in 1931 who showed the importance of CPS for the virulence of *S. pneumoniae* and their role in the serotype specificity of different *S. pneumoniae* strains [9, 10]. The majority of bacterial pathogens carry a dense array of polysaccharides on their surface. In general, CPS are composed of oligosaccharide repeating units and mask other cell surface antigens for preventing immune recognition by the host immune system [11]. CPS differ between different species but they can even vary between *S. pneumoniae* serotypes [12]. Up to date, more than 90 different types of pneumococcal CPS have been identified [13]. In 1930, Francis and Tillett found out that patients infected with *S. pneumoniae* produced CPS-specific antibodies [14, 15]. In order to enhance the immunogenic effect of CPS, Avery and Goebel conjugated pneumococcal CPS to proteins [16]. This idea was based on studies by Landsteiner indicating that small organic molecules ("haptens") that alone do not elicit an immune response induce specific antibody responses when conjugated to an immunogenic carrier protein [17]. However, the first carbohydrate-based vaccines in humans focused on isolated polysaccharides as vaccine candidates.

2   Carbohydrate-Based Vaccines: From the 1930s Till Now

After several tests using isolated pneumococcal polysaccharides, the first two hexavalent CPS-based vaccines against *S. pneumoniae* were approved in the USA in 1947 [16]. At the same time, antibiotics were discovered and were preferably used for the treatment of infectious diseases rather than preventing infection by vaccination. This was the reason why the production of these first polysaccharide vaccines was abandoned in the 1950s [16]. Due to the emergence of numerous antibiotic-resistant bacterial strains during the following decades [18], CPS-based vaccines were reconsidered for preventing bacterial infections [19]. Further studies resulted in a

14-valent CPS-based vaccine produced by Merck Sharp & Dohme against *S. pneumoniae* that was approved in the USA in 1977. Another 14-valent pneumococcal polysaccharide vaccine provided by Lederle Laboratories was licensed in 1979 followed by a 17-valent vaccine in Europe in the 1980s [16]. To provide broader protection against pneumococcal serotypes, a 23-valent CPS-based vaccine produced by both Merck and Lederle was approved in 1983. This vaccine protected against 87 % of pneumococcal diseases in the USA compared to 70–80 % of the 14-valent vaccine [16]. Based on the success of pneumococcal CPS vaccines, meningococcal polysaccharide vaccines were developed and first licensed in the 1970s [20]. Thirteen different types of CPS are known for all *Neisseria meningitidis* strains with five being prevalent and responsible for most of the meningococcal diseases [21]. *N. meningitidis* causes various diseases, mainly meningitis, often associated with a rapid progression of disease and high mortality rates [21]. Thus, there is only a short time window for medical treatment which clearly illustrates the benefit of vaccination [22]. Up to date, bivalent, trivalent, and tetravalent vaccines against *N. meningitidis* have been approved. However, these formulations do not include protection against group B meningococcus due to antigenic similarity of its CPS with polysaccharides of human neuronal tissues [20].

Haemophilus influenzae type b (Hib) represents another Gram-negative bacterium that causes severe infections such as meningitis, sepsis, or pneumonia [23]. In 1985, a polysaccharide vaccine against Hib was licensed in the USA [24]. Typhoid fever, caused by the Gram-negative bacteria *Salmonella typhi*, is also a global health problem with high mortality rates, especially due to the emergence of multidrug-resistant strains [25, 26]. Typhoid fever is an invasive disease manifested by enteric fever, headache, and anorexia [26]. In 1994, a CPS-based vaccine against *S. typhi* was developed and approved (Fig. 1) [26]. However, meanwhile it was observed that CPS do not elicit a protective antibody response in children below the age of 2 years who are at the greatest risk of infection [27, 28]. For instance, invasive Hib as well as *S. typhi* infections mainly occur among young infants [25, 29]. This finding revealed a major drawback of CPS-based vaccines and indicated that new approaches were needed.

3 Polysaccharides Meet the Immune System

The carbohydrate-based vaccine approach is based on the production of CPS-specific antibodies that mediate protection against the respective CPS-carrying pathogen [30]. Typical thymus-dependent (T_D) antigens, i.e., proteins, are generally internalized and processed by antigen-presenting cells (APCs), and antigen-derived

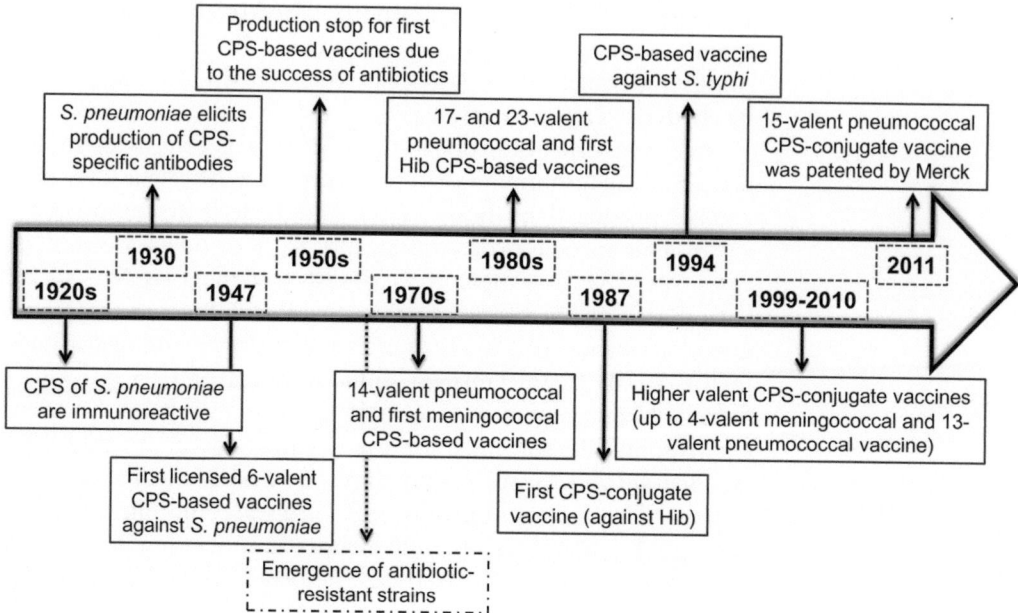

Fig. 1 Historical timeline for the development of carbohydrate-based vaccines

peptides are then presented on MHC-II molecules. Antigen-specific CD4+ T cells bind to the MHC-II-peptide complex via their T cell receptor (TCR) and are activated when additional signals such as co-stimulation and cytokines are provided by APCs [30]. Activated CD4+ T cells subsequently promote B cell proliferation, affinity maturation, isotype switching, as well as immunological memory formation [3].

However, bacterial capsular polysaccharides belong to thymus-independent (T_I) antigens, and thus do not activate T cells [13]. They are not presented on MHC molecules but directly activate B cells [31]. This B cell activation finally results in the production of low-affinity antibodies usually of the IgM isotype due to the absence of affinity maturation and isotype switching [12]. In addition, immunization with T_I antigens does not lead to the generation of T and B cell memory [12]. T_I type 1 antigens, such as bacterial LPS, induce proliferation and differentiation of both naïve and mature B cells whereas T_I type 2 antigens, such as CPS from *S. pneumoniae*, *N. meningitidis*, or *H. influenzae*, only activate mature B cells [13, 31]. Activation likely occurs through cross-linking of cell surface BCRs via the repetitive polysaccharide structures [13, 32]. Consequently, immunization with CPS leads to a T_I immune response and does not involve T cell help. Furthermore, it was observed that T_I type 2 antigens do not elicit protective antibody responses in children below 2 years of age and partially also in the elderly [12, 13].

4   Rediscovery of Glycoconjugate Vaccines

Owing to the lack of protection in young children upon vaccination with pure CPS, studies refocused on the conjugation of glycans to immunogenic carrier proteins to increase vaccine immunogenicity as initially reported by Landsteiner in 1921 [17, 27]. Glycoconjugate vaccines represent a promising strategy to combine the antigenic feature of CPS with the simultaneous induction of T cell help based on carrier protein-derived peptide presentation by MHC-II molecules on APCs. The first CPS-conjugate vaccine was developed against Hib and approved in the USA in 1987 [29, 33]. Further optimization of the vaccine formulation such as the choice of appropriate carrier proteins led to more efficacious glycoconjugate vaccines against Hib [29]. Based on the success of these first Hib conjugate vaccines, a meningococcal conjugate vaccine was introduced in the UK in 1999 [21] followed by a heptavalent glycoconjugate vaccine against *S. pneumoniae* in 2000 [16]. To provide a broader serotype coverage, a tetravalent meningococcal conjugate vaccine was licensed in the USA in 2005 [21] and a 10- and a 13-valent pneumococcal conjugate vaccine followed in 2009 and 2010, respectively [16]. Moreover, Merck has already patented a 15-valent *S. pneumoniae* conjugate vaccine which has shown promising results in preclinical studies (Fig. 1) [34].

Thus, immunogenicity and efficacy of carbohydrate vaccines were successfully enhanced by coupling CPS to proteins. These carrier proteins are typically denatured bacterial toxoids, i.e., diphtheria toxin (DT) or tetanus toxin (TT) [35]. A nontoxic DT mutant, CRM197, is a frequently used carrier protein [35]. Upon vaccination, the glycoconjugates are internalized by APCs, especially by DCs, and are processed to peptides that are subsequently routed to either MHC-I or MHC-II loading. While peptide cross-presentation by MHC-I molecules triggers activation of cytotoxic CD8+ T cells, peptide loading onto MHC-II molecules leads to the activation of CD4+ T cells [36]. Among other effector functions, CD4+ T cells provide T cell help to B cells, thus promoting affinity maturation and isotype switching [37]. B cells with a glycan-specific BCR are activated by these T cells via cytokines as well as CD40/CD40L interaction [12]. Hence, glycoconjugate vaccines induce a protective immune response based on the production of high-affinity glycan-specific antibodies as well as the development of B and T cell memory [37]. In contrast to vaccines based on purified CPS, glycoconjugate vaccines also induce effective protection in children below the age of 2 years [37]. These findings illustrate the major advantage of the glycoconjugate vaccines introduced in the 1980s compared to the formerly available polysaccharide vaccines.

The introduction of pneumococcal conjugate vaccines resulted in a dramatically reduced incidence of invasive pneumococcal disease among young children but also among older people even if they were not vaccinated with the conjugate vaccine [38]. This might be due to the provided "herd immunity" based on a reduced presence of the respective pathogen in the population, thus leading to a decreased transmission [16]. In conclusion, CPS-conjugate vaccines markedly reduced the incidence of Hib, *S. pneumoniae*, and *N. meningitidis* infections. For instance, invasive disease caused by Hib has nearly been eliminated in the industrialized countries since the introduction of the Hib glycoconjugate vaccine [12].

5 Current Focus of Research

The successful story of the first antibacterial carbohydrate-based vaccines stimulated research on preventing other diseases such as cancer or infections caused by viruses, fungi, protozoan parasites, helminths, or further bacteria using similar approaches [39]. For instance, humans exhibit IgG antibodies against polysaccharides of the Gram-positive bacterium *Clostridium difficile* which causes severe diarrheal illness [40, 41]. Hence, current research focuses on the development of carbohydrate-based vaccines against *C. difficile*, particularly triggered by an emerging number of antibiotic-resistant *C. difficile* strains [42, 43].

Polysaccharides such as β-glucans are major components of the fungal cell wall or capsules; thus they represent targets for vaccine design against fungal infections such as *Cryptococcus neoformans*, *Candida* albicans, or *Aspergillus* infections [44]. Some antifungal carbohydrate-based vaccines are currently under development [45].

Glycans are abundantly present on the surface of parasitic protozoans and helminths and are often involved in host invasion and persistence [46]. Parasite-derived carbohydrates are structurally different from mammalian glycans and thus represent promising targets for use in vaccines [47, 48]. Several parasites such as *Leishmania*, *Plasmodium*, or *Schistosoma* cause high morbidity rates in humans because effective treatment is still limited [49, 50]. A number of studies focus on the development of carbohydrate-based vaccines against parasitic infections such as malaria [51, 52] or leishmaniasis [53, 54].

Viral surfaces often include glycoproteins that play crucial roles in virulence and immune evasion [55]. For instance, the HIV-1 envelope glycoprotein gp120 is involved in binding to CD4+ T cells and subsequent virus entry [56]. The difficulty in the development of antiviral carbohydrate-based vaccines is that viral glycans are produced by the glycosylation machinery of the host cell, and thus are usually tolerated by the host immune system. However, several broadly neutralizing antibodies against HIV-1 that specifically bind to oligomannose residues present on the gp120 protein were isolated from infected patients [57–59]. This finding is promising with regard to the development of carbohydrate-based anti-HIV vaccines.

Another interesting approach is the development of carbohydrate-based vaccines against cancer. Tumor cells generally express different glycosylation patterns compared to non-tumor cells and often overexpress certain oligosaccharides called tumor-associated carbohydrate antigens (TACAs) such as the mucin-related (O-linked) Tn antigen, Sialyl Tn, the glycosphingolipid Globo H, or the gangliosides GM2, GD2, GD3, and fucosyl GM1 [60]. Despite the fact that TACAs are self-glycans, they represent attractive targets for vaccines due to their unusually high expression by cancer cells. Several carbohydrate-based anticancer vaccines have been developed and are in preclinical studies or clinical trials [39, 61]. Another approach is based on the synthesis of chemically modified TACAs to avoid immune tolerance [39] (Fig. 2).

It was long thought that carbohydrates only induce a T_I but no T_D immune response. However, several studies revealed that zwitterionic polysaccharides (ZPS) such as polysaccharide A (PSA) from *Bacteroides fragilis* are processed in endosomes and are subsequently presented on MHC-II molecules for activation of CD4$^+$ T helper cells [62]. The alternating positive and negative charges of ZPS mediate their binding to MHC-II molecules via electrostatic

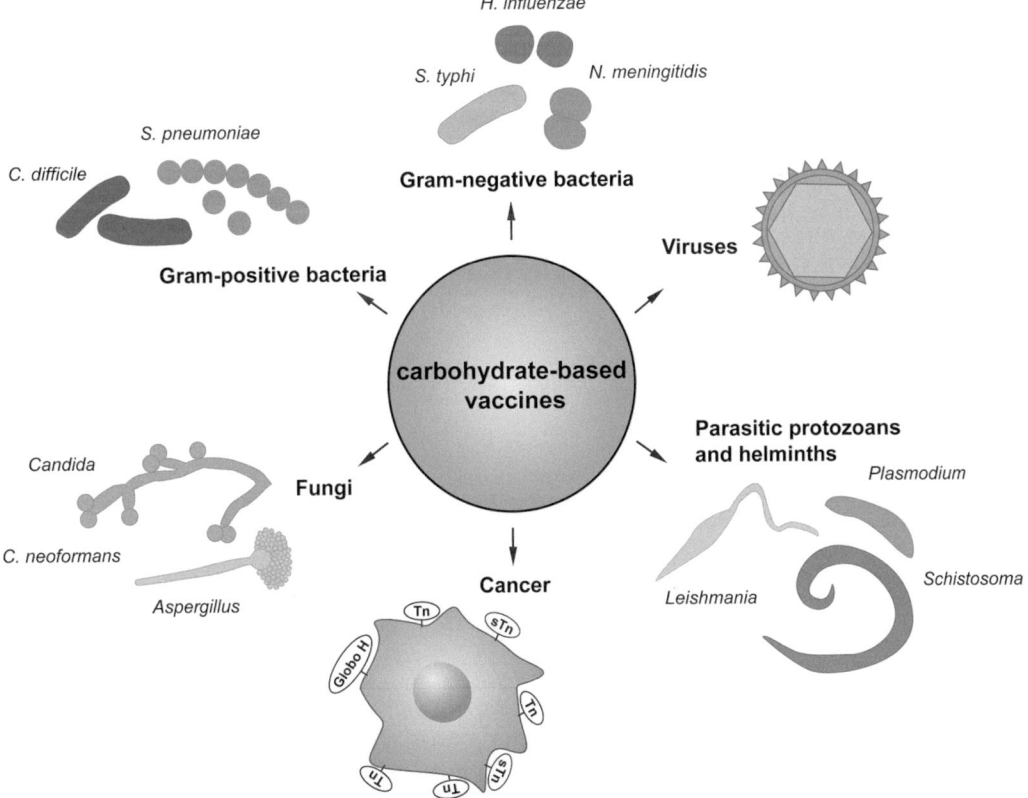

Fig. 2 Current focus of research: carbohydrate-based vaccines against a variety of diseases

interactions [63]. Thus, ZPS promote the activation of glycan-specific T cells [64]. Further examples of ZPS are the polysaccharides of *S. pneumoniae* type 1 and of *Staphylococcus aureus* type 5 and type 8 [64]. In contrast, non-zwitterionic polysaccharides do not bind to MHC-II molecules, and thus cannot be presented to T cells. Importantly, the majority of bacteria do not carry ZPS but uncharged or negatively charged polysaccharides [63]. One promising chemical approach is to introduce positive charges into anionic polysaccharides to enable the carbohydrate presentation on MHC-II molecules for glycan-specific T cell activation [65].

The chemical synthesis of oligosaccharides has tremendously progressed in recent years and represents a promising alternative to the isolation of natural glycans for use in vaccines [66]. The first synthetic glycoconjugate vaccine against Hib was developed in Cuba and was licensed in 2004 [67]. A prerequisite for synthetic carbohydrate-based vaccines is the knowledge of the immunogenic epitope that elicits protective immunity [68]. Recent studies highlight the potential of carbohydrate synthesis as a basis for the development of novel glycoconjugate vaccines [69, 70].

In conclusion, the development of CPS-based vaccines and the design of glycoconjugate vaccines were milestones in the prevention of bacterial infections. Current research deals with the development of carbohydrate-based vaccines against further diseases such as cancer or infections caused by viruses, fungi, protozoan parasites, and helminths.

References

1. Plotkin SA (2005) Vaccines: past, present and future. Nat Med 11:S5–S11
2. Pollard AJ, Perrett KP, Beverley PC (2009) Maintaining protection against invasive bacteria with protein-polysaccharide conjugate vaccines. Nat Rev Immunol 9:213–220
3. Pulendran B, Ahmed R (2011) Immunological mechanisms of vaccination. Nat Immunol 12:509–517
4. Hilleman MR (2000) Vaccines in historic evolution and perspective: a narrative of vaccine discoveries. Vaccine 18:1436–1447
5. Rappuoli R, Black S, Lambert PH (2011) Vaccine discovery and translation of new vaccine technology. Lancet 378:360–368
6. Heidelberger M, Avery OT (1923) The Soluble Specific Substance of Pneumococcus. J Exp Med 38:73–79
7. Heidelberger M, Avery OT (1924) The Soluble Specific Substance of Pneumococcus : Second Paper. J Exp Med 40:301–317
8. Ortqvist A (2001) Pneumococcal vaccination: current and future issues. Eur Respir J 18:184–195

9. Dubos R, Avery OT (1931) Decomposition of the Capsular Polysaccharide of Pneumococcus Type Iii by a Bacterial Enzyme. J Exp Med 54:51–71
10. Avery OT, Dubos R (1931) The Protective Action of a Specific Enzyme against Type Iii Pneumococcus Infection in Mice. J Exp Med 54:73–89
11. Yother J (2011) Capsules of Streptococcus pneumoniae and other bacteria: paradigms for polysaccharide biosynthesis and regulation. Annu Rev Microbiol 65:563–581
12. Mond JJ, Kokai-Kun JF (2008) The multifunctional role of antibodies in the protective response to bacterial T cell-independent antigens. Curr Top Microbiol Immunol 319:17–40
13. Weintraub A (2003) Immunology of bacterial polysaccharide antigens. Carbohyd Res 338:2539–2547
14. Francis T, Tillett WS (1930) Cutaneous reactions in pneumonia. The development of antibodies following the intradermal injection of type-specific polysaccharide. J Exp Med 52:573–585

15. Tillett WS, Francis T (1930) Serological Reactions in Pneumonia with a Non-Protein Somatic Fraction of Pneumococcus. J Exp Med 52:561–571

16. Grabenstein JD, Klugman KP (2012) A century of pneumococcal vaccination research in humans. Clin Microbiol Infect 18(Suppl 5): 15–24

17. Landsteiner K (1921) Über heterogenetisches antigen und hapten. XV. Mitteilung über antigene. Biochem Z 119:294–306

18. Davies J, Davies D (2010) Origins and evolution of antibiotic resistance. Microbiol Mol Biol Rev 74:417–433

19. Vliegenthart JF (2006) Carbohydrate based vaccines. FEBS Lett 580:2945–2950

20. Vipond C, Care R, Feavers IM (2012) History of meningococcal vaccines and their serological correlates of protection. Vaccine 30(Suppl 2): B10–B17

21. Caesar NM, Myers KA, Fan X (2013) Neisseria meningitidis serogroup B vaccine development. Microb Pathog 57:33–40

22. Yogev R, Tan T (2011) Meningococcal disease: the advances and challenges of meningococcal disease prevention. Hum Vaccin 7:828–837

23. Morris SK, Moss WJ, Halsey N (2008) Haemophilus influenzae type b conjugate vaccine use and effectiveness. Lancet Infect Dis 8:435–443

24. Shapiro ED, Berg AT (1990) Protective efficacy of Haemophilus influenzae type b polysaccharide vaccine. Pediatrics 85:643–647

25. Crump JA, Mintz ED (2010) Global trends in typhoid and paratyphoid Fever. Clin Infect Dis 50:241–246

26. Hessel L, Debois H, Fletcher M et al (1999) Experience with Salmonella typhi Vi capsular polysaccharide vaccine. Eur J Clin Microbiol Infect Dis 18:609–620

27. Barrett DJ (1985) Human immune responses to polysaccharide antigens: an analysis of bacterial polysaccharide vaccines in infants. Adv Pediatr 32:139–158

28. Simon R, Levine MM (2012) Glycoconjugate vaccine strategies for protection against invasive Salmonella infections. Hum Vaccin Immunother 8:494–498

29. Adderson EE (2001) Antibody repertoires in infants and adults: effects of T-independent and T-dependent immunizations. Springer Semin Immunopathol 23:387–403

30. Avci FY, Kasper DL (2010) How bacterial carbohydrates influence the adaptive immune system. Annu Rev Immunol 28:107–130

31. Mond JJ, Lees A, Snapper CM (1995) T cell-independent antigens type 2. Annu Rev Immunol 13:655–692

32. Vos Q, Lees A, Wu ZQ et al (2000) B-cell activation by T-cell-independent type 2 antigens as an integral part of the humoral immune response to pathogenic microorganisms. Immunol Rev 176:154–170

33. Goldblatt D (2000) Conjugate vaccines. Clin Exp Immunol 119:1–3

34. Skinner JM, Indrawati L, Cannon J et al (2011) Pre-clinical evaluation of a 15-valent pneumococcal conjugate vaccine (PCV15-CRM197) in an infant-rhesus monkey immunogenicity model. Vaccine 29:8870–8876

35. Knuf M, Kowalzik F, Kieninger D (2011) Comparative effects of carrier proteins on vaccine-induced immune response. Vaccine 29:4881–4890

36. Heath WR, Carbone FR (2001) Cross-presentation in viral immunity and self-tolerance. Nat Rev Immunol 1:126–134

37. Ada G, Isaacs D (2003) Carbohydrate-protein conjugate vaccines. Clin Microbiol Infect 9:79–85

38. Weil-Olivier C, van der Linden M, de Schutter I et al (2012) Prevention of pneumococcal diseases in the post-seven valent vaccine era: a European perspective. BMC Infect Dis 12:207

39. Astronomo RD, Burton DR (2010) Carbohydrate vaccines: developing sweet solutions to sticky situations? Nat Rev Drug Discov 9:308–324

40. Rebeaud F, Bachmann MF (2012) Immunization strategies for Clostridium difficile infections. Expert Rev Vaccines 11:469–479

41. Monteiro MA, Ma Z, Bertolo L et al (2013) Carbohydrate-based Clostridium difficile vaccines. Expert Rev Vaccines 12:421–431

42. Oberli MA, Hecht ML, Bindschädler P et al (2011) A possible oligosaccharide-conjugate vaccine candidate for Clostridium difficile is antigenic and immunogenic. Chem Biol 18: 580–588

43. Martin CE, Broecker F, Oberli MA et al (2013) Immunological evaluation of a synthetic Clostridium difficile oligosaccharide conjugate vaccine candidate and identification of a minimal epitope. J Am Chem Soc 135:9713–9722

44. Cutler JE, Deepe GS Jr, Klein BS (2007) Advances in combating fungal diseases: vaccines on the threshold. Nat Rev Microbiol 5:13–28

45. Cassone A (2013) Development of vaccines for Candida albicans: fighting a skilled transformer. Nat Rev Microbiol 11:884–891

46. Cummings R, Turco S (2009) Parasitic Infections. In: Varki A et al (eds) Essentials of glycobiology, 2nd edn. Cold Spring Harbor (NY), New York

47. Nyame AK, Kawar ZS, Cummings RD (2004) Antigenic glycans in parasitic infections: implications for vaccines and diagnostics. Arch Biochem Biophys 426:182–200

48. van Diepen A, Van der Velden NS, Smit CH et al (2012) Parasite glycans and antibody-mediated immune responses in Schistosoma infection. Parasitology 139:1219–1230

49. Colley DG, Bustinduy AL, Secor WE et al (2014) Human schistosomiasis. Lancet 383: 2253–2264

50. Walker DM, Oghumu S, Gupta G et al (2014) Mechanisms of cellular invasion by intracellular parasites. Cell Mol Life Sci 71:1245–1263

51. Schofield L, Hewitt MC, Evans K et al (2002) Synthetic GPI as a candidate anti-toxic vaccine in a model of malaria. Nature 418:785–789

52. Kwon YU, Soucy RL, Snyder DA et al (2005) Assembly of a series of malarial glycosylphos-phatidylinositol anchor oligosaccharides. Chemistry 11:2493–2504

53. Hewitt MC, Seeberger PH (2001) Solution and solid-support synthesis of a potential leishmaniasis carbohydrate vaccine. J Org Chem 66:4233–4243

54. Liu X, Siegrist S, Amacker M et al (2006) Enhancement of the immunogenicity of synthetic carbohydrates by conjugation to virosomes: a leishmaniasis vaccine candidate. ACS Chem Biol 1:161–164

55. Vigerust DJ, Shepherd VL (2007) Virus glycosylation: role in virulence and immune interactions. Trends Microbiol 15:211–218

56. Wyatt R, Kwong PD, Desjardins E et al (1998) The antigenic structure of the HIV gp120 envelope glycoprotein. Nature 393:705–711

57. Walker LM, Huber M, Doores KJ et al (2011) Broad neutralization coverage of HIV by multiple highly potent antibodies. Nature 477: 466–470

58. Mouquet H, Scharf L, Euler Z et al (2012) Complex-type N-glycan recognition by potent broadly neutralizing HIV antibodies. Proc Natl Acad Sci U S A 109:E3268–E3277

59. Kong L, Lee JH, Doores KJ et al (2013) Supersite of immune vulnerability on the glycosylated face of HIV-1 envelope glycoprotein gp120. Nat Struct Mol Biol 20:796–803

60. Heimburg-Molinaro J, Lum M, Vijay G et al (2011) Cancer vaccines and carbohydrate epitopes. Vaccine 29:8802–8826

61. Hevey R, Ling CC (2012) Recent advances in developing synthetic carbohydrate-based vaccines for cancer immunotherapies. Future Med Chem 4:545–584

62. Cobb BA, Wang Q, Tzianabos AO et al (2004) Polysaccharide processing and presentation by the MHCII pathway. Cell 117:677–687

63. Avci FY, Li X, Tsuji M et al (2013) Carbohydrates and T cells: a sweet twosome. Semin Immunol 25:146–151

64. Cobb BA, Kasper DL (2005) Zwitterionic capsular polysaccharides: the new MHCII-dependent antigens. Cell Microbiol 7: 1398–1403

65. Gallorini S, Berti F, Mancuso G et al (2009) Toll-like receptor 2 dependent immunogenicity of glycoconjugate vaccines containing chemically derived zwitterionic polysaccharides. Proc Natl Acad Sci U S A 106: 17481–17486

66. Lepenies B, Seeberger PH (2010) The promise of glycomics, glycan arrays and carbohydrate-based vaccines. Immunopharmacol Immunotoxicol 32:196–207

67. Verez-Bencomo V, Fernandez-Santana V, Hardy E et al (2004) A synthetic conjugate polysaccharide vaccine against Haemophilus influenzae type b. Science 305:522–525

68. Anish C, Schumann B, Pereira CL et al (2014) Chemical biology approaches to designing defined carbohydrate vaccines. Chem Biol 21:38–50

69. Yang Y, Oishi S, Martin CE et al (2013) Diversity-oriented synthesis of inner core oligosaccharides of the lipopolysaccharide of pathogenic Gram-negative bacteria. J Am Chem Soc 135:6262–6271

70. Cavallari M, Stallforth P, Kalinichenko A et al (2014) A semisynthetic carbohydrate-lipid vaccine that protects against S. pneumoniae in mice. Nat Chem Biol 10:950–956

Chapter 2

Glycans as Vaccine Antigens and Adjuvants: Immunological Considerations

Stephanie Zimmermann and Bernd Lepenies

Abstract

Carbohydrates can be found on the cell surface of nearly every cell ranging from bacteria to fungi right up to mammalian cells. Carbohydrates and their interactions with carbohydrate-binding proteins play crucial roles in multiple biological processes including immunity, homeostasis, cellular communication, cell migration, and the regulation of serum glycoprotein levels. In the last decades, the interest in exploiting the biological activity of glycans as vaccine components has considerably increased. On the one hand, carbohydrates display epitopes to generate protective antibodies against pathogen-derived cell wall structures and on the other hand, glycans have the potential to stimulate the immune system; thus they can act as potent vaccine adjuvants.

An effective vaccine consists of two major components, the vaccine antigen and an adjuvant. The vaccine antigen is an original or modified part of the pathogen that causes the disease. The immune response triggered by vaccination should induce antigen-specific plasma cells secreting protective antibodies as well as the development of memory T and B cells. Carbohydrate structures on pathogens represent an important class of antigens that can activate B cells to produce protective anti-carbohydrate antibodies in adults. A major breakthrough in vaccine development was the design of conjugate vaccines that evoke protective antibody responses against encapsulated bacteria strains such as *Haemophilus influenzae*, *Streptococcus pneumoniae*, or *Neisseria meningitidis* in adults, but also in young children. The first part of this chapter focuses on immune responses triggered by carbohydrate-based vaccines. The second part of the chapter discusses the immunological mechanisms of carbohydrate-based adjuvants to increase the immunogenicity of vaccines.

Key words Adjuvants, Carbohydrate-based vaccines, Conjugate vaccines, Glycans, Immunity, Toll-like receptors, C-type lectin receptors

1 Capsular Polysaccharides as Vaccines

The majority of immunogenic components on the outer cell surface of pathogens, such as bacteria, are glycans or glycoconjugates (*i.e.*, glycoproteins and glycolipids) that often form a thick glycocalyx surrounding the microbe. The cell wall of most bacterial species contains a huge variety of glycoconjugates, such as peptidoglycan, teichoic acid, or lipopolysaccharide (LPS). LPS is a characteristic glycolipid of Gram-negative bacteria that covers approximately

Bernd Lepenies (ed.), *Carbohydrate-Based Vaccines: Methods and Protocols*, Methods in Molecular Biology, vol. 1331, DOI 10.1007/978-1-4939-2874-3_2, © Springer Science+Business Media New York 2015

75 % of the cell surface in *Escherichia coli* and *Salmonella enterica* [1]. Some bacteria possess the ability to completely encapsulate themselves with large molecular-weight surface polysaccharides that enable adherence, prevent them from desiccation, and provide protection against innate defense mechanisms [2]. These capsular polysaccharides (CPS) are proposed to be covalently attached to the outermost layer of the bacterial cell. CPS consist of repeating units with extremely diverse structures. The diversity of CPS is achieved by various determinants such as the length of the repeating units, the combination of different monosaccharides, the stereochemistry of glycosidic linkages, and the presence of branching points or glycan modifications, including sulfation, acetylation, and phosphorylation [3]. More than 90 different capsule structures of *Streptococcus pneumoniae* and more than 80 different capsules or K antigens of *E. coli* have been identified so far [2, 4]. As the outermost surface of encapsulated bacteria, CPS are at the interface between pathogen and host. CPS are essential for the survival of the bacteria in the blood and protecting them against complement deposition. However, certain structures of CPS can be recognized by the host's immune system and uncover the pathogen as foreign invader. Some CPS are major virulence factors and even small differences in their structures influence immunogenicity [5]. Despite the massive abundance of polysaccharides and glycoconjugates on the surface of pathogens, it was believed for a long time that these glycans are not able to activate adaptive immunity and can therefore not induce affinity maturation and class switching of antibodies, or immunologic memory, which is a prerequisite for a successful vaccination [6]. The importance of immunity to CPS was already reported over 80 years ago by Francis and Tillet [7]. However, the introduction of penicillin in the 1940s led to a decreased interest in CPS-based vaccines for a few decades. The persistent high morbidity of pneumococcal diseases finally stimulated the further development of CPS-based vaccines [8].

2   Immunology of Carbohydrates

The human immune system is a complex network of cells, organs, and soluble mediators that has the ability to protect the body against a wide range of pathogens, foreign molecules, or tumor cells. In vertebrates, the immune system is divided into the innate and the adaptive immune system [9, 10]. The innate immune system is evolutionarily ancient and represents a rapid and stereotyped response to a large but limited number of stimuli [11]. In contrast, adaptive immunity requires more time for induction but it is highly antigen-specific. A key function of adaptive immunity is the development of an immunological memory that provides protection from reinfection with the same pathogen. Both arms of immunity strongly cooperate and enable the recognition and destruction of pathogens [10].

Antigen presenting cells (APCs), especially dendritic cells (DCs), bridge innate and adaptive immunity since DCs are able to recognize and to respond to pathogen-associated molecular patterns (PAMPs) or damage-associated molecular patterns (DAMPs) and to initiate adaptive immune responses. The recognition of PAMPs by pattern-recognition receptors (PRRs) expressed by APCs acts as a "danger signal." As a consequence, APCs internalize the pathogen, present antigenic peptides on MHC molecules on their surface, and undergo maturation. During maturation, DCs migrate to the draining lymph nodes, where they prime naïve T cells. For T cell activation, the T cell receptor (TCR) of a naïve T cell has to recognize the antigen peptide-MHC complex on the surface of the APC (Signal 1). In addition, T cell activation by APCs requires co-stimulation, as provided through the interaction of CD80 (B7-1) and CD86 (B7-2) molecules, expressed by mature APCs, with CD28 expressed by T cells (Signal 2). Furthermore, APCs provide polarizing signals such as chemokines and cytokines (Signal 3) [12] (Fig. 1).

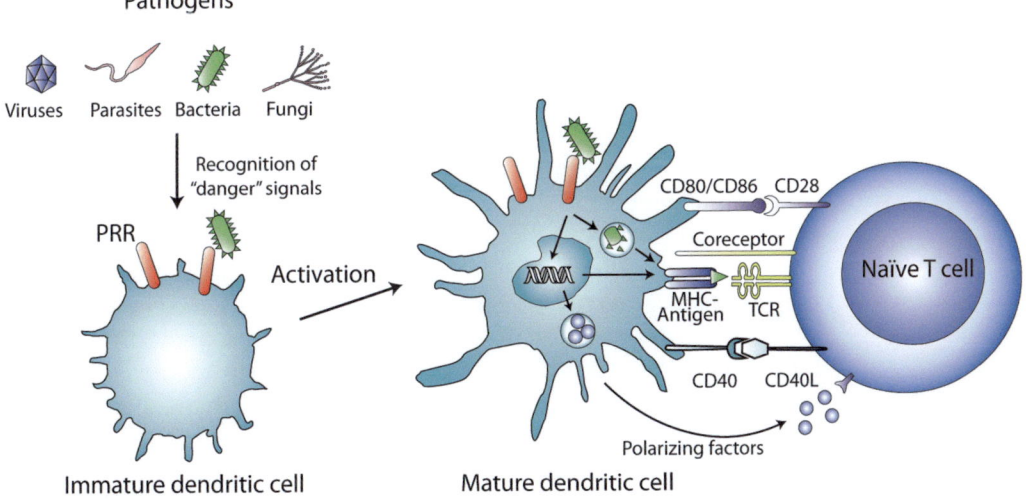

Fig. 1 Recognition of danger signals by APCs and subsequent T cell activation. Immature APCs recognize PAMPs on pathogens such as viruses, bacteria, parasites, and fungi via pattern-recognition receptors (PRRs). Upon pathogen recognition, downstream signaling cascades of PRRs induce APC maturation including the upregulation of co-stimulatory molecules (CD80 and CD86) and major histocompatibility complex (MHC) molecules. Additionally, PRRs induce the expression of cytokines such as IL-12 that drive T cell polarization. Naïve T cells interact with mature APCs that provide *Signal 1* (antigen presentation on MHC-II molecules), *Signal 2* (co-stimulation), and *Signal 3* (secretion of polarizing factors) for antigen-specific T cell activation

Antigen presentation on MHC-I molecules primes naïve $CD8^+$ T cells into cytotoxic T cells that are capable of killing pathogen-infected cells directly. In contrast, antigen presentation on MHC-II molecules primes naïve $CD4^+$ T cells into different T cell subsets including T_H1, T_H2, T_H9, T_H17, T_H22, or follicular helper T (T_{FH}) cells. Secretion of interleukin (IL)-12 by DCs promotes the differentiation of $CD4^+$ T cells into T_H1 cells that produce IFN-γ and IL-2 as signature cytokines [13]. T_H1 cells induce cell-mediated immunity and phagocyte-dependent inflammation and are primarily involved in the eradication of intracellular bacteria and parasites. Secretion of IL-4 in contrast promotes the polarization of naïve $CD4^+$ T cells towards T_H2 cells that produce IL-4, IL-5, and IL-13 as signature cytokines. T_H2 cells provide T cell help to B cells, thus evoke strong antibody responses. In general, T_H2 cells mediate the activation of the humoral immune response and are responsible for the eradication of extracellular pathogens [14]. T_H17 cells that are differentiated in the presence of IL-23 and IL-6 secrete IL-17 and IL-22 and are mainly responsible for neutrophil recruitment and the induction of antimicrobial peptides [15]. While T_H17 cells are involved in host defense against extracellular bacteria and fungi, they also play a critical role in the pathogenesis of autoimmune disease such as rheumatoid arthritis [16]. Key features of adaptive immunity are high-affinity class-switched antibodies and memory B cells. In this context, T_{FH} cells have been shown to mediate the formation of germinal centers and to specifically induce B cell differentiation into antibody secreting plasma cells or memory B cells. T_{FH} cells can be distinguished from other T cell subsets by the expression of IL-21, CXCR5, and the transcription factor Bcl6 [17].

Soluble proteins and peptides are classical thymus-dependent (T_D) antigens. Antibody production against T_D antigens requires T cell-dependent B cell activation resulting in high-affinity antibodies and long-lived B cell memory. For T cell-dependent B cell activation, B cells have to internalize and process the antigen and to present antigen-derived peptides on MHC-II molecules [18]. If the T cell receptor of a T cell is able to recognize the peptide-MHC-II complex, it can provide T cell help to the B cell, thus inducing plasma cell or memory B cell differentiation, affinity maturation, and isotype switching [19]. An efficient activation of B cells requires additional co-stimulatory signals provided by cytokines such as IL-4, IL-6, and IL-21, secreted by the activated T cells, and by the interaction between the receptor CD40 on the B cell and its ligand CD40L expressed by the activated T cell [20].

In contrast to proteins, pure CPS are thymus-independent (T_I) antigens that activate B cells in the absence of T cell help. As a consequence, the immune response to carbohydrates is a primary immune response with the following characteristics: (1) antibodies produced by B cells in response to pure carbohydrates consist mainly of low-affinity IgM and there is no affinity maturation and isotype switching [21]; (2) immune responses are less robust and short-lived, and (3) newborns and children up to 2 years of age fail to induce antibody responses upon vaccination with pure CPS [22, 23]. T_I antigens can be further categorized into T_I type 1 (T_I-1) and T_I type 2 (T_I-2) antigens. T_I type 1 antigens induce a polyclonal proliferation of B cells, whereas T_I-2 antigens—usually polysaccharides of large molecular weight with multiple repeating antigenic epitopes—activate polysaccharide-specific B cells. Typical T_I-2 antigens are CPS from *Streptococcus pneumoniae*, *Haemophilus influenzae* type b, and *Neisseria meningitidis* [24]. These polysaccharides are able to cross-link B cell receptors (BCRs) and active B cells even in the absence of T cell help by providing activation signals through Bruton's tyrosine kinase [25].

CPS-based vaccines, such as Pneumovax23 (Merck) against *S. pneumoniae* or Menomune (Sanofi-Pasteur) against *N. meningitidis*, are currently available on the market and are protective in adults, but fail to induce immunity in newborns and children below 2 years of age. The reason for this unresponsiveness of newborns to T_I-2 antigens can be explained by multiple factors such as an immature B cell population that exhibits a reduced expression of the type 2 complement receptor CD21, low levels of the complement component C3, or a deficiency in the production of certain cytokines by macrophages required for B cell activation [22, 26]. As a consequence, newborns and young children are unable to develop protective antibody responses against encapsulated bacteria, thus are at particularly high risk. The only class of polysaccharides that is capable of eliciting T_D responses is zwitterionic polysaccharides (ZPS), such as polysaccharide A (PSA) from *Bacteroides fragilis* or pneumococcal serotype 1 polysaccharide. ZPS have alternating positive and negative charges and can be internalized and intracellularly processed by B cells [27]. Whereas non-zwitterionic polysaccharides fail to bind to MHC-II molecules, ZPS are loaded on MHC-II molecules and can be presented to CD4$^+$ T cells [28]. In contrast to T_I-2 antigens, T_I-1 antigens are often mitogenic stimuli, such as LPS, CpG DNA, or viral RNA [29]. T_I-1 antigens can synergistically stimulate BCRs and Toll-like receptors (TLRs) [30]. Distinct from T_I-2 antigens, T_I-1 antigens induce B cell activation in both adults and newborns.

To overcome the limitations of T_I-2 antigens, CPS can be covalently linked to carrier proteins. Immunization with these CPS-protein conjugates (glycoconjugate vaccines) leads to T cell-dependent B cell activation and can induce long lasting immunity even in infants [31]. Carrier proteins are immunogenic proteins that are preferably nontoxic, nonreactogenic, and available in high amounts and purity. Carrier proteins that are currently used for conjugate vaccines or that are under development include bacterial products such as tetanus toxoid (TT), diphtheria toxoid (DT), a nontoxic cross-reacting mutant of DT (CRM197), the outer-membrane protein of *Neisseria meningitidis* (OMPC), or keyhole limpet hemocyanin (KLH), a sea snail hemoprotein [32]. Conjugation to carrier proteins may result in unpredictable immunologic interference such as carrier priming or carrier-induced epitopic suppression (CIES) [33]. Carrier priming refers to an enhanced immune response towards the conjugate vaccine due to a prior exposure to the carrier protein. CIES describes the phenomenon that pre-existing immunity against the carrier protein can also inhibit the immune response against the carbohydrate portion of the conjugate vaccine. CIES is a major drawback of conjugate vaccines since it results in an increased immune reaction against the carrier protein whereas the response against the conjugated polysaccharide remains weak [31].

The underlying mechanism of how conjugate vaccines are presented to T cells is still under debate. For a long time, it was thought that the CPS-protein conjugate is recognized and internalized by polysaccharide-specific BCRs of follicular B cells [34]. Subsequently, the protein moiety of the glycoconjugate is processed and presented on MHC-II molecules on the cell surface of the B cell. Recognition of the MHC-II-peptide complex by a peptide-specific T cell then leads to cognate T cell/B cell interactions where the B cell receives activating signals from the T cell (*see* Fig. 2).

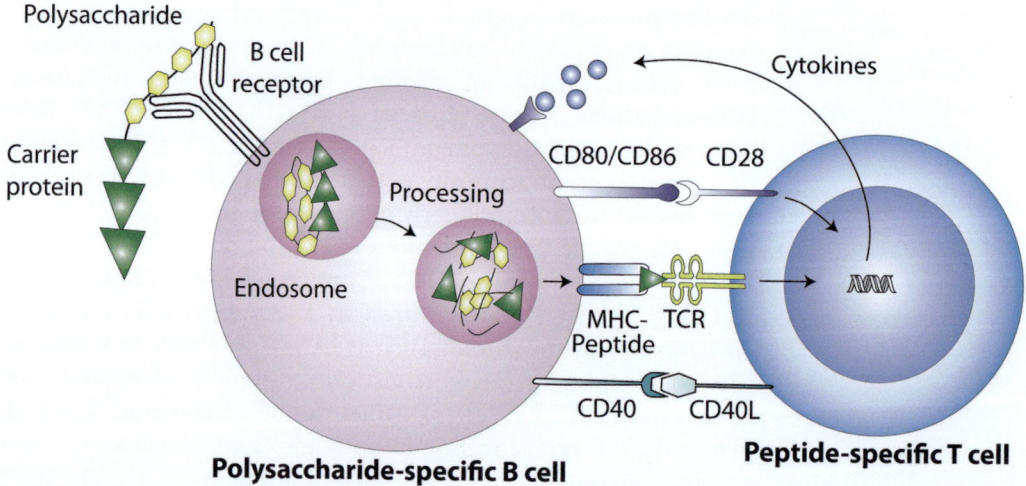

Fig. 2 Mode of action of glycoconjugate vaccines. The B cell receptor (BCR) of a polysaccharide-specific B cell recognizes the sugar portion of the glycoconjugate vaccine. Sugar binding cross-links the BCR and triggers glycoconjugate internalization into endosomes where it is further processed. Peptide fragments of the carrier protein in complex with MHC-II molecules are then presented on the surface of the B cell. The T cell receptor (TCR) of a peptide-specific T cell recognizes the peptide-MHC-II complex. Along with co-stimulation through CD80/CD86, the T cell becomes activated and provides T cell help to the B cell through the secretion of cytokines and CD40-CD40L interaction. This interaction finally leads to B cell activation resulting in differentiation into antibody-secreting plasma cells or memory B cell formation

However, recent studies suggest that after binding to the BCR and subsequent uptake into endosomes, CPS-protein conjugates may also be processed into glycopeptide fragments. The peptide portion can then bind to MHC-II molecules whereas the hydrophilic carbohydrate portion is exposed to the TCR, where it can interact with carbohydrate-specific CD4$^+$ T cells (T$_{carb}$) [35]. Activation of T$_{carb}$ can then mediate carbohydrate-specific T cell responses, independently of the carrier protein. Besides the linker between the carbohydrate antigen and the carrier protein, a number of additional determinants, such as the size of the carbohydrate epitope or the ratio of carbohydrate to carrier protein, impact the initiated immune response, thus may influence vaccine efficacy [36].

3 Advances in Glycoconjugate Vaccine Design

The development of glycoconjugate vaccines was a breakthrough since vaccination protected adults as well as newborns and other high risk groups from infection with encapsulated bacteria. To date, several glycoconjugate vaccines based on CPS are commercially available against bacteria such as *Haemophilus influenzae* type b (Hib), *Neisseria meningitidis*, *Salmonella enterica* serovar *Typhi*, or *Streptococcus pneumoniae* [37]. The first conjugate vaccine was introduced in the late 1980s, where capsular polysaccharides from Hib were covalently linked to diphtheria toxoid. This vaccine was included into large immunization programs and resulted in a dramatic reduction of Hib infections [38].

The success of the Hib glycoconjugate vaccine led to the development of further conjugate vaccines to prevent infections caused by other encapsulated bacteria, such as *N. meningitidis*. Meningococcal diseases caused by *N. meningitidis* serotypes A, B, C, W-135, X, and Y are the leading cause of devastating infections such as meningitis, meningococcemia, or sepsis. The highest risk group includes children below 2 years of age as well as adults older than 65 years. The majority of meningococcal diseases occur in developing countries, such as in the so-called meningitis belt in sub-Saharan Africa, where periodic epidemics are responsible for tens of thousands of deaths [39]. In industrialized countries, the incidence of meningococcal diseases has considerably decreased, particularly after the introduction of the first meningococcal conjugate vaccine in 1999. This monovalent vaccine against *N. meningitides* serotype C, the predominant serotype responsible for outbreaks in the UK, was licensed for the UK, Europe, and Canada and has proven to be highly effective [40]. Today, mono-, di-, and trivalent conjugate vaccines are widely replaced by novel tetravalent meningococcal conjugate vaccines MenACWY-DT (Menactra; Sanofi-Pasteur), MenACWY-CRM$_{197}$ (Menveo, Novartis Vaccines), and MenACWY-TT (Nimenrix, GlaxoSmithKline) that are currently approved in the United States and Europe [41]. In 2010, a campaign initiated by the WHO and the nonprofit organization PATH was started to eliminate meningococcal diseases in Africa, caused by the epidemic serotype A. To this end, a specific conjugate vaccine called MenAfriVac (Ps-A-TT) was especially designed for usage in Africa. Since 2012, more than 100 million people were vaccinated with Ps-A-TT in the meningitis belt and early data already suggest that this vaccine has the potential to eradicate *N. meningitidis* serotype A in Africa [40, 42].

In 2002, the US Food and Drug Administration (FDA) approved the first conjugate vaccine against *S. pneumoniae*, a pathogen that causes a wide range of invasive and noninvasive diseases such as bacteremia, meningitis, pneumonia, otitis media, or sinusitis. This 7-valent-CRM197 conjugate vaccine known as PCV7 (Prevnar, Wyeth Lederle Vaccines) contains CPS from the most frequent serotypes (4, 6B, 9 V, 14, 18C, 19 F, and 23 F) responsible for invasive pneumococcal diseases [43]. In order to provide a broader serotype coverage and to prevent the propagation of the non-included serotypes, a new 10-valent pneumococcal conjugate vaccine, PHiD-CV (Synflorix, GlaxoSmithKline), and a 13-valent pneumococcal conjugate vaccine, PCV13 (Prevnar 13, Wyeth Pharmaceuticals, Inc.), were developed and are currently distributed in more than 40 countries worldwide [44, 45]. In conclusion, the success of antibacterial glycoconjugate vaccines paves the way to the design of "next generation conjugate vaccines" to address other global health issues, such as fungal, parasitic, and viral infections as well as certain types of cancer [37].

4   Carbohydrates as Adjuvants

An adjuvant (from Latin "adjuvare," which means "to help" or "to assist") is any molecule, compound, or complex that enhances innate or adaptive immune responses, thereby helping to elicit effective immune responses to a co-administered antigen. Adjuvants have several desirable properties: First, adjuvants enhance the potency, quality, and longevity of immune responses. Second, they enhance immune responses in poor responder populations (such as newborns, immunocompromised individuals, and elderly patients). Third, vaccine doses and the need for booster immunizations may be reduced in the presence of adjuvants [46].

Adjuvants were first discovered in 1926 when Alexander T. Glenny and colleagues observed that particles of aluminum potassium sulfate enhanced initiated immune responses and triggered a depot effect in vaccine preparations of tetanus and diphtheria toxoids [47]. This finding led to the use of aluminum-based mineral salts as the first adjuvants for human vaccines. More than 80 years later, aluminum hydroxide and aluminum phosphate (alum), and the TLR4-agonist monophosphoryl lipid A formulated in alum (AS04) remain the only adjuvants approved for human vaccines in the United States. In Europe, two additional oil-in-water emulsions (MF59 and AS03) and one liposome-based adjuvant (virosomes) have been licensed as adjuvants for influenza vaccines or hepatitis A vaccines, respectively [48]. Although alum has been used for billions of injections worldwide, the mechanism of action is complex and still not fully understood. Only recently, mechanistic studies shed light on the molecular and cellular events triggered by alum [49]. These studies show that alum activates the nucleotide binding domain-like receptor protein 3 (Nalp3) inflammasome, resulting in the secretion of IL-1β and IL-18 by macrophages [50]. It was further observed that alum favors T_H2 immune responses, but the exact mechanism of how alum-mediated Nalp3 activation favors T_H2 cell polarization remains to be analyzed [51].

Recent advances in the development of recombinant or synthetic vaccines created the need for novel adjuvants. One reason is that modern vaccines are much purer and therefore less immunogenic than vaccines that were composed of live, attenuated or inactivated pathogens [52]. Although there are other immune stimulatory components that are more potent than alum, the majority of these substances have failed in the use for human applications since they often display higher levels of toxicity and are less well tolerated than alum. In addition, the mode of action of many adjuvant formulations is still not fully understood. However, research in the last decades has provided a deeper insight into innate immunity and antigen presentation and has led to the design of several novel adjuvants that are currently in preclinical and clinical development [53]. The enhancement of immune responses by adjuvants may be based on one or more of the following mechanisms: (1) activation of the inflammasome, (2) sustained release of the antigen at the site of injection (depot effect), (3) recruitment of immune cells to the site of injection, (4) enhanced antigen uptake by APCs, (5) binding of adjuvants to PRRs on APCs which in turn leads to APC maturation [54]. As a consequence, novel adjuvants can be designed and selected on the basis of their molecular interactions. A new class of effective and safe vaccine adjuvants is based on PRR ligands that specifically activate Toll-like receptor (TLR) signaling [55] or C-type lectin receptor (CLR) signaling in APCs [56]. Owing to their high expression by pathogens, numerous glycan structures have been identified to be recognized by PRRs. Therefore, glycan-based compounds have the potency to enhance antigen uptake, processing, and presentation by APCs and to stimulate subsequent T cell responses.

5 Glycan Recognition by Pattern-Recognition Receptors (PRRs)

APCs express a number of germline-encoded PRRs enabling APCs to bind and to recognize highly conserved PAMPs that are exclusively found on pathogens but not on host cells. Binding of PAMPs to PRRs can induce pathogen uptake as well as intracellular signaling cascades resulting in APC maturation and subsequent T cell activation [57]. Since the surface of nearly every pathogen is covered with specific glycoconjugates such as glycoproteins or glycolipids, glycans represent an important class of PAMPs. Recognition of these carbohydrate-containing PAMPs by cells of the innate immune system is mediated by several glycan-binding proteins that function as PRRs.

6 C-Type Lectin Receptors

Lectins are glycan-binding proteins that are ubiquitously expressed by viruses, bacteria, insects, crustaceans, plants, and animals [58]. C-type lectins, a family of lectins that share a common carbohydrate recognition domain (CRD), often recognize carbohydrates in a Ca^{2+}-dependent manner. Myeloid C-type lectins are expressed by APCs such as macrophages and DCs and serve as PRRs. They recognize glycans such as high mannose and fucose carbohydrate structures, present on the cell surface of pathogens including viruses, bacteria, fungi, and parasites [59, 60]. The main function of APC-expressed CLRs is to internalize pathogens and to mediate their subsequent degradation in lysosomal compartments. Thereby, CLRs are able to enhance antigen processing and presentation by APCs [61]. Additionally, CLR engagement leads to the induction of intracellular signaling cascades activating innate response genes responsible for the production of reactive oxygen species and pro-inflammatory cytokines and chemokines [62]. Thus, CLR targeting using carbohydrate ligands is a promising approach to shape immune responses [63]. CLR-triggered responses may be diverse and strongly depend on the nature of the ligand, cross-talk with other PRRs such as TLRs and the expression pattern of CLRs on APCs [60]. The specificity of CLRs for certain carbohydrates is defined by conserved amino acid motifs in the CRD. In general, CLRs can be divided into receptors that recognize mannose and/or fucose-terminated glycans and receptors that recognize galactose-terminated or N-acetylgalactosamine-terminated glycans on pathogens [64]. Targeting of some CLRs, such as Dendritic Cell-Specific Intercellular adhesion molecule-3-grabbing non-integrin (DC-SIGN), has been shown to enhance antigen cross-presentation, thereby promoting antigen-specific CD8+ T cell responses [65].

Examples of glycoconjugates as potent stimulators of innate immunity that trigger signaling via CLRs are trehalose 6,6'-dimycolate (TDM), a glycolipid present in the cell wall of mycobacteria ("cord factor"), and β-glucan, a polysaccharide composed of β-1,3-linked glucose found in the cell wall of many fungal species. TDM, recognized by the CLRs Mincle and MCL [66], and β-glucan, recognized by the CLR Dectin-1 in collaboration with TLR2, were both shown to activate innate response genes in APCs and to facilitate T cell differentiation into T_H1 and T_H17 [67]. Thus, TDM and β-glucan formulations may have potential as human vaccine adjuvants [68].

7   Toll-Like Receptors (TLRs)

Toll-like receptors are classical PRRs that play a central role in innate immunity. TLR ligands are potent activators of innate immune responses that trigger DC maturation, finally leading to the induction of adaptive immune responses. Thus, TLR agonists represent a new class of vaccine adjuvants that currently undergo testing for safety and efficacy in humans [69]. TLRs are characterized by varying numbers of extracellular leucine-rich repeat (LRR) motifs and an intracellular Toll/IL 1 receptor like (TIR) domain, crucial for the recruitment of downstream signaling molecules [70]. TLR1, TLR2, TLR4, and TLR6 are transmembrane receptors that recognize PAMPs displayed on the surface of pathogens. In contrast, TLR3, TLR7, TLR8, and TLR9 are localized in endosomes and are involved in the recognition of pathogen-derived structures released after internalization and degradation. TLR2 and TLR4 are involved in the recognition of pathogen-derived polysaccharides and glycoconjugates [71, 72].

Various TLR agonists have been identified and are currently tested in clinical trials but have not yet been licensed [49]. One of the most prominent targets of the TLR family is TLR4 that recognizes LPS from Gram-negative bacteria. Recognition of LPS by the TLR4-MD-2 complex leads to APC activation that is important for early antibacterial responses [73]. Ligand binding by TLRs induces receptor dimerization and subsequent conformational changes leading to the association with intracellular adaptor molecules such as the Myeloid differentiation primary-response protein 88 (MyD88), the TIR-domain-containing adaptor protein (TIRAP), the TIR-domain-containing adaptor protein inducing IFN-β (TRIF), and the TRIF-related adaptor molecule (TRAM). Downstream signaling activates interferon regulatory factors (IRFs), mitogen-activated protein kinases (MAPKs), and the transcription factor NF-κB [70]. Thus, TLR agonists may elicit the expression of pro-inflammatory cytokines, MHC molecules, and co-stimulatory molecules.

One major advantage of pure carbohydrates as adjuvants is that they are usually well tolerated, nontoxic, and can be easily metabolized [74]. One candidate that has already entered phase II clinical trials is Advax [75]. Advax is based on β-D-(2-1) polyfructofuranosyl-α-D-glucose (delta inulin) that is extracted from the plant roots of the *Compositae* family or can be chemically synthesized from sucrose. Delta inulin was shown to enhance humoral and cellular immunity by triggering monocyte activation [76]. Unlike alum that mainly evokes T_H2-mediated responses inulin-based adjuvants seem to induce both T_H1 and T_H2 responses [74].

An additional mechanism of how carbohydrate-based adjuvants can enhance innate immune responses is the activation of the Nalp3 inflammasome. Zymosan and mannan purified from fungal cell walls were shown to induce caspase-1 activation and IL-1β secretion [77]. Another class of carbohydrate derivatives with strong adjuvant activity is saponins, secondary metabolites found in plants such as the *Rhamnaceae*, *Araliaceae*, *Polygalaceae*, or *Fabaceae* family [68]. Saponins are natural steroid or triterpene glycosides with a great variability of structures and properties including immune stimulatory activity. The saponin QS-21 isolated from the bark of the *Quillaja saponaria* (QS) tree is one of the most potent adjuvants and is currently under investigation in vaccines against infectious diseases such as tuberculosis, malaria, acquired immunodeficiency syndrome, hepatitis, as well as cancer [78]. The adjuvant activity of saponins may be attributed to their pore-forming properties which may trigger cell activation [68]. Furthermore, it has been suggested that QS saponins can provide co-stimulation for T cell activation through binding to amino groups on T cell surface receptors [79]. However, the exact mode of action of saponins remains unclear [78]. QS-21 enhances B cell responses and also induces robust T_H1 and T_H2 responses. Since natural QS-21 is toxic and unstable, it has not yet been licensed for human use. Approaches to overcome these challenges are to chemically modify natural QS-21, to synthesize less toxic analogues, or to incorporate QS-21 into lipid particles [80]. Examples of adjuvant formulations that contain QS-21 are AS01 and AS02 that are currently tested as adjuvant candidates in human malaria vaccines [81].

In conclusion, carbohydrate-based adjuvants act through multiple immunological mechanisms as also indicated in Fig. 3. At present, only few adjuvants have been applied to clinic since the discovery and approval of alum in the 1930s. However, the examples shown here highlight the utility of carbohydrate-based adjuvants to increase the efficacy of vaccines. Further studies are needed to fully understand their mode of action and to exploit their potential for clinical applications.

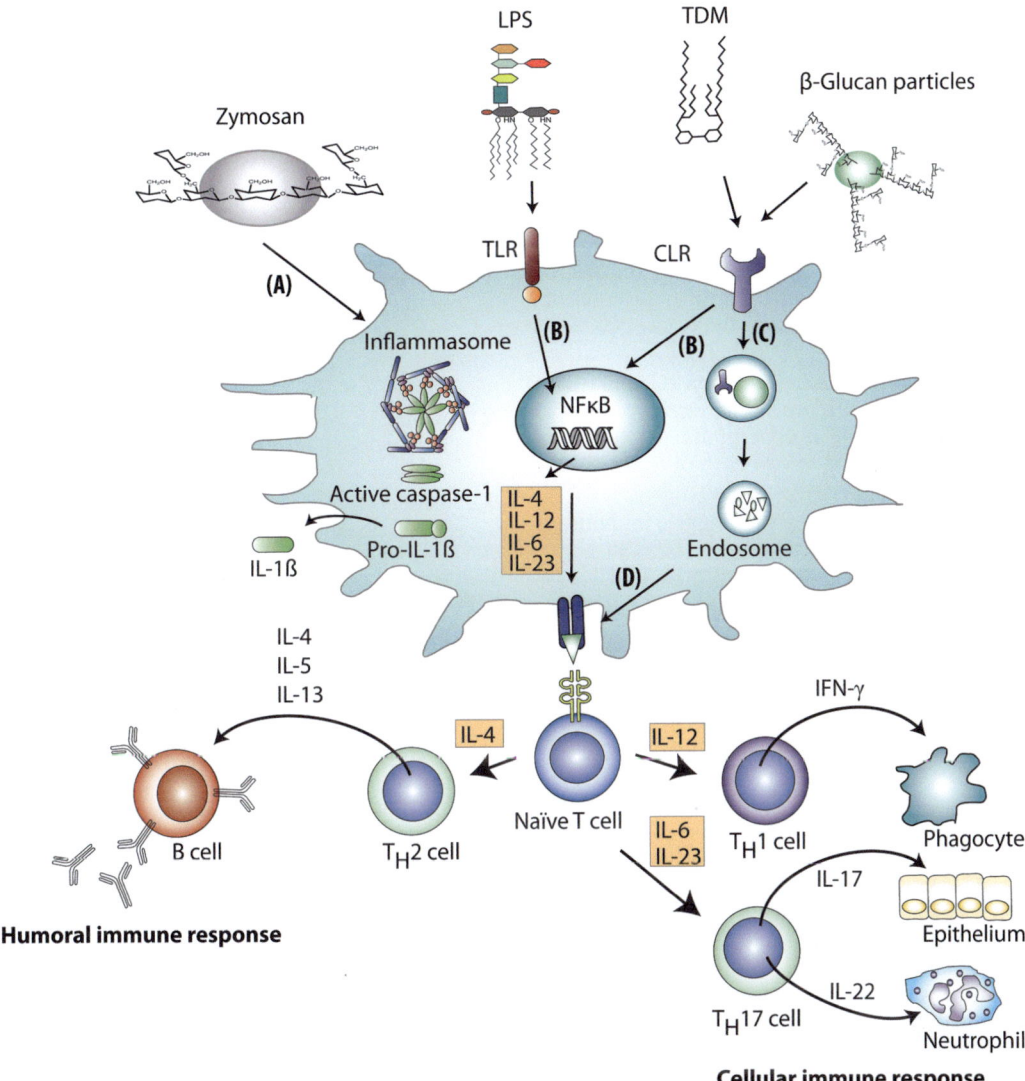

Fig. 3 Mode of action of carbohydrate-based adjuvants. Some polysaccharide compounds such as zymosan activate the inflammasome leading to the production of IL-1β (A). Other carbohydrate-based adjuvants such as LPS, TDM, or β-glucan particles are recognized by PRRs such as TLRs and CLRs, respectively. Activation of TLR and CLR signaling leads to the expression of cytokines (B), enhanced antigen internalization (C), and presentation on MHC-II molecules (D). Cytokines and chemokines produced in response to carbohydrate-based adjuvants influence the recruitment of immune cell subsets to the site of infection and may direct T cell responses towards T_H1, T_H2, or T_H17

Acknowledgements

This work was supported by the International Max Planck Research School (IMPRS) on Multiscale Bio-Systems. The authors would like to thank Benjamin Schumann for critical reading of the manuscript.

References

1. Whitfield C, Trent MS (2014) Biosynthesis and export of bacterial lipopolysaccharides. Annu Rev Biochem 83:99–128

2. Roberts IS (1996) The biochemistry and genetics of capsular polysaccharide production in bacteria. Annu Rev Microbiol 50: 285–315

3. Marino K, Bones J, Kattla JJ et al (2010) A systematic approach to protein glycosylation analysis: a path through the maze. Nat Chem Biol 6:713–723

4. Kadioglu A, Weiser JN, Paton JC et al (2008) The role of Streptococcus pneumoniae virulence factors in host respiratory colonization and disease. Nat Rev Microbiol 6:288–301

5. Avci FY, Kasper DL (2010) How bacterial carbohydrates influence the adaptive immune system. Annu Rev Immunol 28:107–130

6. Cobb BA, Kasper DL (2005) Coming of age: carbohydrates and immunity. Eur J Immunol 35:352–356

7. Francis T, Tillett WS (1930) Cutaneous reactions in pneumonia. The development of antibodies following the intradermal injection of type-specific polysaccharide. J Exp Med 52:573–585

8. Schumann B, Anish C, Pereira CL et al (2014) Chemical biology approaches to designing defined carbohydrate vaccines. Chem Biol 21:38–50

9. Medzhitov R, Janeway CA Jr (1997) Innate immunity: impact on the adaptive immune response. Curr Opin Immunol 9:4–9

10. Palm NW, Medzhitov R (2009) Pattern recognition receptors and control of adaptive immunity. Immunol Rev 227:221–233

11. Cruvinel WD, Mesquita D, Araujo JAP et al (2010) Immune system—Part I Fundamentals of innate immunity with emphasis on molecular and cellular mechanisms of inflammatory response. Rev Bras Reumatol 50:434–461

12. Kapsenberg ML (2003) Dendritic-cell control of pathogen-driven T-cell polarization. Nat Rev Immunol 3:984–993

13. Romagnani S (2000) T-cell subsets (Th1 versus Th2). Ann Allergy Asthma Immunol 85:9–21

14. Zhu J, Yamane H, Paul WE (2010) Differentiation of Effector CD4 T Cell Populations. Annu Rev Immunol 28:445–489

15. O'Connor W, Zenewicz LA, Flavell RA (2010) The dual nature of TH17 cells: shifting the focus to function. Nat Immunol 11:471–476

16. Singh RP, Hasan S, Sharma S et al (2014) Th17 cells in inflammation and autoimmunity. Autoimmun Rev 13:1174–1181

17. Crotty S (2011) Follicular Helper CD4 T Cells (TFH). Annu Rev Immunol 29:621–663

18. Parker DC (1993) T cell-dependent B cell activation. Annu Rev Immunol 11:331–360

19. Murphy K, Travers P, Walport M, Janeway CA (2012) Janeway's immunobiology. Garland Science, New York

20. Guttormsen H-K, Sharpe AH, Chandraker AK et al (1999) Cognate stimulatory B-cell–T-cell interactions are critical for T-cell help recruited by glycoconjugate vaccines. Infect Immun 67:6375–6384

21. Mond JJ, Lees A, Snapper CM (1995) T cell-independent antigens type 2. Annu Rev Immunol 13:655–692

22. Landers C, Chelvarajan RL, Bondada S (2005) The role of B cells and accessory cells in the neonatal response to TI-2 antigens. Immunol Res 31:25–36

23. Weintraub A (2003) Immunology of bacterial polysaccharide antigens. Carbohydr Res 338: 2539–2547

24. Vos Q, Lees A, Wu ZQ et al (2000) B-cell activation by T-cell-independent type 2 antigens as an integral part of the humoral immune response to pathogenic microorganisms. Immunol Rev 176:154–170

25. López-Herrera G, Vargas-Hernández A, González-Serrano ME et al (2014) Bruton's tyrosine kinase—an integral protein of B cell development that also has an essential role in the innate immune system. J Leukoc Biol 95:243–250

26. Bondada S, Chelvarajan RL (2004) Neonatal immunity to polysaccharide antigens: role of B cells versus macrophages. Nat Rev Immunol. doi:10.1038/nri1394-c1

27. Velez CD, Lewis CJ, Kasper DL et al (2009) Type I Streptococcus pneumoniae carbohydrate utilizes a nitric oxide and MHC II-dependent pathway for antigen presentation. Immunology 127:73–82

28. Avci FY, Li X, Tsuji M et al (2013) Carbohydrates and T cells: a sweet twosome. Semin Immunol 25:146–151

29. Vinuesa CG, Chang P-P (2013) Innate B cell helpers reveal novel types of antibody responses. Nat Immunol 14:119–126

30. Alugupalli KR, Akira S, Lien E et al (2007) MyD88- and Bruton's tyrosine kinase-mediated signals are essential for T cell-independent pathogen-specific IgM responses. J Immunol 178:3740–3749

31. Dagan R, Poolman J, Siegrist C-A (2010) Glycoconjugate vaccines and immune interference: a review. Vaccine 28:5513–5523

32. Adamo R, Nilo A, Castagner B et al (2013) Synthetically defined glycoprotein vaccines: current status and future directions. Chem Sci 4:2995–3008

33. Pobre K, Tashani M, Ridda I et al (2014) Carrier priming or suppression: understanding carrier priming enhancement of anti-polysaccharide antibody response to conjugate vaccines. Vaccine 32:1423–1430

34. Lucas AH, Apicella MA, Taylor CE (2005) Carbohydrate moieties as vaccine candidates. Clin Infect Dis 41:705–712

35. Avci FY, Li X, Tsuji M et al (2011) A mechanism for glycoconjugate vaccine activation of the adaptive immune system and its implications for vaccine design. Nat Med 17:1602–1609

36. Kuberan B, Lindhardt RJ (2000) Carbohydrate based vaccines. Curr Org Chem 4:653–677

37. Astronomo RD, Burton DR (2010) Carbohydrate vaccines: developing sweet solutions to sticky situations? Nat Rev Drug Discov 9:308–324

38. Makela PH (2003) Conjugate vaccines-a breakthrough in vaccine development. Southeast Asian J Trop Med Public Health 34:249–253

39. Irving TJ, Blyuss KB, Colijn C et al (2012) Modelling meningococcal meningitis in the African meningitis belt. Epidemiol Infect 140:897–905

40. Cohn A, Harrison L (2013) Meningococcal vaccines: current issues and future strategies. Drugs 73:1147–1155

41. Hedari CP, Khinkarly RW, Dbaibo GS (2014) Meningococcal serogroups A, C, W-135, and Y tetanus toxoid conjugate vaccine: a new conjugate vaccine against invasive meningococcal disease. Infect Drug Resist 7:85–99

42. Daugla DM, Gami JP, Gamougam K et al (2014) Effect of a serogroup A meningococcal conjugate vaccine (PsA-TT) on serogroup A meningococcal meningitis and carriage in Chad: a community study [corrected]. Lancet 383:40–47

43. Arguedas A, Soley C, Abdelnour A (2011) Prevenar experience. Vaccine 29:C26–C34

44. Prymula R, Schuerman L (2009) 10-valent pneumococcal nontypeable Haemophilus influenzae PD conjugate vaccine: Synflorix™. Expert Rev Vaccines 8:1479–1500

45. Gruber WC, Scott DA, Emini EA (2012) Development and clinical evaluation of Prevnar 13, a 13-valent pneumococcal CRM197 conjugate vaccine. Ann NY Acad Sci 1263:15–26

46. Alving CR, Peachman KK, Rao M et al (2012) Adjuvants for human vaccines. Curr Opin Immunol 24:310–315

47. Marrack P, McKee AS, Munks MW (2009) Towards an understanding of the adjuvant action of aluminium. Nat Rev Immunol 9:287–293

48. Rappuoli R, Mandl CW, Black S et al (2011) Vaccines for the twenty-first century society. Nat Rev Immunol 11:865–872

49. De Gregorio E, Tritto E, Rappuoli R (2008) Alum adjuvanticity: unraveling a century old mystery. Eur J Immunol 38:2068–2071

50. Eisenbarth SC, Colegio OR, O'Connor W et al (2008) Crucial role for the Nalp3 inflammasome in the immunostimulatory properties of aluminium adjuvants. Nature 453:1122–1126

51. Dostert C, Ludigs K, Guarda G (2013) Innate and adaptive effects of inflammasomes on T cell responses. Curr Opin Immunol 25:359–365

52. Mohan T, Verma P, Rao D (2013) Novel adjuvants & delivery vehicles for vaccines development: a road ahead. Indian J Med Res 138:779–795

53. Mbow ML, De Gregorio E, Valiante NM et al (2010) New adjuvants for human vaccines. Curr Opin Immunol 22:411–416

54. Awate S, Babiuk LA, Mutwiri G (2013) Mechanisms of action of adjuvants. Front Immunol. doi:4 10.3389/fimmu.2013.00114

55. Duthie MS, Windish HP, Fox CB et al (2011) Use of defined TLR ligands as adjuvants within human vaccines. Immunol Rev 239:178–196

56. Lang R, Schoenen H, Desel C (2011) Targeting Syk-Card9-activating C-type lectin receptors by vaccine adjuvants: Findings, implications and open questions. Immunobiology 216:1184–1191

57. Janeway CA, Medzhitov R (2002) Innate immune recognition. Annu Rev Immunol 20:197–216

58. Cummings RD, McEver RP (2009) C-type lectins. In: Varki A, Cummings RD, Esko JD, Freeze HH, Stanley P, Bertozzi CR, Hart GW, Etzler ME (eds) Essentials of glycobiology, 2nd edn. Cold Spring Harbor, New York, NY

59. Sancho D, Reis e Sousa C (2012) Signaling by myeloid C-type lectin receptors in immunity and homeostasis. Annu Rev Immunol 30:491–529

60. Geijtenbeek TBH, Gringhuis SI (2009) Signalling through C-type lectin receptors: shaping immune responses. Nat Rev Immunol 9:465–479

61. van Kooyk Y, Geijtenbeek TBH (2003) DC-SIGN: escape mechanism for pathogens. Nat Rev Immunol 3:697–709

62. Drummond RA, Brown GD (2013) Signalling C-type lectins in antimicrobial immunity. PLoS Pathog 9, e1003417

63. Lepenies B, Lee J, Sonkaria S (2013) Targeting C-type lectin receptors with multivalent carbohydrate ligands. Adv Drug Deliv Rev 65: 1271–1281

64. Drickamer K (1992) Engineering galactose-binding activity into a C-type mannose-binding protein. Nature 360:183–186

65. van Kooyk Y, Unger WWJ, Fehres CM et al (2013) Glycan-based DC-SIGN targeting vaccines to enhance antigen cross-presentation. Mol Immunol 55:143–145

66. Miyake Y, Toyonaga K, Mori D et al (2013) C-type lectin MCL is an FcRγ-coupled receptor that mediates the adjuvanticity of mycobacterial cord factor. Immunity 38:1050–1062

67. Carvalho A, Giovannini G, De Luca A et al (2012) Dectin-1 isoforms contribute to distinct Th1/Th17 cell activation in mucosal candidiasis. Cell Mol Immunol 9:276–286

68. Petrovsky N, Cooper PD (2011) Carbohydrate-based immune adjuvants. Expert Rev Vaccines 10:523–537

69. Egli A, Santer D, Barakat K et al (2014) Vaccine adjuvants—understanding molecular mechanisms to improve vaccines. Swiss Med Wkly 144:w13940

70. Akira S, Takeda K (2004) Toll-like receptor signalling. Nat Rev Immunol 4:499–511

71. Zughaier SM (2011) Neisseria meningitidis capsular polysaccharides induce inflammatory responses via TLR2 and TLR4-MD-2. J Leukoc Biol 89:469–480

72. Akira S, Uematsu S, Takeuchi O (2006) Pathogen recognition and innate immunity. Cell 124:783–801

73. Park BS, Lee J-O (2013) Recognition of lipopolysaccharide pattern by TLR4 complexes. Exp Mol Med 45, e66

74. Petrovsky N, Aguilar JC (2004) Vaccine adjuvants: current state and future trends. Immunol Cell Biol 82:488–496

75. Gordon DL, Sajkov D, Woodman RJ et al (2012) Randomized clinical trial of immunogenicity and safety of a recombinant H1N1/2009 pandemic influenza vaccine containing Advax™ polysaccharide adjuvant. Vaccine 30: 5407–5416

76. Cooper PD, Petrovsky N (2011) Delta inulin: a novel, immunologically active, stable packing structure comprising β-d-[2 → 1] poly(fructofuranosyl) α-d-glucose polymers. Glycobiology 21:595–606

77. Lamkanfi M, Malireddi RKS, Kanneganti T-D (2009) Fungal zymosan and mannan activate the cryopyrin inflammasome. J Biol Chem 284:20574–20581

78. Fernández-Tejada A, Chea EK, George C et al (2014) Development of a minimal saponin vaccine adjuvant based on QS-21. Nat Chem 6:635–643

79. Marciani DJ (2003) Vaccine adjuvants: role and mechanisms of action in vaccine immunogenicity. Drug Discov Today 8:934–943

80. Ragupathi G, Gardner JR, Livingston PO et al (2011) Natural and synthetic saponin adjuvant QS-21 for vaccines against cancer. Expert Rev Vaccines 10:463–470

81. Agnandji ST, Fendel R, Mestré M et al (2011) Induction of plasmodium falciparum-specific CD4+ T cells and memory B cells in Gabonese children vaccinated with RTS, S/AS01(E) and RTS, S/AS02(D). PLoS One 6, e18559

Chapter 3

The Glycan Array Platform as a Tool to Identify Carbohydrate Antigens

Li Xia and Jeffrey C. Gildersleeve

Abstract

Carbohydrate antigens are important targets for the immune system, but identification of key glycan antigens is challenging. Direct analysis of glycomes by mass spectrometry is difficult, and detection reagents, such as monoclonal antibodies and lectins, are only available for a small subset of glycans. An alternative approach involves profiling serum anti-glycan antibody populations to identify unique antibodies or changes in antibody subpopulations. Glycan microarray technology allows rapid evaluation of hundreds to thousands of antigen-antibody interactions in a single experiment. This high-throughput format is particularly useful in profiling complex anti-glycan antibodies in serum. Here we elaborate the use of this technology to explore clinically relevant carbohydrate antigens by profiling serum anti-glycan antibodies. Detailed protocols from glycan microarray fabrication to microarray binding assays and analysis of microarray data are presented.

Key words Glycan microarray, Carbohydrate array, Anti-glycan antibody, Carbohydrate antigen, Tumor antigen, Serum antibody profile

1 Introduction

Carbohydrates are abundantly present on all cell surfaces, including mammalian cells and pathogen cells. Some of the glycans expressed on tumor cells and pathogens are structurally distinct from normal healthy human glycans [1]. As a result, they can stimulate an immune response which can be harnessed in the diagnosis and treatment of many diseases including cancers and pathogen infections [2, 3]. For example, bacterial carbohydrates that stimulate a neutralizing response can inform vaccine design. However, identification of carbohydrate antigens is extremely challenging due to the complexity of diverse glycan structures in nature, a dearth of structural information on those glycans, and a lack of detecting tools [4, 5].

Antigens, in general, are often identified indirectly by profiling antibody and cellular responses [6]. For example, protein arrays have been used frequently to compare antibody populations before and after infection or vaccination [7]. When antibodies to a particular

Bernd Lepenies (ed.), *Carbohydrate-Based Vaccines: Methods and Protocols*, Methods in Molecular Biology, vol. 1331, DOI 10.1007/978-1-4939-2874-3_3, © Springer Science+Business Media New York 2015

peptide are detected after immune stimulation, this information is then used to trace the response back to the original antigen. Glycan microarray technology allows analogous evaluations of anti-glycan immune response. On the microarray, a large number of structurally distinct glycans derived from either natural or synthetic sources are immobilized on a glass slide in a spatially defined pattern [8–10]. The source of glycans can be from human, bacteria, virus, or other organisms, and only tiny amounts of material are required. This miniaturized format allows high-throughput screening of hundreds of carbohydrate-protein interactions on a single slide. This technology has been used in many research areas including functional glycomics, drug discovery, and diagnosis [11–14]. One of the applications in vaccine development is discovery of clinically relevant biomarkers by profiling serum anti-glycan antibodies [15]. For example, one can study ligand specificities of the isolated monoclonal antibodies produced in vaccinated or pathogen-infected animals [16, 17]. One can also compare antibody populations of diseased subjects to a group of healthy control individuals to discover disease-specific antigens [for some recent examples, see [18–21]]. Another approach is to evaluate antibody changes in individuals before and after stimulation (e.g., vaccination, pathogen infections) to discover antigens on vaccines or pathogens [for some recent examples, see [22–24]].

The general approach is relatively straightforward. A slide is first incubated with the sample of interest (e.g., vaccinated or infected sera, monoclonal antibodies). After washing off unbounded samples, the slide is incubated with fluorophore or streptavidin-labeled detection reagents (e.g., fluorophore-labeled anti-human IgG and IgM antibodies) and the captured antibodies on the array are detected with a fluorescent scanner (Fig. 1). Since it is often advantageous to profile many different samples and/or to profile individual samples multiple times under different conditions, many groups use a slide format in which multiple copies of the array are printed on each slide (e.g., 16 arrays/slide, Fig. 1). After physically separating the replicate arrays using a well module, one can carry out multiple array assays on each slide. The protocol described here covers procedures for microarray fabrication, microarray binding assay, and data analysis. In addition, technical challenges and potential pitfalls are also discussed.

2   Materials

2.1   Reagents and Materials

2.1.1   For Glycan Microarray Construction

1. Microarray printing pins (Arrayit, SMP2).
2. SuperEpoxy 2 microarray substrate slides (Arrayit).
3. 384-well low profile microplate, V-bottom.
4. Thermowell aluminum sealing tape.

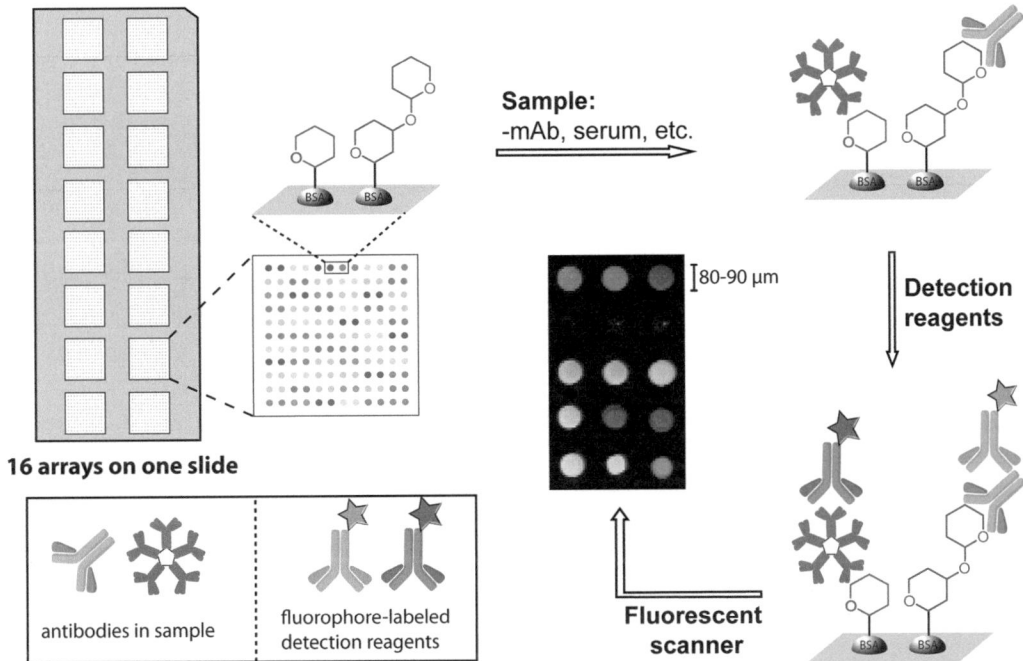

Fig. 1 *Glycan microarray binding assay.* 16 arrays are printed on a single slide with hundreds of BSA-modified neoglycoproteins on one array. Prior to the assay, the slide is fitted with a 16-well module that physically separates the individual arrays. In the binding assay, the slide is first blocked to deactivate reactive functional groups on surface. After blocking, it is incubated with sample of interest, and then the captured antibodies are detected with fluorophore-labeled secondary reagents. Binding is quantitated by a fluorescent scanner. In the example shown, IgG and IgM antibodies are detected with secondary reagents labeled with different dyes

5. Print dye: Alexa Fluor 555 azide.

6. Glycans for microarray: Glycans used in our lab are either purchased from commercial sources or chemically synthesized in the lab. There are several ways to immobilize glycans onto a solid support. The protocol described here uses neoglyco-proteins (proteins in which glycans are covalently linked via a nonnatural linkage) as substrates that are immobilized onto epoxide-modified glass slides. Many natural or synthetic gly-cans with a free reducing end can be coupled to BSA or HSA in a single step via reductive amination [25]. Our most recent array contains over 500 glycoconjugates, encompassing human glycans, non-human glycans, glycopeptides, and glycoproteins. The glycans are stored at −20 °C until use.

2.1.2   For Glycan Array Profiling of Serum Antibodies

1. ProPlate multi-array slide module.

2. ProPlate adhesive seal-strips.

3. Aluminum foil.

4. Detecting antibodies: DyLight 549-conjugated Goat Anti-Human IgG and DyLight 649-conjugated Goat Anti-Human IgM.

5. Reference serum: pooled serum samples from multiple healthy donors to serve as a control.

6. Serum samples: For case-control study, sera from a group of diseased subjects and a group of healthy controls are recommended; for profiling the same subject at different time points, pre- and post-stimulation (e.g., vaccination, pathogen infection) sera from individual subjects are recommended.

2.2 Equipment

1. Microarray printer (Biorobotics, MicroGrid II).

2. Microarray Fluorescent scanner (Molecular Devices, GenePix 4000B).

3. FLx800 microplate fluorescence reader (BioTek Instruments).

4. Microscope.

5. Centrifuge.

6. Incubator shaker.

7. Sonicator.

2.3 Software

1. GenePix Pro 6.0 (Molecular Devices).

2. Excel.

2.4 Buffer Solutions

Buffer solutions 2–4 should be prepared fresh on the day of experiment.

1. Phosphate buffered saline (PBS): 10 mM $NaHPO_4$, 1.8 mM KH_2PO_4, 2.7 mM KCl, 137 mM NaCl, pH 7.4.

2. Print buffer: $1 \times PBS$ buffer with 2.5 % (v/v) glycerol, 0.0005 % (v/v) Triton-X 100, 0.7 μg/mL print dye (Alexa Fluor 555 azide). Store in the dark to protect the dye.

3. Blocking buffer: PBS with 3 % (w/v) bovine serum albumin (BSA, globulin free).

4. Washing buffer: PBS with 0.05 % (v/v) Tween 20 (PBST).

5. Serum incubation buffer: PBST with 3 % (w/v) BSA and 1 % (w/v) human serum albumin (HSA, globulin free).

6. Antibody incubation buffer: PBS with 1 % (w/v) BSA and 3 % (w/v) HSA.

3   Methods

3.1   Construction of Glycan Antigen Microarray

3.1.1   Preparing Source Plates for Printing

1. Dilute glycan antigens with printing buffer containing print dye Alexa 555 (*see* **Note 1**) to a final concentration of 125 µg/mL.

2. Place 8 µL/well (*see* **Note 2**) diluted glycan solution into 384-well plates. The layout of 384-well plate depends on the number and configuration of pins used for printing (*see* **step 1** in Subheading 3.1.2 for detail coordination of source plate with pins).

3. Scan the finished 384-well plates in FLx800 Microplate Fluorescence Reader for any missing samples.

4. Seal plates with aluminum seal to prevent evaporation and photobleaching of the dye, then cover with lids and store at 4 °C till use.

5. Spin down the source plate at 1000 rpm for 5 min before printing.

3.1.2   Set Up Robotic Microarray Printer

1. Program the printer so that each slide accommodates 16 arrays and each glycan is printed in duplicate (Fig. 2).

2. Sonicate printer pins for 15 min in Milli-Q water and blow-dry with argon. Handle with care so as not to touch the pin tips. Tips are delicate, and the quality of pins directly affects the printing quality.

3. Inspect pins under a microscope for any physical damages, and make sure no debris is trapped inside of pin channels.

4. Load pins into pin tool.

5. Load epoxide-coated slides and source plates in the printer.

6. Humidify the whole printer until humidity reaches 50–60 %. Maintain this humidity during the whole printing.

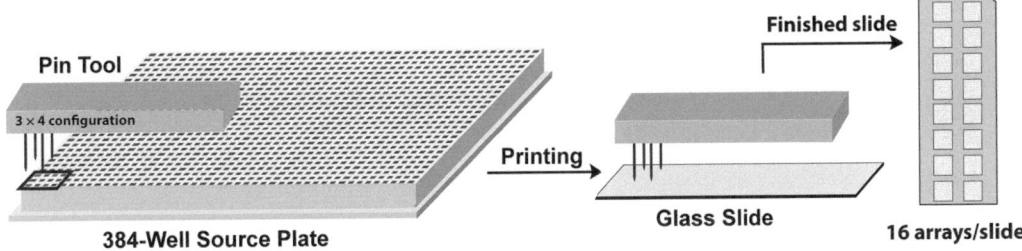

Fig. 2 *Coordination of pin configuration with 384-well source plate.* Four pins are arranged in 3 × 4 configuration. Correspondingly, samples in 384-well source plate are arranged so that each sample is distributed in four wells matching with four pins. During printing, four pins are dipped into 384-well source plate, and then deliver sample onto glass slide for printing. The four-pin system allows four arrays to be printed simultaneously. A total of 16 arrays are printed on a single slide

3.1.3 Printing Microarray

1. Start printing. If possible, printing should be carried out in the dark to minimize photobleaching of the print dye.

2. During printing, frequently check humidity and pins. Make sure pins are not stuck in the pin tool (*see* **Note 3**).

3. After a print run is complete, scan slides in a fluorescent microarray scanner for any smear, merging spots, and missing spots. Keep record of any defects.

4. Collect slides in a covered box, and store it in a vacuum-sealed bag at –20 °C. Slides can be stored in this manner for ≥ 6 months without loss of binding capacity or integrity.

3.1.4 Quality Control

In addition to scanning the slides after a print run to detect smears, missing spots, and merged spots, quality of the printed slides could be evaluated by incubating with lectins with known binding specificities (*see* **Note 4**). We typically profile a set of 4 lectins on at least one slide from every print batch to assess quality and reproducibility. New data is compared to all previous batches and to known binding preferences.

3.2 Glycan Array Profiling of Serum Antibodies

1. Remove vacuum-sealed slides from –20 °C freezer and let them warm up to room temperature before opening the package.

3.2.1 Preparing Slide Module and Microarray

2. Inspect slides for any defects (smear, non-uniformed spots, missing spots, etc.) with fluorescent scanner at PMT 600. Save this pre-scan image as .tiff file for future data analysis. Choose slides with minimum defects for high quality results.

3. Assemble slide module over the slide to separate 16 printed arrays. The printed area of each array should be aligned in the center of each well. Be careful not to smudge or smear the array surface.

3.2.2 Blocking Glycan Microarray

This process serves dual roles: blocking excess epoxide on slide surface and washing off print dye.

1. Prepare a fresh solution of blocking buffer: PBS with 3 % BSA (*see* **Note 5**).

2. Gently pipet 200 µL blocking buffer into each well. Make sure not to drop the liquid directly on the printed area. Slowly pipet solution against the corner of well while holding the slide slightly tilted (*see* **Note 6**).

3. Seal the slide module with an adhesive strip to prevent evaporation, and incubate without shaking at room temperature for 2 h, or at 4 °C for overnight (*see* **Note 7**).

4. Discard the blocking solution after removing the strip, and wash the array with 6 × 200 µL/well washing buffer. After each wash cycle, invert the module and tap vigorously on paper towel to remove excess liquid.

3.2.3 Preparing Serum Samples and Reference Sample

All the following steps should be handled in class II biosafety cabinet to prevent bloodborne pathogen infection from serum samples.

1. Prepare a fresh serum incubation buffer: PBST with 3 % BSA and 1 % HSA (*see* **Note 8**).

2. Dilute serum samples with incubation buffer to an appropriate concentration (1:50 dilution is recommend. However, best condition needs to be optimized for different samples). Mix well by gently tapping the tube rather than vortex.

3. Prepare a reference sample (1:50 dilution) to serve as a control on each slide. Pooled serum sample could be used as reference. To avoid thaw-freeze cycles, make aliquots of the reference and store at −80 °C. Use a fresh aliquot each time (*see* **Note 9**).

3.2.4 Incubating Microarray with Serum Samples

Perform the following steps in class II biosafety cabinet.

1. Design layout of the array. Take following factors into consideration to minimize technical errors introduced into experiment:

    (a) Each slide includes a reference control sample;

    (b) Samples are duplicated in wells printed by different pins;

    (c) Samples from the same patient are performed on the same slides.

2. Pipet 100 μL serum samples and 100 μL reference into their assigned wells. Use care to prevent bubble formation.

3. Seal the slides and incubate in an orbital shaker at 100 rpm and 37 °C for 4 h.

4. Discard serum samples and immediately rinse with 200 μL/ well washing buffer for three times (*see* **Note 10**). Repeat the wash cycle three times, let it sit for 2 min before discarding the solution. After each wash cycle, invert the module and tap vigorously on paper towel to remove excess liquid.

3.2.5 Detecting with Fluorophore-Labeled Secondary Antibodies

1. Prepare a fresh antibody incubation buffer: PBS with 1 % BSA and 3 % HSA.

2. Mix DyLight 549 anti-human IgG and DyLight 649 anti-human IgM (1:500 dilutions for both antibodies) in buffer and cover with aluminum foil. Depending on the scanner, two or more detecting antibodies could be used simultaneously (*see* **Note 11**).

3. Pipet 100 μL detecting reagent into each well and seal the module with adhesive strip.

4. Incubate in an orbital shaker at 100 rpm and 37 °C for 2 h. Cover the module with aluminum foil to prevent photobleaching.

5. Discard the solution and wash with 7×200 μL/well washing buffer. Let the solution sit for 2 min before discarding the solution for the last three wash cycles. After each wash cycle, invert the module and tap vigorously on paper towel to remove excess liquid.

3.2.6 Preparing Slides for Scanning

1. Carefully remove the slide from the slide module, transfer it into a bath containing washing buffer, and let it soak for 5 min. Make sure the printed surface is facing up. Only hold the slide by the edges to avoid smudge of the printed array surface.

2. Remove the slide from bath and place it into a 50 mL tube.

3. Dry the slide by centrifuging at 1000 rpm for 5 min.

4. Scan the slide in fluorescent scanner at both high and low PMT voltage settings. The scanning resolution is set to 10 μm or finer. Save the generated .tiff files for future data analysis (*see* **Note 12**).

3.3 Quantification of Microarray Data

3.3.1 Create a Template for Microarray Quantification and Align It with Microarray Image

1. Open the image-processing software, GenePix 6.0, and create a GenePix Array List (GAL) file that encodes the size and position of each printed glycan. This will generate a template mask with circular boundaries.

2. Open the pre-scan image from **step 2** in Subheading 3.2.1 and adjust the GAL file so that the mask aligns with the actual spots on the slide. Inspect the whole slide and flag spots as "bad" that are either missing or contain defects in printing. These glycans will be excluded from further analysis. Save these modified settings as a GenePix Setting (GPS) file.

3.3.2 Calculate Signal Intensity for Individual Spot

1. Open image of the high PMT scan from **step 4** of Subheading 3.2.6, and load the GPS file from the previous step. Finely adjust the size and position of the each circle mask so that it fits with individual spot size perfectly (*see* **Note 13**).

2. Use the software to calculate median pixel intensity (MI) and local background pixel intensity (BI) for all of the spots. The signal intensity (SI) of each spot is defined as MI–BI, representing background-corrected signal. Save the result as a GenePix Result (GPR) file and export the data as a .txt file that can be opened in Excel.

3. Open image of the low PMT scan from **step 4** of Subheading 3.2.6, and analyze as above.

4. In Excel, import .txt result files for both high and low PMT settings. The primary data are derived from the high PMT scan. The low PMT scan is only used to correct those spots that are saturated in the high PMT scan. Use the following equations to calculate signal intensities for each spot (*see* **Note 14** for Correction Factor (CF) calculation):

For unsaturated spots (RFU < 50,000): $SI = SI_{\text{high PMT}}$

For saturated spots (RFU >= 50,000): $SI = \text{Correction Factor (CF)} \times SI_{\text{low PMT}}$

3.3.3 Determine the Average Signal Intensity for Each Glycan

1. Average the signal intensity of duplicate spots for each glycan on the same array.

2. Determine a floor value and adjust all signals below this value to floor value. A floor value 0.5–1.0 times higher than background is recommended so that signal noise close to background is excluded from consideration (*see* **Note 15**).

3.3.4 Normalize Slides Using a Reference Serum Sample

If experiments are carried out on different days, the slides used are printed at different times, or different detecting reagents are used, signal intensities may vary between slides. Therefore, signal normalization is recommended when comparing data across different slides. This is done by comparing a reference serum sample on each slide. In our protocol, we selected several glycans on the array that universally bind to serum antibodies (IgG, IgM, and IgA) with mid-intensities (*e.g.*, blood group antigens, alpha-Gal antigens). The median signal for these selected glycans on each slide is adjusted to a standard value. For example, we use 50,000 RFU for IgM and 15,000 RFU for IgG. These values are for comparison purpose only and have no physical meanings.

1. For the reference sample on each slide, determine the median intensity for the selected glycans. Calculate the ratio of this median value to the standard value. This ratio is used as normalization factor (NF) for this slide.

2. Normalize all signal intensities on this slide by dividing by NF.

3. Normalize all slides in the same manner.

3.3.5 Average Signals for Duplicate Serum Samples

In a single experiment, each serum sample is measured in two wells printed by different pins, and each pin prints glycans in duplicate. That is, the same antigen-antibody interaction is measured four times per experiment. This accounts for any printing variabilities within and between pins. Final data are Log-transformed to facilitate analysis of both high signals and low signals on the same scale.

1. Average each glycan signal resulting from duplicate serum samples.

2. Apply a Log (base 2) transformation to the final data.

3.4 Post-assay Data Analysis

There are three general strategies used to identify carbohydrate antigens using glycan arrays: identifying glycans bound by monoclonal antibodies, comparing antibody profiles in case subjects versus control subjects, and profiling changes within an individual before and after a stimulus. The approach for evaluating the glycan array data varies based on the type of experiment.

For monoclonal antibodies, one typically compares the data of the antibody of interest to a negative control, such as non-carbohydrate-specific isotype control or buffer. The positive binders for the antibody but not the control provide a list of potential antigen targets [26]. Potential antigens should then be confirmed using other methods.

Case-control studies are used to compare antibody profiles in two different groups to identify specific antibody populations that are unique to one of the groups. This approach is most frequently used to identify disease-specific antigens, although they could be used to identify responses to vaccination or other stimuli [for some recent examples, *see* [18–21]]. When comparing two groups of people, natural variation between individuals can lead to differences between groups by chance. Therefore, it is important to have enough subjects in each group to identify statistically significant differences and to properly matched subjects. For example, factors such as race, age, gender, and blood type can influence anti-glycan antibody profiles [27–30], and numerous other factors that have not been extensively studied, such as diet, smoking, and season of blood draw, may also affect anti-glycan antibodies. Care should be taken to avoid misinterpretation of the data [31]. Finally, technical variability in the measurement can also affect the results. To minimize the effects of day-to-day technical variability, one should randomize the order in which cases and controls are assayed.

Carbohydrate antigens can also be identified by monitoring changes within individuals over time [for some recent examples, *see* [22–24]]. For example, one can compare antibody profiles before and after vaccination or acquisition of a disease (*see* Fig. 3). Since the samples being compared come from the same person, differences due to factors such as genetics, smoking, and diet are typically much smaller than when comparing profiles between individuals. However, technical variability is compounded since one is comparing two different measurements, each with their own error. In addition, one must define the natural biological fluctuations in antibody levels over the relevant time frame. One generally focuses on changes in antibody levels that are larger than the natural variation over time. It is important to note that the immune responses to a stimulus can vary from one patient to another. Hence, during data analysis it may be necessary to evaluate patients in subgroups according to their age, gender, or blood types to identify clinically relevant antigens [32].

The identified antibody responses provide insights into the corresponding antigen repertoires that trigger antibody response. However, one has to be aware that the structures of glycans on the array may not represent the true antigens present on the cell surface of vaccine or pathogen samples. Fortunately, diversity of glycans on the array can be harnessed to identify the specific motifs interacting with the bound antibodies. A comparison of binding to an antigen of interest to its structurally similar analogs on the array can facilitate rapid identification of epitopes that induce the

Pre-vaccination   **Post-vaccination**

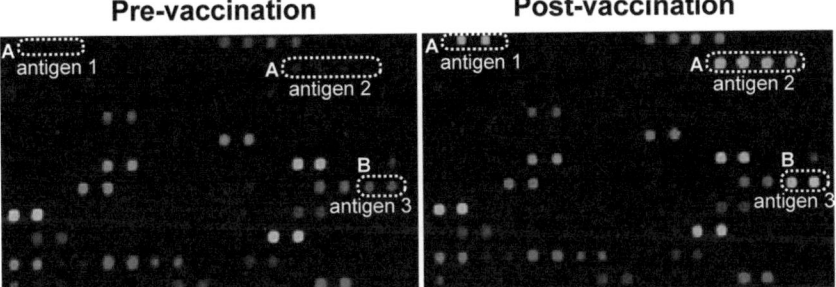

Fig. 3 Example of different types of antibody responses observed before and after vaccination in the same patient. (**a**) New antibodies to antigen 1 and 2 are produced (from no binding to binding); (**b**) Existing antibodies to antigen 3 increase in signal (from weak binding to strong binding. IgG signals are shown in *green* and IgM signals are shown in *red*

specific antibody response. For instance, antibody binding to both trisaccharide (Rhaα1-3-Glcβ1-4Glc) and disaccharide (Rhaα1-3-Glc) but not to monosaccharide (Rha) suggests the minimal binding epitope is a disaccharide unit [33]. Once the antigen repertoires are identified, additional follow-up experiments are needed to trace their source of origin and confirm their true presentations on the cell surface.

3.5 Technical Challenges

Several factors can give rise to technical challenges when comparing samples [20, 29, 34]. The most important factor is the quality and reproducibility of the printed slides. Batches printed by different pins or at different times may introduce variation in the measurements. In addition, variation can also arise when using different detection reagents. When multi-well array slides are used, a reference sample can be included in one well on each slide and used to normalize all data on that slide. In this way, signals from different slides can be normalized to the same reference allowing for comparison between different slides. The use of a standard reference sample also allows one to more readily detect printing problems. To further minimize technical variability, serum samples can be run on multiple slides or wells (*e.g.*, samples are assayed in different wells printed by different pins in order to minimize variation between pins). The process used for quantification can also affect the results of measurements. Spot boundaries in the imaging software must be positioned correctly and sized appropriately for precise quantification.

Finally, statistical analysis is an essential tool to determine whether a response is random or specific. Since many individual comparisons are carried out in microarray experiments, one frequently observes statistically significant differences/changes by chance. It is recommended to perform analysis in two stages. First, the analysis method is applied to a training set of patients to identify specific hits. These hits are then verified using a second, separate patient population. This approach helps to minimize false positive results arising from the data analysis [35].

4   Notes

1. This dye is to help visualize the printed spots. Any dye can be used as long as it meets the following criteria: excitation and emission wavelengths match with scanner; water soluble; compatible with glycans; can be easily washed off; minimum photobleaching over time.

2. We routinely use 8 μL per sample to print five batches of 30 slides. Less sample volume can be used depending on the number of slides, as long as evaporation problem is minimized.

3. Pause printing if pins are clogged. Use a forceps to move the pin up and down 20–30 times until pin moves freely. Then resume printing.

4. The protocol for quality check of microarray slides has been published [25].

5. Use globulin-free BSA and HSA to avoid high background binding to the slide.

6. Rapid drop of blocking buffer will smear the array and create comet tail. Use of gel-loading pipet tips is recommended for this step.

7. After adding blocking buffer, keep the module still without shaking. Disturbing the module during blocking will disrupt printed array.

8. Addition of HSA and BSA serves two purposes: prevent binding of anti-albumin antibodies to the carrier protein core of the neoglycoproteins that can be present in human serum; block nonspecific binding to the array.

9. Concentration of serum samples needs to be optimized for different samples. However, reference sample is always diluted to 1:50. It is used as a control across different slides.

10. This quick wash step is to minimize cross-contamination between wells.

11. Cross-reactivity of the secondary antibody with glycans on the array should be checked by incubating the secondary antibody on the array without any prior serum incubation.

12. We routinely use combinations of PMT voltages of 520/430. However, these settings need to be adjusted for different scan-

ners. The aim is to detect low intensity signals at high PMT scan, and then acquire data for high intensity signals at low PMT (signal that are saturated at high PMT). At low PMT scan, no spots should be saturated.

13. This step is critical in data analysis as the size of the spot boundary can dramatically change measured signal intensity, hence, influences experiment results. Only allow spots to be resized to 20 % below the actual print size.

14. Identify those spots with mid-range intensities (at 30–50 % saturation) at high PMT. For these spots, calculate the intensity ratios at high PMT vs. low PMT. The average ratio is then used as a correction factor (CF). This approach is to extend the dynamic range of the microarray scanner [36].

15. This step is mainly to adjust for negative values resulting from background correction. Negative values interfere with future mathematic calculations.

Acknowledgement

This work was supported by Intramural Research Program in National Institutes of Health.

References

1. Dennis JW, Granovsky M, Warren CE (1999) Glycoprotein glycosylation and cancer progression. Biochim Biophys Acta 1473:21–34

2. Roy R (2004) New trends in carbohydrate-based vaccines. Drug Discov Today Technol 1:327–336

3. Lucas AH, Apicella MA, Taylor CE (2005) Carbohydrate moieties as vaccine candidates. Clin Infect Dis 41:705–712

4. Cummings RD (2009) The repertoire of glycan determinants in the human glycome. Mol Biosyst 5:1087–1104

5. Cummings RD, Pierce JM (2014) The challenge and promise of glycomics. Chem Biol 21:1–15

6. Robinson WH (2006) Antigen arrays for antibody profiling. Curr Opin Chem Biol 10:67–72

7. Mattoon D, Michaud G, Merkel J et al (2005) Biomarker discovery using protein microarray technology platforms: Antibody-antigen complex profiling. Expert Rev Proteomics 2: 879–889

8. Horlacher T, Seeberger PH (2008) Carbohydrate arrays as tools for research and diagnostics. Chem Soc Rev 37:1414–1422

9. Liu Y, Palma AS, Feizi T (2009) Carbohydrate microarrays: Key developments in glycobiology. Biol Chem 390:647–656

10. Park S, Gildersleeve JC, Blixt O et al (2013) Carbohydrate microarrays. Chem Soc Rev 42:4310–4326

11. Shin I, Park S, Lee MR (2005) Carbohydrate microarrays: An advanced technology for functional studies of glycans. Chemistry 11: 2894–2901

12. Rillahan CD, Paulson JC (2011) Glycan microarrays for decoding the glycome. Annu Rev Biochem 80:797–823

13. Donczo B, Kerekgyarto J, Szurmai Z et al (2014) Glycan microarrays: New angles and new strategies. Analyst 139:2650–2657

14. Geissner A, Anish C, Seeberger PH (2014) Glycan arrays as tools for infectious disease research. Curr Opin Chem Biol 18:38–45

15. Oyelaran O, Gildersleeve JC (2007) Application of carbohydrate array technology to antigen discovery and vaccine development. Expert Rev Vaccines 6:957–969

16. Gildersleeve JC, Wang B, Achilefu S et al (2012) Glycan array analysis of the antigen rep-

ertoire targeted by tumor-binding antibodies. Bioorg Med Chem Lett 22:6839–6843

17. Hong X, Ma MZ, Gildersleeve JC et al (2013) Sugar-binding proteins from fish: Selection of high affinity "lambodies" that recognize biomedically relevant glycans. ACS Chem Biol 8:152–160

18. Wang CC, Huang YL, Ren CT et al (2008) Glycan microarray of Globo H and related structures for quantitative analysis of breast cancer. Proc Natl Acad Sci U S A 105:11661–11666

19. Padler-Karavani V, Hurtado-Ziola N, Pu MY et al (2011) Human xeno-autoantibodies against a non-human sialic acid serve as novel serum biomarkers and immunotherapeutics in cancer. Cancer Res 71:3352–3363

20. Jacob F, Goldstein DR, Bovin NV et al (2012) Serum antiglycan antibody detection of nonmucinous ovarian cancers by using a printed glycan array. Int J Cancer 130:138–146

21. Chen K, Gentry-Maharaj A, Burnell M et al (2013) Microarray glycoprofiling of ca125 improves differential diagnosis of ovarian cancer. J Proteome Res 12:1408–1418

22. Liao SF, Liang CH, Ho MY et al (2013) Immunization of fucose-containing polysaccharides from Reishi mushroom induces antibodies to tumor-associated Globo H-series epitopes. Proc Natl Acad Sci U S A 110:13809–13814

23. Luyai AE, Heimburg-Molinaro J, Prasanphanich NS et al (2014) Differential expression of anti-glycan antibodies in schistosome-infected humans, rhesus monkeys and mice. Glycobiology 24:602–618

24. Campbell CT, Gulley JL, Oyelaran O et al (2014) Humoral response to a viral glycan correlates with survival on PROSTVAC-VF. Proc Natl Acad Sci U S A 111:E1749–E1758

25. Zhang Y, Gildersleeve JC (2012) General procedure for the synthesis of neoglycoproteins

and immobilization on epoxide-modified glass slides. Methods Mol Biol 808:155–165

26. Di Cristina M, Nunziangeli L, Giubilei MA et al (2010) An antigen microarray immunoassay for multiplex screening of mouse monoclonal antibodies. Nat Protoc 5:1932–1944

27. Dotan N, Altstock RT, Schwarz M et al (2006) Anti-glycan antibodies as biomarkers for diagnosis and prognosis. Lupus 15:442–450

28. Huflejt ME, Vuskovic M, Vasiliu D et al (2009) Anti-carbohydrate antibodies of normal sera: Findings, surprises and challenges. Mol Immunol 46:3037–3049

29. Oyelaran O, McShane LM, Dodd L et al (2009) Profiling human serum antibodies with a carbohydrate antigen microarray. J Proteome Res 8:4301–4310

30. Muthana SM, Gildersleeve JC (2014) Glycan microarrays: Powerful tools for biomarker discovery. Cancer Biomark 14:29–41

31. Obukhova P, Piskarev V, Severov V et al (2011) Profiling of serum antibodies with printed glycan array: Room for data misinterpretation. Glycoconj J 28:501–505

32. Campbell CT, Gulley JL, Oyelaran O et al (2013) Serum antibodies to blood group a predict survival on prostvac-vf. Clin Cancer Res 19:1290–1299

33. Martin CE, Broecker F, Oberli MA et al (2013) Immunological evaluation of a synthetic clostridium difficile oligosaccharide conjugate vaccine candidate and identification of a minimal epitope. J Am Chem Soc 135:9713–9722

34. Wang L, Cummings RD, Smith DF et al (2014) Cross-platform comparison of glycan microarray formats. Glycobiology 24:507–517

35. Taylor JM, Ankerst DP, Andridge RR (2008) Validation of biomarker-based risk prediction models. Clin Cancer Res 14:5977–5983

36. Lyng H, Badiee A, Svendsrud DH et al (2004) Profound influence of microarray scanner characteristics on gene expression ratios: Analysis and procedure for correction. BMC Genomics 5:10

Antibody-Carbohydrate Recognition from Docked Ensembles Using the AutoMap Procedure

Tamir Dingjan, Mark Agostino, Paul A. Ramsland, and Elizabeth Yuriev

Abstract

Carbohydrate-protein recognition is vital to many processes in health and disease. In particular, elucidation of the structural basis of carbohydrate binding is important to the development of oligosaccharides and oligosaccharide mimetics as vaccines for infectious diseases and cancer. Computational structural techniques are valuable for the study of carbohydrate-protein recognition due to the challenges associated with experimental determination of carbohydrate-protein complexes. AutoMap is a computer program that we have developed to study protein-ligand recognition. AutoMap determines the interactions taking place in a set of highly ranked poses obtained from molecular docking and processes these to identify the protein residues most likely to be involved in interactions. In this protocol, we describe the use of AutoMap and illustrate its suitability for studying antibody recognition of the Lewis Y tetrasaccharide, which is a potential cancer vaccine antigen.

Key words Site mapping, Protein-ligand interactions, Carbohydrate recognition, Ligand docking

1 Introduction

Recognition of carbohydrates by proteins underpins many processes in host-pathogen interactions [1]. Viral infection relies on attachment to and subsequent invasion of human cells; this initial step is often mediated by carbohydrate-protein recognition [2] (such as in influenza hemagglutinin recognition of sialic acid [3] or in noroviral binding to blood group antigens [4]). Similarly, cancer cells aberrantly display antigens, such as Lewis blood group carbohydrates, on their surface [5]. These disease-associated carbohydrate markers are potential candidates for development of antibody-based therapies and vaccines.

Carbohydrate-based vaccines rely on the recognition of oligosaccharide antigens by immune system proteins. The rational design of such vaccines can be aided with specific knowledge of

Bernd Lepenies (ed.), *Carbohydrate-Based Vaccines: Methods and Protocols*, Methods in Molecular Biology, vol. 1331, DOI 10.1007/978-1-4939-2874-3_4, © Springer Science+Business Media New York 2015

how the binding site residues recognize the antigen, and to what extent each residue contributes to the overall recognition event. The knowledge of carbohydrate-protein recognition is particularly valuable in designing vaccines based on carbohydrate mimetics [6].

Structural analyses are accomplished experimentally by X-ray crystallography and nuclear magnetic resonance spectroscopy. Computational structural approaches complement these methods and are usually less expensive and more rapid to deploy. To facilitate the computational study of carbohydrate-protein interactions, we developed a site mapping approach, which is based on our earlier work with antibody-binding peptides identified by combinatorial peptide chemistry approaches [7, 8]. Following this early work, we have implemented site mapping for carbohydrates [9, 10] and since then have extensively validated it for the study of a wide variety of protein-carbohydrate systems [11–15]. The automation of our method is implemented in AutoMap, the program demonstrated here. For accompanying information and in-depth discussion regarding AutoMap's design and function, please refer to Agostino et al. [16].

AutoMap is a software package that can be used to analyze protein-ligand complexes. The software identifies binding site amino acids that contribute to recognition, and to what extent. This data can be used to construct a per-residue rationale for protein-ligand recognition. AutoMap is designed to take as its input molecular structure files, such as might be generated by a docking program or obtained from the Protein Data Bank [17]. AutoMap works by tallying and normalizing the hydrogen bonding and van der Waals (vdW) interactions determined in an ensemble of docked poses for a given protein-ligand system. The normalized tallies are used to identify the key protein residues involved in ligand recognition and are included into site maps based on cumulative sum cutoffs, which are optimized to reproduce experimentally observed interactions in specific protein-ligand systems, e.g., in antibody-carbohydrate complexes. A summary of the protocol is provided in Fig. 1.

In this protocol, we have provided step-by-step instructions to show how AutoMap can be applied to study a protein-carbohydrate complex. Many of the steps shown here are also described in the AutoMap manual, which is included in the downloadable package.

To exemplify the use of AutoMap, we have selected the complex of hu3S193 Fab with the tetrasaccharide Lewis Y antigen, a system of interest to the development of epithelial carcinoma vaccines. While the 1.9 Å resolution crystal structure of this complex is available (PDB ID: 1S3K) [18], the AutoMap method is applicable to cases where only the crystal structure of a native protein is available or a homology model is generated.

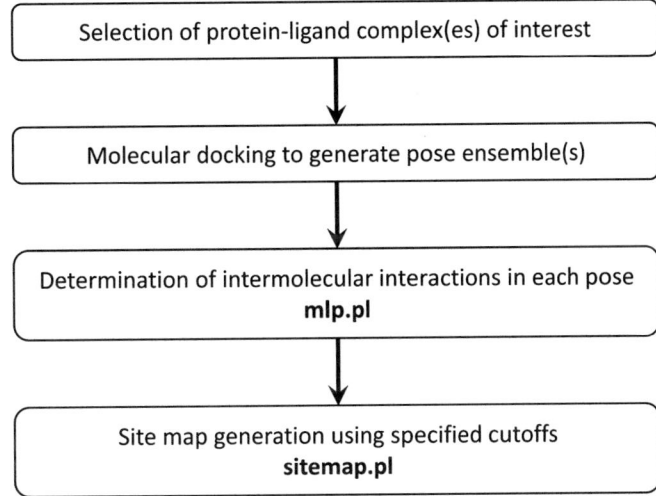

Fig. 1 A flowchart describing the AutoMap process. *See* Subheadings 3.2, 3.3, and 3.5

2   Software Requirements

AutoMap is written in Perl and requires some supporting software utilities to run. These software utilities and the operating environment requirements for AutoMap are detailed below. AutoMap is available from http://sourceforge.net/projects/ligmap/.

2.1   Operating Environment

AutoMap has been tested in openSUSE 10.x, 11.x and recent editions of Ubuntu (12.04 onwards), but should run in any modern Linux distribution. AutoMap can also be used in the Windows environment. To use AutoMap in Windows, Cygwin is required (www.cygwin.com), with the following additional packages also required to be installed within Cygwin:

- dos2unix. a utility used to convert text files from DOS or Mac formats to UNIX formats
- tcsh. a UNIX shell
- perl: the Perl interpreter

2.2   Required Utilities

2.2.1   LigPlot+

LigPlot+ [19] is used by AutoMap to generate interaction data from the molecular structure files. The package is freely available to academic users following registration from https://www.ebi.ac.uk/thornton-srv/software/LigPlus/. To enable LigPlot+ to identify a wide variety of functional groups and residues, it is recommended to obtain the Het Group Dictionary from the PDB, available at http://rcsb-deposit.rutgers.edu/het_dictionary.txt.

The Silico toolkit [20] is used by AutoMap to convert molecular structure file formats and perform other common analysis tasks. The package is available from http://silico.sourceforge.net/. Follow the instructions included with the Silico toolkit to install it.

2.3   Recommended Utilities

2.3.1   3D Visualization Software

While not required to run AutoMap, it is recommended that the user has a molecular visualization program installed. We suggest the use of either Maestro [21] or PyMOL [22], which are freely available to academic users:

- Maestro: http://www.schrodinger.com/freemaestro/.

- PyMOL Molecular Graphics System: http://sourceforge.net/projects/pymol/, Linux users may also install this via their package manager.

In the example, steps requiring molecular visualization software are carried out with PyMOL.

2.3.2   Software for Manipulation of 2D Binding Site Interaction Diagrams

In addition to generating interaction data, LigPlot+ can produce 2D binding site diagrams showing intermolecular interactions between the ligand and receptor. These diagrams are saved into two files, a PostScript file for immediate viewing (**ligplot.ps**) and a second file that can be edited using the LigPlot+ graphical user interface (**ligplot.drw**). Java is required to use the LigPlot+ graphical user interface (contained in the LigPlot+ package as **LigPlus.jar**). PostScript files may be further edited using a range of graphical processing tools, such as Inkscape, available from www.inkscape.org (or within the package manager of common Linux distributions).

2.4   Molecular Docking Output

AutoMap requires as input a receptor coordinate file and the coordinates of multiple docked ligands. Therefore, the user should have at least one molecular docking program available. We have tested [15, 16] AutoMap with input from Glide [23], GOLD [24], AutoDock [25, 26], DOCK [27], and FRED [28]. However, AutoMap should be compatible with output from any molecular docking program that writes molecular structures as its output. The output from some programs requires additional processing steps before it can be used correctly with AutoMap; the steps to achieve this are described in the AutoMap manual. In the example, the use of pose input from Glide is described.

3   Methods

3.1   Installation

1. Download AutoMap from the link above. Save the file as **automap.zip**.

2. Extract the contents of the archive file to a folder containing the protein receptor and docked ligand coordinate files (the working directory)

3. Download LigPlot+ from the link above. Save the file as **ligplus.zip** (*see* **Note 1**).

4. Extract the contents of the archive file to a temporary folder. The executable files for LigPlot+ are in the /lib folder of the archive, categorized by operating system. Identify the folder appropriate to your system and copy all the files in it to the working directory.

5. Copy the LigPlot+ parameter file to the working directory (located in /lib/params/ligplot.prm of the LigPlot+ package).

6. Optionally, download the Het Group Dictionary from the link above and save it in the working directory (**het_dictionary.txt**).

3.2 Preparing Receptor Input Files

The protein and docked ligand ensemble must be prepared as separate files. Prepare the protein file using the following procedure:

1. In Maestro, ensure that your protein has chain identifiers and that the residue numbers do not exceed 999. If the structure does not feature chain identifiers, add these. For proteins with residues numbered greater than 999, renumber these, using multiple chain identifiers to break up the structure as required.

2. Save the protein coordinates used in the docking simulation as a PDB file.

3. Open the saved PDB file in a text editor.

4. Remove all noncoordinate lines from the file (lines beginning with anything other than "ATOM"). Ensure all lines beginning with "TER", "ENDMDL", or "END" are removed in this step.

5. Replace all "HETATM" line headers with "ATOM ". Note the extra two spaces after "ATOM", required to preserve column widths in the PDB format.

6. Add a final "TER" line to the file. This line should contain the "TER" header and the atom number one greater than the final "ATOM" line. For example, if the final "ATOM" line is numbered 3491, the added "TER" line should read: "TER 3492". This "TER" line marks the end of the file.

7. Save the resulting file as **<protein>.pdb**, where <protein> is the name of the protein. *See* Fig. 2a for an example of a prepared protein file for the Lewis Y-hu3S193 Fab complex.

3.3 Preparing Ligand Input Files

Docking output files from Glide, GOLD, DOCK, AutoDock, and FRED are all compatible with AutoMap. It is highly recommended that, prior to docking, the ligand is set up with appropriately defined residues; this entails numbering the residues starting from 1 and ensuring unique residue names for each residue. For example, trimannose (MAN-MAN-MAN) should be given different residue names for each mannose unit (such as MAA, MAB, and MAC).

a

```
ATOM      1  N   GLU H   1     -27.713  20.568  15.957  1.00 33.77        N1+
ATOM      2  CA  GLU H   1     -27.049  21.879  15.711  1.00 32.60        C
ATOM      3  C   GLU H   1     -27.676  22.594  14.523  1.00 29.93        C
                                    . . . . .
ATOM   3502 HG21 THR L 112       6.701  40.890  13.340  1.00  0.00        H
ATOM   3503 HG22 THR L 112       6.977  39.205  13.697  1.00  0.00        H
ATOM   3504 HG23 THR L 112       6.812  39.734  12.020  1.00  0.00        H
TER    3505
```

b

```
HEADER    1S3K
COMPND    1S3K
REMARK       SILICO DATA
REMARK       SCLR 'entry_id' '2'
                              . . . . .
REMARK       SCLR 'PDB_EXPDTA_TEMPERATURE' '100'
REMARK       END SILICO DATA
ATOM      1  C1  FUC   223    -36.572  46.259  15.173  0.00  0.00         C
ATOM      2  C2  FUC   223    -37.666  47.292  15.505  0.00  0.00         C
ATOM      3  C3  FUC   223    -38.854  46.588  16.175  0.00  0.00         C
                              . . . . .
ATOM     90  H6  FUC   226    -30.670  39.053  13.810  0.00  0.00         H
ATOM     91  H7  FUC   226    -30.160  40.864  14.625  0.00  0.00         H
CONECT    1    2   10   11   28
CONECT    2    1    3    7   12
                              . . . . .
CONECT   91   79
END
```

Fig. 2 An example of the processed structure files. (**a**) The hu3S193 (PDB ID: 13SK) protein structure file (**<protein>.pdb**) showing the first and last few lines of the final output. Note that the processed file begins with an ATOM line and ends with a TER line. (**b**) The Lewis Y tetrasaccharide structure file (**<ensemble>_lib. pdb**), as generated by the Silico toolset [20], showing samples of each section of a particular ligand entry. Note that this file contains structural information for multiple ligand conformations. The file commences with a HEADER line, while each ligand entry begins with a COMPND line. If additional information about the ligand is present in the file supplied for conversion, this is retained in the resulting file in a series of REMARK lines. ATOM lines list the atomic coordinates and CONECT lines detail chemical bonds. Each ligand entry is terminated by an END line. Only the COMPND and ATOM lines are used by the interaction analysis script; all other lines are ignored. COMPND lines indicate the commencement of each ligand entry in the ligand structure file and thus allow the file to be separated for LigPlot+ analysis, while ATOM lines contain the atomic coordinates describing each ligand structure

Chain identifiers must also be removed from the ligand. Additional preparatory steps are required to use AutoDock files (*see* **Note 2**).

An ensemble of poses for a specific ligand must be saved as a separate file. Site mapping can be done for a specific ligand (as in the example presented here) or for a range of related ligands (e.g., major carbohydrate xenoantigens all bearing terminal Galα1-3Gal epitopes [11]).

Prepare the ligand ensemble using the following procedure:

1. Import the docked ligand ensemble into Maestro (*see* **Note 3** for alternative procedure if Maestro is unavailable).

2. Select the desired docked ligand entries in the Project Table and export as a single uncompressed Maestro file **<ligands>. mae**, where <ligands> is the name of the ligand(s).

3. Open a terminal in the directory containing the ligand ensemble and execute:

> read_write_mol2 **<ligands>.mae**

where <ligands> is the name of the ligand ensemble file. This step creates a file called **<ligands>_new.mol2**.

4. Next, execute the following:

> read_write_pdb **<ligands>_new.mol2**

This creates a file called **<ligands>_new_new.pdb**.

5. Rename **<ligands>_new_new.pdb** to **<ensemble>_lib.pdb**, where <ensemble> is the name of the ligand, or any other identifying label for the ligand ensemble. The "**_lib**" suffix must be present as it is used by AutoMap to identify the ligand file.

6. Remove any underscores from ligand names.

See Fig. 2b for an example of a prepared ligand file for the Lewis Y-hu3S193 Fab complex.

A further processing step is required if using docked ligand output from GOLD (*see* **Note 4**)

If the ligand structure input files have been generated using Windows, the Dos2Unix utility must be used to convert the files into a Linux-readable format. This can be executed in Cygwin using the command:

> dos2unix <file>

Run this command on the **<ensemble>_lib.pdb** and **<protein>. pdb** files.

3.4 Optimizing Interaction Cutoffs

This part of the protocol is optional but highly advisable. If it is skipped, the default cutoffs of 80/80 for hydrogen bonding and vdW interactions (*see* **Note 5**) could be used for running AutoMap and visualizing site maps (Subheadings 3.5, **step 2**, and 3.6 below). A summary of the cutoff optimization procedure is provided in Fig. 3. If the user wants to run AutoMap without cutoff optimization, please proceed directly to Subheading 3.5.

To optimize the cutoff values, the AutoMap package includes a script called **optcutoff.pl**. For detailed justification and discussion of cutoffs, refer to Agostino et al. [16]. The following procedure describes the process of optimization:

1. For a given protein-ligand system (e.g., antibody-carbohydrate system), select a range of high resolution crystal structures of protein-ligand complexes.

2. For each protein-ligand complex to be used for cutoff optimization, prepare an ensemble of docked ligand poses. Ligands should be docked into the protein file extracted from the crystal

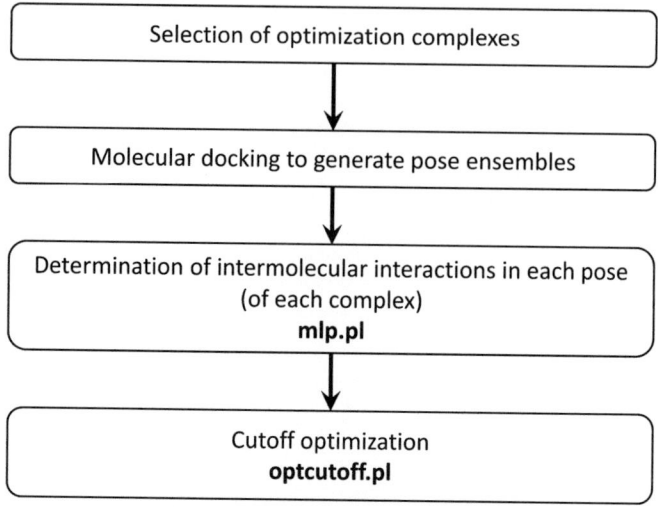

Fig. 3 A flowchart detailing the cutoff optimization process. *See* Subheading 3.4

structure of the protein-ligand complex (referred to below as **<protein-ligand>.pdb**).

3. Prepare the protein and ligand files as described above (Subheadings 3.2 and 3.3).

4. For each protein-ligand complex, execute the interaction analysis script:

    > ./mlp.pl ./**<protein>.pdb**

    Before executing the interaction analysis script on each subsequent protein-ligand complex, rename the resulting **allsum.csv** file to **<protein>_allsum_VAL.csv**.

5. For each protein-ligand complex, a list of interacting residues in the crystal structure of the complex must be generated. The file must be named **<protein>.lst** and be placed in the working directory. To create this file:

    (a) Execute LigPlot+ on the complex (the original PDB file of the crystal structure is suggested):

        > ligplot **<protein-ligand>.pdb** *aaa bbb*

        where *aaa* and *bbb* are the first and last ligand residue numbers. This step lists the interacting residues in the file **ligplot.sum**

    (b) Open **ligplot.sum** in a text editor. Create a new file named **<protein>.lst**. Note that each interaction features a protein and ligand residue. For each interaction residue, copy the protein residue information to the **<protein>.lst** file. The file should contain as many entries as there are interactions. No distinction between residues involved in hydrogen bonding or vdW contacts is made. Each entry must be formatted as below:

RESxxxAC

where RES is the three-letter residue ID, xxx is the residue number, A is a letter that may appear after the number, and C is the chain identifier. Note that the spacing of these elements should be maintained even if some fields are left blank. Some examples of this format:

ASP 5 A: Aspartate 5 of chain A

GLY 27 L: Glycine 27 of chain L

TRP100AH: Tryptophan 100A of chain H

6. After performing **steps 1–5** above for each protein-ligand complex, used for cutoff optimization, execute the following:

> ./optcutoff.pl <resolution>

where <resolution> is the interval at which cutoff values will be tested (*see* **Note 6**).

Once this process is complete, the optimal cutoff values will be displayed on-screen. Record these values; these are to be used when generating site maps for this type of protein-ligand system (*see* Subheading 3.5, step 2, below). These values are only applicable to protein-ligand systems similar to the optimization system.

3.5 Running AutoMap

For a brief summary of AutoMap's operation, *see* **Note 7**.

Ensure that all of the LigPlot+ and AutoMap files have the executable attribute set (**mlp.pl, optcutoff.pl, sitemap.pl**) (*see* **Note 8**).

To run AutoMap, use the following procedure:

1. Open a terminal and execute (for further options, *see* **Note 9**):

> ./mlp.pl ./**<protein>.pdb**

2. The second step of AutoMap implementation identifies protein residues involved in ligand recognition and their relative contributions to hydrogen bonding and vdW interactions (*see* **Note 10** for a brief description of this process). This step outputs a text file listing these residues (option (a) below) as well as additional protein PDB files, which could be used for visualizing site maps (option (b) below).

(a) Once **step 1** is complete, execute:

./sitemap.pl <hb-cutoff><vdw-cutoff>

In this command, <hb-cutoff> and <vdw-cutoff> are the cutoff values determined above (Subheading 3.4, **step 6**). This step generates an output called **site.map**; *see* Fig. 4 for an example output file generated for the Lewis Y-hu3S193 Fab complex.

TYR 35	H	33.3333333333333	12.3893805309735
SER104	H	20.5128205128205	3.53982300884956
ARG101	H	10.2564102564103	7.9646017699115
THR100	H	10.2564102564103	5.30973451327434
TYR 37	L	10.2564102564103	7.9646017699115
TYR 33	H	5.12820512820513	13.2743362831858
TRP105	H	0	10.6194690265487
HIS 31	L	5.12820512820513	7.9646017699115
GLY103	H	0	7.07964601769912
TYR 50	H	0	6.19469026548673
PHE101	L	0	5.30973451327434
ASP102	H	0	3.53982300884956

Fig. 4 An example site map file of the crystallographic Lewis Y-hu3S193 Fab complex showing interacting amino acids, and their relative contributions to hydrogen bonding (middle column) and van der Waals interactions (last column). The values in this file are normalized adjustments to the B-factor values of the protein receptor file, which will yield a visualized site map when applied in Subheading 3.6, **step 2**

(b) Once **step 1** is complete, execute:

>./sitemap.pl <hb-cutoff> <vdw-cutoff>./<**protein**>. **pdb**

This step creates the files <**protein**>_**hbd.pdb** and <**protein**>_**vdw.pdb**, in which the B-factor column values are modified to represent the relative contribution to recognition interactions (*see* **Note 11**).

3.6 Visualizing Site Maps

Most molecular visualization programs can be used to visualize site maps by coloring residues by B-factor/temperature factor. The specific procedure for PyMOL is as follows:

1. Open the two site map files (<**protein**>_**hbd.pdb** and <**protein**>_ **vdw.pdb**) in PyMOL.

2. Execute the following command from the PyMOL terminal:

cmd.spectrum("b","blue_white_red",selection=" <selection-name>",minimum=0,maximum=73)

<selection-name> is the name of the desired entry in PyMOL. This same command can be used for either the hydrogen bonding or vdW site map.

3. Show the surface representation of the colored proteins by applying the surface render.

See Fig. 5 for site maps generated for the Lewis Y-hu3S193 Fab complex. This figure illustrates correspondence between the experimental (Fig. 6) and computational delineation of intermolecular interactions involved in the recognition of the Lewis Y antigen by the hu3S193 monoclonal antibody. Similar mapping of antibody recognition of chemically modified Lewis Y antigens may be of future use in the rational design of carbohydrate-based cancer vaccines.

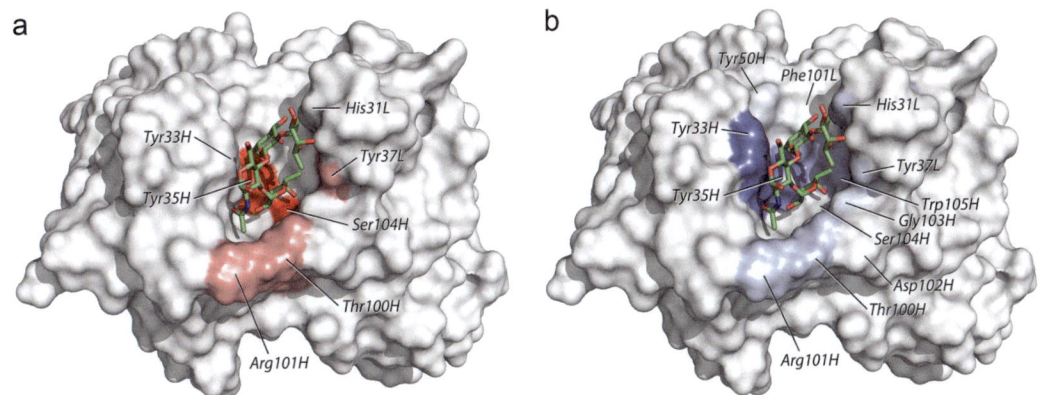

Fig. 5 Application of AutoMap to Lewis Y recognition by the hu3S193 antibody. (**a**) A hydrogen-bonding site map. Residues contributing to hydrogen-bonding interactions are colored in *red*, with more intense color corresponding to more frequent involvement across the ensemble of docked ligand poses. Interacting residues (with n ≥ 7 %): Tyr35H, Thr100H, Arg101H, Ser104H, Tyr37L (*see* **Note 11** for the definition of n). (**b**) A van der Waals site map. Residues contributing to van der Waals interactions are colored in blue, with more intense color corresponding to more frequent involvement across the ensemble of docked ligand poses. Interacting residues (with n ≥ 7 %): Tyr33H, Tyr35H, Arg101H, Gly103H, Trp105H, His31L, Tyr37L. In panels (**a**) and (**b**), the crystallographic structure of Lewis Y bound to hu3S193 Fab is displayed as a stick model (carbon, *green*; oxygen, *red*; nitrogen, *blue*)

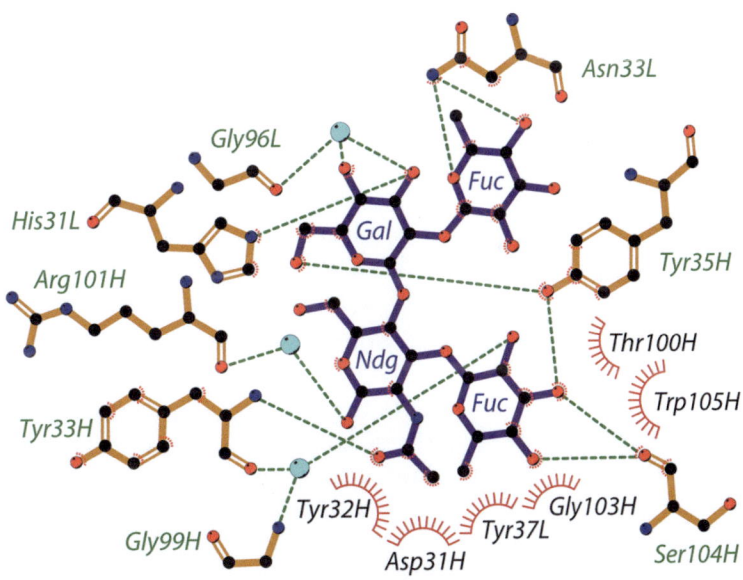

Fig. 6 A LigPlot+ diagram showing the interactions between Lewis Y antigen and antibody residues in the crystallographic complex (PDB ID: 1S3K). *Green lines* represent hydrogen bonds; *red arcs* represent vdW interactions; *cyan circles* represent water molecules. Only hydrogen bonding-mediating water molecules are shown. Carbohydrate residues are labeled with their 3-letter codes from the PDB file: Gal, galactose; Fuc, fucose; Ndg, *N*-acetylglucosamine. Comparison between these experimentally determined intermolecular interactions and the computationally delineated interacting residues shown in Fig. 5a, b shows the correspondence between the binding interactions identified by the AutoMap procedure and those found in the crystallographic complex

4   Notes

1. To download LigPlot+, go to the website listed in the Subheading 2.2.1. Click "Download" in the right-hand margin menu. You will be required to register for a free academic license prior to downloading the software; click on "academic license" and enter your details. Once you have received an access code by email, you will be able to download LigPlot+.

2. To use docking output from AutoDock, additional preparatory steps are required. The AutoMap manual included in the downloadable package contains detailed steps for the preparation of AutoDock-generated poses.

3. Our procedure for the preparation of the ligand ensemble makes use of Maestro and the Silico toolkit. **Steps 1–3** consist of file format conversions to MOL2 format, and can be replaced by equivalent procedures if Maestro is not available. Most molecular visualization programs are able to read and write structures in the MOL2 format. **Steps 4–5** convert the file to the PDB format with the necessary headers, and are highly recommended.

4. To use docking output from GOLD, a single additional step is required. In the ligand ensemble file, the data in the COMPND field should be simplified before proceeding. The COMPND field typically resembles: 1SL5|1sl5_lig|mol2|1|dock1. The pipe symbols in this field will affect AutoMap, so it is necessary to remove them. This can be quickly done by opening the ensemble file in a text editor and replacing all the COMPND field values (i.e., the characters "1SL5|1sl5_lig|mol2|1|dock1") with a noninterfering label such as "ligand".

5. Previously, we have developed site maps for antibody-carbohydrate [11], antibody-ganglioside [13], lectin-carbohydrate [12], and antibody-peptide [29] systems using the default 80/80 cutoffs for the hydrogen bonding and vdW interactions. In all these systems, multiple crystal structures were available to optimize the hydrogen bonding and vdW cutoffs: $80/90, 90/40, 90/-$, and $60/90$ for the four systems, respectively [16]. Recently, we optimized the hydrogen bonding and vdW cutoffs for the protein-glycosaminoglycan system: $90/50$ [15]. In the example, presented here, the $80/90$ cutoffs were used.

6. Cutoff values are tested in the range of 0–100 %. If <resolution> is not specified, a default value of 10 % is used (i.e., a resolution of 10 causes the optimization script to evaluate cutoff values ranging from 0 to 100 in increments of 10). Resolution values should be selected such that 100 divided by the resolution still results in an integer (e.g., 10, 5, 2.5, etc.).

Larger resolution values may return inaccurate cutoff values, and smaller resolutions will increase the time taken for the optimization. In practice, a resolution of 10 takes about 5 min to complete the optimization process, and provides a suitable level of accuracy. The optimization step need only be run once for each protein-ligand system, so it is worthwhile to perform an optimization calculation and use the accurate values produced.

7. AutoMap operates in two main stages; first, the script analyzes the protein-ligand system and tallies all hydrogen bonding and vdW interactions from each docked ligand pose. Second, the site mapping script transfers the recorded interaction information to a site map file. This map can be imprinted on a protein structure file, allowing the interacting residues to be colored according to their contribution to the overall binding interaction.

8. In Linux, files can be made executable using the chmod command: "chmod 755 <file>" will assign execution permissions to the file owner, but not to the group or general access.

9. This script tallies up interactions between the protein and each ligand pose in the docked ligand ensemble(s). All the interactions are saved to a single file for each type of interaction (hydrogen bonds—**allhhb.csv**; vdW—**allnnb.csv**). A summary file containing all the interactions in these two files is also written (**allsum.csv**). To collect the interactions for each protein-ligand complex in separate files, create the following folders in the directory containing the data and AutoMap scripts prior to execution:

Pics: To collect the visual output of LigPlot+ for each complex (**.ps** files).

hhb: To collect the individual **ligplot.hhb** files for each complex (hydrogen-bonding interactions only).

nnb: To collect the individual **ligplot.nnb** files for each complex (vdW interactions only).

sum: To collect the individual **ligplot.sum** files for each complex (summary of hydrogen bonding and vdW interactions).

This information can add several hundred megabytes to the output data, and so is not stored by default.

10. The sitemap.pl script tallies up the interactions occurring in each pose and assigns a percentage contribution to each protein residue (calculated by dividing the number of interactions featuring that residue by the total number of interactions). These normalized percentages are then sequentially added, beginning from the largest. Once the summed total meets the specified cutoff for that interaction type (hydrogen bond or

vdW), all residues already included in the cumulative value are considered significant for recognition.

11. The normalized tallies are converted to "color values" used to visualize B-factor values, allowing more intense colors to be applied to more relevant residues. Namely, atoms with B factors in the 5.0–37.0 range are rendered by most molecular modeling programs according to a blue-to-white ramp, and those in the 37.0–73.0 range according to a white-to-red ramp. This feature is used in the conversion of the normalized tallies to "B factors" for color mapping. Residues which account for ≤5 % interactions are assigned a B factor of 37.0 (i.e., rendered white). Residues which account for ≥20 % hydrogen bonds are assigned a B factor of 73.0 (i.e., rendered red). Residues which account for ≥20 % vdW interactions are assigned a B factor of 5.0 (i.e., rendered blue). For residues that contribute 5–20 % of the total hydrogen bonding or vdW interactions, the B factors are calculated according to the formulas below, where n is the normalized tally for a given residue in percentage terms:

$$B_{HB} = \frac{12 \times n}{5} + 25$$

$$B_{vdW} = \frac{-32 \times n}{15} + \frac{143}{3}$$

Acknowledgements

TD is supported by an Australian Postgraduate Award. MA is a recipient of an NHMRC Early Career Fellowship (GNT1054245). The authors gratefully acknowledge the contribution toward this study from the Victorian Operational Infrastructure Support Program received by the Burnet Institute. We would also like to acknowledge computational resources provided by the Victorian Life Sciences Computational Infrastructure through grant VR0250.

References

1. Gabius H (ed) (2009) The sugar code. Fundamentals of glycosciences. Wiley-VCH, Weinheim

2. Air GM (2011) The role of carbohydrates in viral infections. In: Binghe W, Boons G-J (eds) Carbohydrate recognition. John Wiley & Sons, Inc, New Jersey, pp 65–91

3. Weis W, Brown JH, Cusack S et al (1988) Structure of the influenza virus haemagglutinin complexed with its receptor, sialic acid. Nature 333(6172):426–431

4. Hutson AM, Atmar RL, Estes MK (2004) Norovirus disease: changing epidemiology and host susceptibility factors. Trends Microbiol 12(6):279–287

5. Yuriev E, Farrugia W, Scott AM et al (2005) Three-dimensional structures of carbohydrate determinants of Lewis system antigens: implications for effective antibody targeting of cancer. Immunol Cell Biol 83(6):709–717

6. Agostino M, Sandrin MS, Thompson PE et al (2011) Carbohydrate-mimetic peptides: structural

aspects of mimicry and therapeutic implications. Expert Opin Biol Ther 11(2):211–224

7. Yuriev E, Ramsland PA, Edmundson AB (2001) Docking of combinatorial peptide libraries into a broadly cross-reactive human IgM. J Mol Recognit 14(3):172–184

8. Yuriev E, Ramsland PA, Edmundson AB (2002) Recognition of IgG-derived peptides by a human IgM with an unusual combining site. Scand J Immunol 55(3):242–253

9. Milland J, Yuriev E, Xing PX et al (2007) Carbohydrate residues downstream of the terminal Galα(1,3)Gal epitope modulate the specificity of xenoreactive antibodies. Immunol Cell Biol 85:623–632

10. Yuriev E, Sandrin MS, Ramsland P (2008) Antibody-ligand docking: insights into peptide-carbohydrate mimicry. Mol Simulat 34:461–468

11. Agostino M, Sandrin MS, Thompson PE et al (2009) In silico analysis of antibody-carbohydrate interactions and its application to xenoreactive antibodies. Mol Immunol 47:233–246

12. Agostino M, Yuriev E, Ramsland P (2011) A computational approach for exploring carbohydrate recognition by lectins in innate immunity. Front Immunol 2:23

13. Agostino M, Yuriev E, Ramsland PA (2012) Antibody recognition of cancer-related gangliosides and their mimics investigated using in silico site mapping. PLoS One 7(4), e35457

14. Agostino M, Sandrin MS, Thompson PE et al (2010) Identification of preferred carbohydrate binding modes in xenoreactive antibodies by combining conformational filters and binding site maps. Glycobiology 20:724–735

15. Agostino M, Gandhi NS, Mancera RL (2014) Development and application of site mapping methods for the design of glycosaminoglycans. Glycobiology 24:840–851

16. Agostino M, Mancera RL, Ramsland PA et al (2013) AutoMap: A tool for analyzing protein-ligand recognition using multiple ligand binding modes. J Mol Graph Model 40:80–90

17. Berman HM, Westbrook J, Feng Z et al (2000) The Protein Data Bank. Nucleic Acids Res 28(1):235–242

18. Ramsland PA, Farrugia W, Bradford TM et al (2004) Structural convergence of antibody binding of carbohydrate determinants in Lewis Y tumor antigens. J Mol Biol 340(4):809–818

19. Laskowski RA, Swindells MB (2011) LigPlot+: multiple ligand-protein interaction diagrams for drug discovery. J Chem Inf Model 51(10): 2778–2786

20. Chalmers DK, Roberts BP (2011) Silico—a Perl molecular modelling toolkit.

21. Maestro, version 9.7 (2014). Schrödinger, LLC, New York

22. The PyMOL Molecular Graphics System, Version 1.5.0.4. Schrodinger, LLC, New York.

23. Friesner RA, Banks JL, Murphy RB et al (2004) Glide: a new approach for rapid, accurate docking and scoring. 1. Method and assessment of docking accuracy. J Med Chem 47(7):1739–1749

24. Verdonk ML, Cole JC, Hartshorn MJ et al (2003) Improved protein-ligand docking using GOLD. Proteins 52:609–623

25. Morris GM, Goodsell DS, Halliday RS et al (1998) Automated docking using a Lamarckian genetic algorithm and an empirical binding free energy function. J Comput Chem 19(14): 1639–1662

26. Huey R, Morris GM, Olson AJ et al (2007) A semiempirical free energy force field with charge-based desolvation. J Comput Chem 28(6):1145–1152

27. Lang PT, Brozell SR, Mukherjee S et al (2009) DOCK 6: combining techniques to model RNA-small molecule complexes. RNA 15(6): 1219–1230

28. McGann M (2011) FRED pose prediction and virtual screening accuracy. J Chem Inf Model 51(3):578–596

29. Agostino M, Sandrin MS, Thompson PE et al (2011) Peptide inhibitors of xenoreactive antibodies mimic the interaction profile of the native carbohydrate antigens. Biopolymers 96: 193–206

Chapter 5

Generation of Monoclonal Antibodies against Defined Oligosaccharide Antigens

Felix Broecker, Chakkumkal Anish, and Peter H. Seeberger

Abstract

Unique carbohydrate antigens are expressed on the surface of various pathogens, including bacteria, parasites, and viruses, and aberrant glycosylation is a frequent feature of cancer cells. Antibodies recognizing such carbohydrate antigens may be used for the specific detection of potentially harmful cells, immunohisto-chemistry, and diagnostic and therapeutic applications. The generation of specific and strongly binding antibodies against defined carbohydrate epitopes is challenging, since isolated carbohydrates often suffer from low purity, usually have limited immunogenicity, and induce antibodies of low affinity. We describe a protocol to generate highly affine monoclonal antibodies (mAbs) against pure and defined synthetic carbohydrate antigens. First, an oligosaccharide is covalently coupled to an immunogenic carrier protein to obtain a glycoconjugate. This glycoconjugate is used to raise oligosaccharide specific antibodies in mice, and splenocytes are fused with myeloma cells to form hybridomas. Hybridoma clones producing oligosaccharide-specific mAbs are selected by glycan microarray screening. Selected clones are expanded and mAbs are purified from the cell culture supernatant. This protocol is suitable to procure carbohydrate-specific mAbs of high specificity, selectivity, and affinity that may be useful for a variety of biochemical and medical applications.

Key words Antigen, Carbohydrate, Epitope, Glycan, Glycoconjugate, Hybridoma, mAb, Monoclonal antibody, Oligosaccharide, Pathogen detection

1 Introduction

Unique surface carbohydrate antigens expressed by bacteria, cancer cells, or viruses are valuable targets for monoclonal antibodies (mAbs) that can be used for the specific detection of bacteria [1–4] or cancer cells [5], and show promise as therapeutic agents against various diseases such as cancer [6] and viral infections [7]. However, the generation of strongly binding mAbs against defined carbohydrate antigens is challenging, as isolated carbohydrates often suffer from low purity and limited immunogenicity and usually induce antibodies of low affinity directed against poorly defined epitopes. Generation of anti-carbohydrate mAbs benefits from recent advances in the chemical synthesis of pure and well-defined

Bernd Lepenies (ed.), *Carbohydrate-Based Vaccines: Methods and Protocols*, Methods in Molecular Biology, vol. 1331,
DOI 10.1007/978-1-4939-2874-3_5, © Springer Science+Business Media New York 2015

complex oligosaccharide antigens that are not available through isolation from natural sources [8]. We have successfully employed synthetic oligosaccharides to produce highly affine, specific and selective mAbs against different cell surface carbohydrate antigens, for instance, of *Bacillus anthracis* and *Yersinia pestis* bacteria that bind with nanomolar equilibrium constants [9, 10].

Here, we describe a protocol to generate and characterize highly affine mAbs against defined synthetic oligosaccharide antigens. To overcome the weak immunogenicity of free carbohydrates, a synthetic oligosaccharide antigen is first covalently coupled to an immunogenic carrier protein to form a glycoconjugate [11]. Orientation-specific attachment of oligosaccharides is achieved through a pre-installed primary amine function and a spacer molecule, di(*N*-succinimidyl) adipate (DSAP), yielding well-defined glycoconjugates (Fig. 1). Various carrier proteins suitable for glycoconjugate synthesis have been described, including keyhole limpet hemocyanin (KLH) [12], diphtheria (DT), and tetanus toxoids (TT) [13]. Following the herein described procedure, we have used KLH [1], the nontoxic diphtheria toxin variant CRM_{197} [14–16] as well as ovalbumin (OVA) [17] to prepare glycoconjugates with a variety of oligosaccharide antigens. The glycoconjugate is characterized by sodium dodecyl sulfate-polyacrylamide gel electrophoresis (SDS-PAGE) and matrix-assisted laser desorption/ionization time-of-flight mass spectrometry (MALDI-TOF-MS). The conjugation ratio, i.e., the number of oligosaccharides bound to one molecule of the carrier protein, is calculated by comparing the masses of the glycoconjugate and the nonconjugated carrier protein. Then, mice are immunized with a defined amount of carrier protein-bound oligosaccharide antigen. We have used C57BL/6 [4, 14, 16] and BALB/c mice [15] to generate oligosaccharide-specific antibodies. Although non-adjuvanted glycoconjugates can induce oligosaccharide-specific immune responses [16], we recommend the use of an adjuvant, such as Freund's Adjuvant. Mice are immunized with an immunization regime containing 3 μg

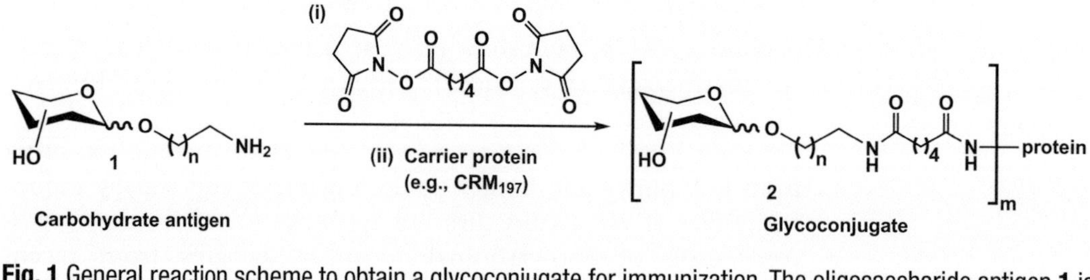

Fig. 1 General reaction scheme to obtain a glycoconjugate for immunization. The oligosaccharide antigen **1** is procured by chemical synthesis and bears an aliphatic amine-terminal linker moiety. This oligosaccharide is reacted in a one-pot procedure first with an excess of a spacer molecule, Di(*N*-succinimidyl) adipate (DSAP), and subsequently with a carrier protein to obtain glycoconjugate **2**. Reagents and conditions: (i) DSAP, Et₃N DMSO; (ii) Carrier protein, 100 mM sodium phosphate, pH 7.4

oligosaccharide antigen per dose with two immunizations in two-week intervals. The immune response in terms of serum antibody production is analyzed using glycan microarrays. Thereby, the mouse with the highest oligosaccharide-specific serum antibody response is selected. This mouse is subjected to a third immunization and antibody-producing splenocytes are fused with myeloma cells to form hybridoma fusion cells. Hybridoma clones that produce antibodies with the desired specificities are identified using glycan microarrays. Finally, mAbs are purified from the cell culture supernatant of expanded hybridoma clones.

Binding characteristics of purified mAbs may be assessed by glycan microarray, surface plasmon resonance (SPR), and saturation transfer difference-NMR, as we have described recently [9, 10]. Depending on the oligosaccharide antigen employed, the obtained mAbs may be useful, for instance, for cell type-specific immunolabeling, pathogen detection and identification, and antigen-specific targeting approaches.

2 Materials

Mention of trade names or commercial products in this procedure is solely for the purpose of providing specific information and does not imply recommendation or endorsement.

2.1 Preparation and Characterization of a Glycoconjugate for Immunization

1. Carrier protein, ~1 mg (*see* **Note 1**).

2. Conjugation buffer: Prepare a 100 mM sodium phosphate monobasic (NaH_2PO_4) stock solution by dissolving 1.2 g anhydrous NaH_2PO_4 in 100 mL double-distilled water (ddH_2O). Prepare a 100 mM sodium phosphate dibasic (Na_2HPO_4) stock solution by dissolving 1.42 g anhydrous Na_2HPO_4 in 100 mL ddH_2O. To prepare the conjugation buffer, add sodium phosphate monobasic stock solution to 30 mL sodium phosphate dibasic stock solution, until pH 7.4 is reached while stirring. Approximately 10 mL of sodium phosphate monobasic stock solution will be added. Stock solutions and conjugation buffer can be stored up to 6 months at room temperature (RT).

3. Oligosaccharide antigen with primary amine linker, ~3–4 nmol (*see* **Note 2**).

4. Anhydrous (*see* **Note 3**) dimethyl sulfoxide (DMSO). Store at RT.

5. Di(*N*-succinimidyl) adipate (DSAP) spacer (*see* **Note 4**). Store desiccated at RT.

6. Centrifugal filter device with 10,000 Da exclusion size and ~4 mL starting volume, e.g., Amicon Ultra-4 Ultracel with 10 kDa NMWL (Millipore).

7. 6× Polyacrylamide separation gel buffer: For 1 L, dissolve 181.65 g Tris base in 800 mL ddH$_2$O, adjust to pH 8.8 with concentrated HCl while stirring, and add ddH$_2$O to a final volume of 1 L to yield 1.5 M Tris–HCl. Store at RT up to 6 months.

8. Acrylamide/bis-acrylamide (29:1) solution, 30 % (w/v) (*see* **Note 5**). Store at RT.

9. 10 % Ammonium persulfate (APS) solution (w/v): to 1 g of ammonium persulfate, add 10 mL ddH$_2$O (*see* **Note 6**). Store at 4 °C up to 1 month.

10. TEMED (*N*,*N*,*N'*,*N'*-tetramethyl-ethylenediamine) reagent. Store at 4 °C and protect from light.

11. 4× Polyacrylamide stacking gel buffer: For 1 L, dissolve 60.55 g Tris base in 800 mL ddH$_2$O, adjust to pH 6.8 with concentrated HCl while stirring, and add ddH$_2$O to a final volume of 1 L to yield 0.5 M Tris–HCl. Store at RT up to 6 months.

12. Protein determination kit, e.g., Micro BCA Protein Assay Kit (Thermo). Store at RT.

13. 4× SDS-PAGE loading buffer: Prepare 100 mL of a 1 M Tris–HCl stock solution (pH 6.8) by dissolving 12.1 g of Tris base in 70 mL ddH$_2$O, adjusting to pH 6.8 with concentrated HCl while stirring, and adding ddH$_2$O to a final volume of 100 mL. Prepare 100 mL of a 0.5 M EDTA stock solution (pH 8) by dissolving 18.6 g EDTA (disodium, dihydrate) in 80 mL ddH$_2$O. Add concentrated NaOH while stirring to bring to pH 8. Store both stock solutions at 4 °C up to 1 month. Prepare 10 mL of 4× SDS-PAGE loading buffer by mixing 2.6 mL ddH$_2$O, 2 mL 1 M Tris–HCl stock solution (pH 6.8), 1 mL 0.5 M EDTA stock solution (pH 8), 4 mL glycerol, 0.4 mL 2-mercaptoethanol, 0.8 g sodium dodecyl sulfate (SDS), and 8 mg bromophenol blue. Aliquots stored at –20 °C are stable for several months.

14. SDS-PAGE running buffer: Prepare a 10× stock solution by dissolving 30.5 g Tris–HCl, 144 g glycine, 10 g sodium dodecyl sulfate in 1 L ddH$_2$O. To prepare 1× SDS-PAGE running buffer, add 900 mL ddH$_2$O to 100 mL of the 10× stock solution. Store at RT.

15. PageRuler Plus Prestained Protein Ladder (Thermo) or a similar protein size marker suitable for SDS-PAGE. Use ~3 μL of this marker in one pocket of a polyacrylamide gel of about 8×8×0.1 cm. Store at –20 °C.

16. Coomassie gel staining solution: For 1 L, mix 500 mL methanol, 430 mL ddH$_2$O, 70 mL glacial acetic acid, and 2.5 g Coomassie Brilliant Blue R-250. Store at RT.

17. Destaining solution: For 1 L, mix 500 mL methanol, 400 mL ddH$_2$O, and 100 mL glacial acetic acid. Store at RT.

18. DHAP (2,5-dihydroxyacetophenone) matrix: Suspend 7.6 mg (50 μmol) 2,5-dihydroxyacetophenone in 375 μL ethanol, then add 125 μL (10 μmol) of diammonium hydrogen citrate (stock solution: 27 mg in 1.5 mL ddH$_2$O) [18]. Store protected from light at 4 °C up to 1 month.

19. MALDI-MS-TOF instrument and matching target plate. We use the Autoflex Speed instrument and a ground steel MTP 384 target plate (Bruker).

2.2 Generation and Purification of Oligosaccharide-Specific Monoclonal Antibodies in Mice

1. Coupling buffer: Prepare a 6× stock solution (300 mM sodium phosphate, pH 8.5) by dissolving 0.41 g anhydrous NaH$_2$PO$_4$ and 3.785 g anhydrous Na$_2$HPO$_4$ in 90 mL ddH$_2$O. Adjust to pH 8.5 with concentrated HCl or NaOH while stirring. Bring the final volume to 100 mL with ddH$_2$O. To prepare 1× coupling buffer, dilute the 6× stock solution 1:6 with ddH$_2$O. Pass through a filter with 0.2 μm pore size. Store 6× and 1× coupling buffers at 4 °C.

2. Piezoelectric microarray spotting device. We use the sciFLEX-ARRAYER S3 (Scienion). This noncontact printer can be equipped with up to eight piezo dispense capillaries, holds up to 20 microarray slides and one microtiter source plate with 96 or 384 wells. Other types of microarray printers may be used according to the manufacturer's recommendations.

3. Surface-modified NHS ester-activated microarray glass slides. We use CodeLink Activated Slides (SurModics). Store desiccated at RT.

4. FlexWell 64 incubation chambers (Grace Bio-Labs) and matching stainless steel spring clips.

5. Humidified chamber. It can be easily prepared by placing paper towels into a sealable box (it should fit at least one microarray slide storage box) and soaking the paper towels with ddH$_2$O.

6. Microarray quenching buffer: Dissolve 7.1 g anhydrous Na$_2$HPO$_4$ and 6.1 g (6 mL) ethanolamine in 800 mL ddH$_2$O. Adjust to pH 9 with concentrated HCl while stirring. Bring the final volume to 1 L with ddH$_2$O to yield 100 nM ethanolamine and 50 nM sodium phosphate. Store protected from light at RT up to 6 months.

7. 6–8 Weeks old female mice. Use C57BL/6 or BALB/c mice.

8. Sterile disposable needles with Luer taper suitable for s.c. immunizations in mice.

9. Complete Freund's Adjuvant (CFA) and Incomplete Freund's Adjuvant (IFA). Store at 4 °C. Agitate CFA before use to distribute inactivated mycobacteria contained in this adjuvant.

10. Sterile phosphate-buffered saline (PBS). You may use premade PBS, such as Dulbecco's phosphate-buffered saline

(DPBS) or prepare by yourself as follows. Dissolve 8 g NaCl, 0.2 g KCl, 1.44 g anhydrous Na_2HPO_4, 0.24 g anhydrous potassium phosphate monobasic (KH_2PO_4) in 800 mL ddH_2O. Adjust to pH 7.4 with concentrated HCl while stirring and bring the final volume to 1000 mL with ddH_2O. Sterilize by autoclaving or by filtering through a 0.2 μm filter. Store at RT for up to 6 months.

11. 1 mL glass syringes with Luer taper and Luer taper adapter suitable to connect two glass syringes.

12. Microarray sample buffer: Weigh 0.5 g (1 % (w/v)) BSA (lyophilized) in a conical 50 mL centrifuge tube and add 50 mL PBS (*see* **item 10**). The PBS does not have to be sterile at this point. Add 5 μL (0.01 % (v/v)) Tween-20. Place the tube on a rocking plate for at least 10 min at RT or shake vigorously. Do so until the BSA has dissolved completely. Store at 4 °C up to 1 week.

13. Microarray washing buffer: To 500 mL of PBS, add 0.5 mL (0.1 % (v/v)) Tween-20. The PBS does not have to be sterile at this point. Mix using a magnetic stirrer and/or vigorously shaking until the Tween-20 has completely dissolved. Store at RT for up to 1 month.

14. Fluorescence-labeled anti-mouse IgG detection antibody. Alexa Fluor 635 Goat Anti-Mouse IgG (H + L) (Life Technologies). Protect from light and store at 4 °C. This antibody is also available with a range of other fluorescence labels. Use this detection antibody at 1:400 in microarray sample buffer (*see* **item 12**).

15. Microarray hybridization covers (22 × 74 mm). HybriSlip (Grace Bio-Labs).

16. Microarray scanner. We use the Genepix 4300A microarray scanner equipped with four lasers; 488, 532, 594, and 635 nm (Molecular Devices).

17. P3X63Ag8.653 myeloma cells (ATCC cat. no. CRL-1580). Frozen stocks are kept in liquid nitrogen vapor phase.

18. Myeloma growth medium: Under sterile conditions, add the following to a fresh bottle of RPMI 1640 medium (500 mL): 50 mL heat-inactivated fetal calf serum (FCS), L-glutamine to a final concentration of 2 mM (if not included in the RPMI 1640 medium), sodium pyruvate to a final concentration of 1 mM, 5 mL of a 100× nonessential amino acid (NEA) concentrate, 5 mL of a 100× penicillin-streptomycin concentrate, gentamycin to a final concentration of 50 μg/mL, 2-mercaptoethanol to a final concentration of 50 μM, and 10 mL of a 50× hypoxanthine/thymidine concentrate. Mix well and heat to 37 °C prior to use. Store at 4 °C for up to 6 months.

19. 50 % Polyethylene glycol (PEG) 1500 (w/v) suitable for cell culture. Store at 4 °C.

20. RPMI 1640 medium without supplements. Store at 4 °C.

21. Heat-inactivated fetal calf serum (FCS). Store at –20 °C.

22. Hybridoma selection medium: Prepare alike myeloma growth medium (*see* **item 18**), except for the following changes. Instead of adding 10 mL of a 50× hypoxanthine/thymidine (HT) concentrate, add 10 mL of a 50× hypoxanthine/aminopterin/thymidine (HAT) concentrate. Add BM Condimed H1 supplement to a final concentration of 10 % (v/v). Mix well and heat to 37 °C prior to use. Use RPMI 1640 medium containing phenol red pH indicator. Store at 4 °C for up to 6 months.

23. DMSO cell culture grade. Store at RT.

24. Serum-free hybridoma medium: Under sterile conditions, add 10 mL of a 100× penicillin-streptomycin concentrate and gentamycin to a final concentration of 50 µg/mL to 1000 mL ISF-1 medium (Biochrom). Store at 4 °C for up to 6 months.

25. Centrifugal filter device with 50,000 Da exclusion size and ~15 mL starting volume, e.g., Amicon Ultra-15 Ultracel with 50 kDa NMWL (Millipore).

26. Proteus Protein G Antibody Purification Midi Kit (AbD Serotec).

3   Methods

3.1   Preparation and Characterization of a Glycoconjugate for Immunization

3.1.1   Conjugation of an Oligosaccharide to a Carrier Protein

1. Dissolve the carrier protein in conjugation buffer. We recommend using 1 mg carrier protein in a volume of 1 mL (i.e., 1 mg/mL). Prepare this solution in a 5 mL glass sample vial and add a magnetic stir bar. Keep this sample at RT.

2. The oligosaccharide antigen should be a pure lyophilized powder. We recommend using approximately 3–4 nmol oligosaccharide (for a typical oligosaccharide, such as the Lewis X trisaccharide of ~500 Da [17], this corresponds to 1.5–2 mg) for 1 mg carrier protein.

3. Dissolve the oligosaccharide in 0.1 mL anhydrous DMSO.

4. If you observe non-dissolved oligosaccharide, add more anhydrous DMSO and mix by pipetting up and down until complete dissolution. Addition of 10 µL triethylamine or heating to 37 °C may facilitate dissolution. Keep this solution at RT. Try to keep the final volume as low as possible.

5. Use DSAP spacer molecule in ~tenfold molar excess to the oligosaccharide. This is crucial to avoid oligosaccharide dimer formation (two oligosaccharide molecules linked by one spacer molecule). E.g., when using 4 nmol oligosaccharide, use 40 nmol (~14 mg) of DSAP (MW = 340.3 Da). Weigh in the appropriate amount of DSAP in a 2 mL glass sample vial using a precision scale. Add a magnetic stir bar.

6. Add 0.1 mL anhydrous DMSO, then 10 µL triethylamine to the vial. Place vial on a magnetic stirrer at 200 rpm to dissolve DSAP.

7. If you observe non-dissolved DSAP, add more anhydrous DMSO until complete dissolution and/or heat to 37 °C. Try to keep the final volume as low as possible.

8. While stirring at 200 rpm, slowly add the oligosaccharide solution from **step 4** to the vial over 30 min (e.g., if the volume of the oligosaccharide solution is 0.1 mL, slowly add five times 20 µL every 6 min).

9. Let the reaction incubate at RT for additional 90 min to complete the reaction, while stirring at 200 rpm.

10. After 90 min, non-reacted DSAP spacer molecules are extracted from the reaction mixture with chloroform (*see* **Note 7**). First, transfer the reaction mixture to a conical 15 mL centrifuge tube. Rinse the glass vial twice with 0.2 mL conjugation buffer (the buffer should turn turbid) and transfer to the same 15 mL tube. From now on, it is important to work quickly to minimize hydrolysis of the *N*-hydroxysuccinimidyl group of monoester intermediates.

11. Carefully pipet 10 mL chloroform to the 15 mL tube and screw the cap on. Shake the tube vigorously for 10 s. Immediately centrifuge at top speed in a table-top centrifuge for 1 min.

12. Carefully remove the tube from the centrifuge; two phases should be visible, an upper aqueous phase (~0.4 mL) and a lower chloroform phase on the bottom (~10 mL). Carefully remove the upper phase and transfer to a new conical 15 mL centrifuge tube. Repeat the extraction procedure from **steps 11 to 12** twice, but in the last step transfer the upper phase to the 5 mL glass sample vial containing the carrier protein solution of **step 1**. Transfer of residual chloroform should be avoided. Discard the chloroform phases.

13. Place the glass sample vial on a magnetic stirrer. Let the reaction stir at 200 rpm overnight (12–18 h) at RT.

3.1.2 Desalting of the Glycoconjugate

1. In this step, non-reacted monoesters are removed and the conjugation buffer is replaced with ddH_2O, which is required for MALDI-TOF MS analysis later on.

2. Transfer the reaction mixture to the upper chamber of a centrifugal filter device with 10,000 Da exclusion size. Fill the upper chamber up with ddH_2O.

3. Centrifuge the filter device at 3000 to $4000 \times g$ in a table-top centrifuge until the remaining volume in the upper chamber is 1 mL or less. The required centrifugation time depends on the type of centrifugal filter device, the sample, and the centrifugation speed. For the centrifugal filter devices we use, 10–15 min at $3000 \times g$ is usually sufficient.

4. Discard the flow-through.

5. Fill the upper chamber up with ddH$_2$O. Repeat **steps 3–5** twice, but in the last step, instead of filling the upper chamber up with ddH$_2$O, transfer the content of the upper chamber to a new 1.5 mL reaction tube. Keep the glycoconjugate solution at 4 °C (*see* **Note 8**).

3.1.3 Characterization of the Glycoconjugate by SDS-PAGE

1. To prepare two 10 % resolving gels of about $8 \times 8 \times 0.1$ cm, prepare appropriate gel casting systems. Mix 4.9 mL ddH$_2$O, 1.7 mL 6× polyacrylamide separation gel buffer, 3.3 mL 30 % acrylamide/bis-acrylamide. Just before casting, add 0.1 mL 10 % APS and 10 μL TEMED to induce polymerization. Mix the solution well and immediately pour the gels with a micropipette, filling each chamber to about three quarters. Gently overlay with isopropanol.

2. Allow the resolving gels to polymerize for 30 min.

3. After 30 min, pour the isopropanol and remove residual liquid with Whatman filter paper.

4. To prepare two 4 % stacking gels, mix 3 mL ddH$_2$O, 1.25 mL 4× polyacrylamide stacking gel buffer, 0.65 mL 30 % acrylamide/bis-acrylamide. Just before casting, add 0.1 mL 10 % APS and 10 μL TEMED to induce polymerization. Mix the solution well and immediately pour the gels with a micropipette on top of the separation gels, filling the chamber completely. Apply combs.

5. Allow stacking gels to polymerize for 40 min.

6. After 40 min, carefully remove the combs. The gels may be used directly, or can be stored up to 1 week at 4 °C wrapped in water-soaked paper towels and placed in plastic bags.

7. Determine the protein concentration of the desalted glycoconjugate solution from Subheading 3.1.2, **step 5** with a protein determination kit. Do not use direct UV/Vis spectroscopy to determine the protein concentration of glycoconjugate solutions (*see* **Note 9**). Typically, about 90 % of the original amount of carrier protein will be recovered as glycoconjugate.

8. In a new 0.5 mL reaction tube, dilute ~2 μg of the glycoconjugate with ddH$_2$O to a volume of 12 μL, then add 4 μL 4× SDS-PAGE loading buffer. This amount of protein will result in clearly visible protein bands in $8 \times 8 \times 0.1$ cm gels after Coomassie staining.

9. In another 0.5 mL reaction tube, dilute ~2 μg of unconjugated carrier protein with ddH$_2$O to a volume of 12 μL, then add 4 μL 4× SDS-PAGE loading buffer.

10. Incubate both reaction tubes for 5 min at 95 °C.

11. Briefly spin down the reaction tubes in a table-top centrifuge.

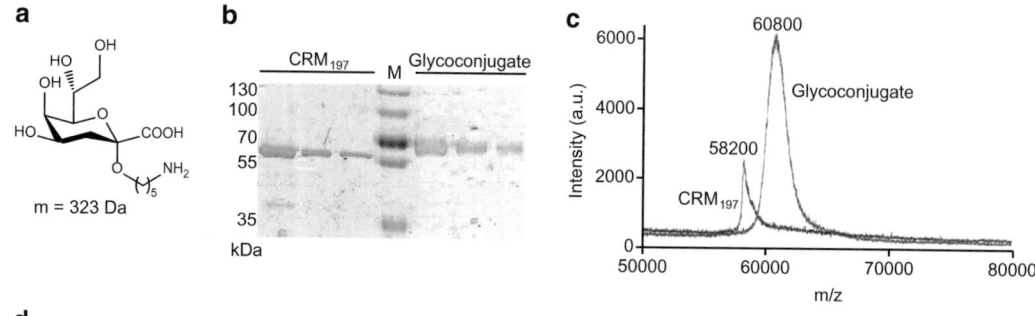

a

m = 323 Da

b

CRM₁₉₇   M   Glycoconjugate

130
100
70
55

35

kDa

c

60800

Glycoconjugate

58200

CRM₁₉₇

Intensity (a.u.)

50000   60000   70000   80000

m/z

d

$$\text{conjugation ratio} = \frac{^m/_z\,(\text{glycoconjugate}) - {}^m/_z\,(\text{carrier protein})}{m(\text{oligosaccharide}) + m(\text{adipoyl})} = \frac{60800 - 58200}{323 + 114} = 6 \qquad \text{(Eq. 1)}$$

$$c(\text{oligosaccharide}) = \frac{m(\text{oligosaccharide})}{m(\text{protein})} \cdot \text{conjugation ratio} \cdot c(\text{protein}) = \frac{323}{58200} \cdot 6 \cdot 3\,{}^{\mu g}/_{\mu L} = 0.1\,{}^{\mu g}/_{\mu L} \qquad \text{(Eq. 2)}$$

Fig. 2 Characterization of glycoconjugates. (**a**) Structure of a model oligosaccharide antigen, the monosaccharide 3-deoxy-α-D-*manno*-oct-2-ulosonic acid (Kdo) equipped with an aminopentyl linker at the reducing end [19] is shown. (**b**) SDS-PAGE analysis of a glycoconjugate of the Kdo antigen shown in (**a**) and the CRM₁₉₇ carrier protein obtained with the conjugation chemistry shown in Fig. 1. Both the nonconjugated CRM₁₉₇ carrier protein and the glycoconjugate were loaded in three different amounts; 2.5, 1, and 0.5 μg (from *left* to *right*, respectively). M denotes the protein size marker, PageRuler Plus Prestained Protein Ladder (Thermo Scientific). (**c**) MALDI-TOF-MS analysis of CRM₁₉₇ carrier protein and glycoconjugate. Peaks corresponding to singly charged ions ($z = 1$) are shown. (**d**) Exemplary calculation of the conjugation ratio Eq. 1 and oligosaccharide concentration Eq. 2 assuming a protein concentration of the glycoconjugate solution of 3 μg/μL

12. Install an SDS-PAGE gel into an appropriate running chamber filled with SDS-PAGE running buffer. Make sure that the wells are completely covered by buffer and remove any air bubbles.

13. Pipet the samples of **steps 8–11** into individual pockets of the gel, add an appropriate protein size marker into an adjacent pocket.

14. Let the gel run at 120 V until the blue dye has reached the bottom.

15. Remove the gel from the frame, discard the stacking gel, and soak the gel in Coomassie gel staining solution for 15 min.

16. After 15 min, carefully remove the staining solution (*see* **Note 10**). Destain the gel by soaking in destaining solution. Complete destaining takes several hours. To accelerate destaining, replace the destaining solution once it has turned blue.

17. Scan the gel. The protein band of the glycoconjugate should be broader and running at higher masses than the nonconjugated carrier protein (Fig. 2).

3.1.4 Characterization of the Glycoconjugate by MALDI-TOF-MS and Estimation of the Conjugation Ratio

1. The nonconjugated carrier protein is used as a reference to estimate the oligosaccharide loading of the glycoconjugate. For MALDI-TOF-MS analysis it should be dissolved in salt-free ddH₂O, as salts, detergents, and contaminants interfere with the ionization process. If necessary, desalt the nonconjugated carrier protein as described in Subheading 3.1.2.

2. Prepare two 0.5 mL reaction tubes by first pipetting 2 μL of DHAP matrix and then adding 2 μL of 2 % trifluoroacetic acid (TFA) in ddH$_2$O (v/v). To one of the reaction tubes, add 2 μL of the desalted glycoconjugate and mix gently for 10 s by swirling with the pipette tip. To the other reaction tube, add 2 μL of the desalted nonconjugated carrier protein. Mix gently for 10 s by swirling with the pipette tip.

3. Add 1 μL of each mixture to one spot on a 384-well MALDI-MS target plate. Wait until the spots have completely dried before inserting into the MALDI-TOF-MS instrument.

4. Acquire mass spectra, using linear positive ion mode and detection within an m/z range that is appropriate for the carrier protein and the glycoconjugate (*see* **Note 11**).

5. Compare mean m/z ratios of glycoconjugate and nonconjugated carrier protein. The glycoconjugate should show higher mean m/z ratios and broader peaks (*see* **Note 12**). Calculate the average conjugation ratio by dividing the m/z difference by the molecular mass of the oligosaccharide-adipoyl moiety according to Eq. 1, in which the charge z is equal to 1 (representing the singly charged ions) while the molecular mass of the adipoyl moiety, m(adipoyl), is 114 Da.

$$\text{conjugation ratio} = \frac{{}^{m}\!/\!_{z}\left(\text{glycoconjugate}\right) - {}^{m}\!/\!_{z}\left(\text{carrier protein}\right)}{m\left(\text{oligosaccharide}\right) + m\left(\text{adipoyl}\right)} \quad (1)$$

6. Finally, calculate the oligosaccharide concentration, c(oligosaccharide), of the glycoconjugate solution from Subheading 3.1.3, **step 7** according to Eq. 2. An exemplary calculation with a model antigen, the monosaccharide α-3-deoxy-D-*manno*-oct-2-ulosonic acid (Kdo) equipped with an aminopentyl linker at the reducing end [19] can be found in Fig. 2.

$$c\left(\text{oligosaccharide}\right) = \frac{m\left(\text{oligosaccharide}\right)}{m\left(\text{protein}\right)} \bullet \text{conjugation ratio} \bullet c\left(\text{protein}\right) \quad (2)$$

3.2 Generation and Purification of Oligosaccharide-Specific Monoclonal Antibodies in Mice

Microarray technology serves to evaluate the immune response in mice and to identify oligosaccharide-specific hybridoma clones, as described below in Subheadings 3.2.2 and 3.2.4, respectively (Fig. 3). Therefore, the oligosaccharide antigen is immobilized on the microarray surface via its primary amine function.

3.2.1 Preparation of Glycan Microarray Slides

1. Dissolve the oligosaccharide antigen in coupling buffer. We recommend oligosaccharide concentrations ranging from 0.1 to 1 mM to achieve good binding signals. Around 50 μL of such oligosaccharide solutions is sufficient to produce a sufficient number of microarray slides.

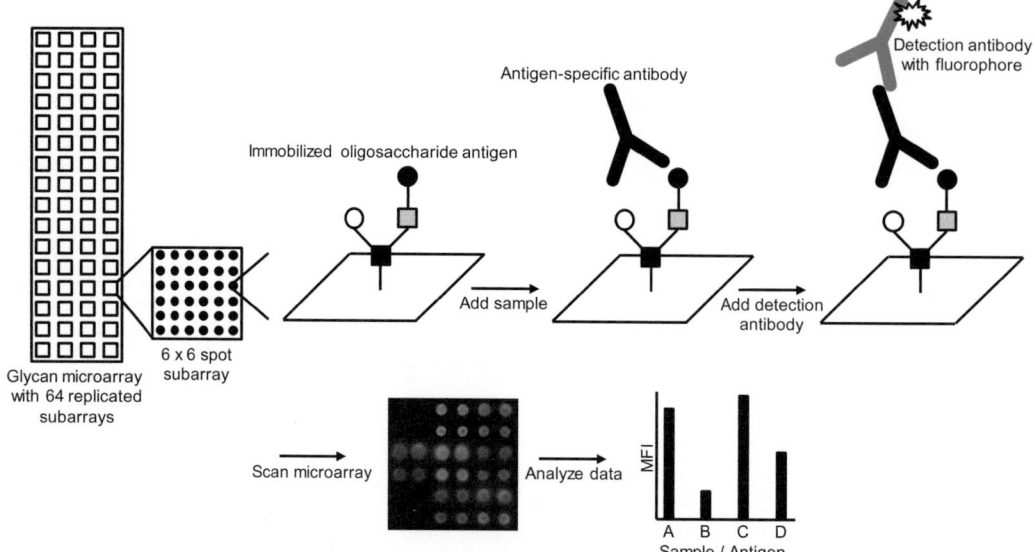

Fig. 3 Principle of detecting oligosaccharide-specific antibodies by glycan microarray. A printing pattern can be programmed such that 64 identical subarrays, each containing 6 × 6 spots, are printed on one microarray slide. If a sample (e.g., serum or hybridoma supernatant) contains antibodies that bind to the immobilized oligosaccharide antigen, they can be visualized with a secondary fluorescence-labeled detection antibody. After scanning of the microarray slides, binding signals (for instance, expressed as MFI for mean fluorescence intensity) can be quantified

2. Pipet 20–50 μL of oligosaccharide solutions per well of a 384-well V-shaped microplate (Genetix). We recommend preparing solutions of different concentrations (e.g., 0.1, 0.5, and 1 mM) that will allow the detection of concentration-dependent binding signals later on. Use coupling buffer as negative control. You may also include the nonconjugated carrier protein and a dummy glycoconjugate prepared with a different carrier protein, e.g., bovine serum albumin (BSA), and a non-related monosaccharide or oligosaccharide bearing an identical linker moiety (in the herein described example, an aminopentyl moiety, *see* Fig. 2a) prepared and characterized according to Subheadings 3.1.1 through 3.1.4 and dissolved in coupling buffer (recommended concentration is 1 μM for proteins). This will allow the detection of antibodies directed against the carrier protein and the generic spacer moiety (consisting of adipoyl and pentyl groups in this case), respectively (*see* **Note 13**). You may also include one or more control oligosaccharides (each bearing an amine-terminal linker) at 1 mM dissolved in coupling buffer.

3. Print microarray slides using a piezoelectric spotting device at about 60 % relative humidity. For printing amine-functionalized oligosaccharides, we use NHS ester-activated glass microarray slides. When using CodeLink amine binding slides, note that

only one side of the glass slides is functionalized. This side is up when the "CodeLink" lettering is readable. We recommend a printing pattern that allows the analysis of multiple samples on one microarray slide. For example, printing of 64 identical square fields (each field will fit at least 6×6 spots with a pitch of 400 μm) will allow the use of 64-well incubation chambers later on for parallel analysis of multiple samples on one slide (Fig. 3). In each of the 64 identical fields, we recommend including replicates of each substance. Printing of 20 identical microarray slides will usually be sufficient for assessing the immune response in mice and the subsequent analysis and selection of hybridoma clones. If printing duplicated spots on 64 fields, set the uptake volume of the capillary to 3 μL (*see* **Note 14**).

4. After printing is completed, remove microarray slides from spotting device and place them into a slide storage box (*see* **Note 15**). Set the box open with the printed side of the slides upwards into a humidified chamber and incubate at RT overnight to complete the coupling reaction. Seal the 384-well microplate with sealing film and store at –20 °C up to 6 months.

5. After the overnight incubation, place microarray slides into microarray quenching buffer. Incubate at 50 °C for 1 h to quench the remaining free NHS ester groups on the microarray surface, then wash three times with ddH$_2$O. In each washing step, spin down residual water by centrifuging at $1200 \times g$ for 5 min. For this, you may place individual microarray slides into conical 50 mL centrifuge tubes or utilize a slide adapter for swing-bucket centrifuge rotors (e.g., the CombiSlide system by Eppendorf).

6. After washing, keep the slides in sealed slide storage boxes at 4 °C.

3.2.2 Immunization of Mice and Glycan Microarray-Assisted Evaluation of the Immune Response

Generating an oligosaccharide-specific IgG response in animals is the first step towards the production of mAbs. This protocol describes the generation of murine mAbs. Both C57BL/6 and BALB/c mouse strains can be used. For some oligosaccharide antigens we have observed that either strain is superior over the other in producing antigen-specific IgGs. Three consecutive subcutaneous (s.c.) immunizations in 2-week intervals with an antigen dose of 3 μg per immunization formulated with Freund's Adjuvant are usually suitable to generate sufficient numbers of highly proliferating antigen-producing B cells in the mouse spleens used for hybridoma fusion later on. In some cases it may be necessary to perform additional boosting immunizations or to choose other immunization routes, such as intraperitoneal (i.p.) injections. This is a trial-and-error process that depends on the nature of the oligosaccharide antigen, the type of carrier protein, the conjugation ratio, and the mouse strain. Depending on the investigator's objectives other animals, such as rats or rabbits, may be used [20].

1. Immunize at least three mice with an amount of glycoconjugate corresponding to 3 μg oligosaccharide antigen via the s.c. route (*see* **Note 16**). For example, use 30 μL of a glycoconjugate solution with c(oligosaccharide) = 0.1 μg/μL (*see* calculation in Fig. 2d) per animal. The total injection volume should be 100 μL per mouse, administered with a sterile disposable needle. Just prior to each immunization, an emulsion with CFA or IFA at 1:1 ratio is prepared. Dilute the glycoconjugate preparation accordingly with sterile PBS. For this example, add 20 μL of sterile PBS to 30 μL glycoconjugate solution for each mouse (yielding 50 μL), then prepare an emulsion with 50 μL CFA or IFA. Prepare a solution with an amount of glycoconjugate sufficient for the total number of mice and add one additional volume to account for the dead volume of syringes and needles. In this example, using three mice, mix 120 μL glycoconjugate solution with 80 μL sterile PBS (yielding 200 μL) and prepare an emulsion with 200 μL CFA or IFA. Use the same glass syringe and needle to immunize all mice. The emulsions are prepared using two identical 1 mL glass syringes connected by an appropriate Luer taper adapter through firmly pushing the mixture from one side to the other at least 20 times. Alternatively, prepare the emulsion in a 1.5 mL reaction tube by vigorously pipetting up and down at least 20 times with a micropipette. The emulsion should be a uniform, white and viscous liquid. Always use clean and autoclaved glass syringes and adapters. We recommend the following immunization regime:

   Day 0—primary immunization using CFA, blood collection

   Day 14—first boosting using IFA

   Day 21—blood collection and evaluation of the immune response by glycan microarray

   Day 28—second boosting using IFA (one selected mouse)

   Day 35—harvest of splenocytes, cell fusion

2. After immunization, wash the syringes and adapters ten times each with methanol, ethanol, and ddH$_2$O. Afterwards, disassemble syringes and adapters and sterilize by autoclaving.

3. You may collect blood weekly, but most importantly at days 0 and 21. To collect blood, nick the tail vein with a needle or lancet and draw blood into a glass microcapillary tube, then transfer into a 1.5 mL reaction tube with an appropriate pipetting device. Alternatively, puncture the submandibular vein with a lancet and let blood drip directly into 1.5 mL reaction tubes. 50 μL blood per animal is sufficient to evaluate the immune response.

4. Prepare sera by letting whole blood samples clot at RT for 30 min. Remove blood clots by spinning down at 1000–2000×g for 10 min. Pipet the sera into new 0.5 mL reaction

tubes. Discard the 1.5 mL reaction tubes containing the blood clots. Store sera at −20 °C or use them directly for microarray analysis. Avoid repeated freeze-thaw cycles.

5. Block one printed and quenched microarray slide prepared according to Subheading 3.2.1, by incubating the slide in microarray sample buffer at RT for 1 h. You can also perform this blocking step overnight at 4 °C. Wash the slide three times with PBS similar to Subheading 3.2.1, **step 5**.

6. Add a 64-well incubation chamber on top of the functionalized side of the blocked microarray slide using two stainless steel spring clips.

7. Dilute individual sera 1:100 (v/v) in microarray sample buffer. When using a 64-well incubation chamber, a volume of 30 µL per serum dilution is sufficient. Duplicates of each sample are recommended. You may include further dilutions (e.g., 1:200, 1:500, 1:1000, or higher) to better compare serum antibody levels later on.

8. Place diluted serum samples into individual wells of the incubation chamber attached to the microarray slide with a micropipette. Incubate at RT for 1 h or overnight at 4 °C in a humidified chamber.

9. Keeping the incubation chamber attached, wash the microarray slide three times with microarray washing buffer. Remove the samples by decanting and add 50 µL microarray washing buffer per well with an 8-channel micropipette (place an appropriate volume of microarray washing buffer in a disposable reagent reservoir). Repeat three times. Finally, wash once with PBS. Decant the liquid and carefully remove the incubation chamber. Spin down residual fluid as described in Subheading 3.2.1, **step 5**.

10. Prepare the detection antibody by diluting fluorescence-labeled anti-mouse IgG antibody in microarray sample buffer. Avoid exposure to light.

11. Place the microarray slide in a Petri dish (printed side upwards). Drop about 100–200 µL of the diluted detection antibody onto the microarray slide with a micropipette. Carefully add a microarray hybridization cover with the help of a forceps and incubate at RT for 1 h in a humidified chamber. From now on, protect the microarray slide from light. Therefore, line the humidified chamber with aluminum foil.

12. Carefully remove the microarray hybridization cover with the help of forceps and wash three times by pipetting about 10 mL microarray washing buffer into the Petri dish. Avoid pipetting directly onto the slide. In each washing step, incubate the buffer-covered slide at RT for 10 min in a humidified chamber placed on a rocking platform, then decant the liquid. The slide will attach to the Petri dish surface by adhesive forces.

Finally, rinse the microarray slide once with ddH_2O for 30 s and decant the liquid. Remove residual fluid by centrifugation as described in Subheading 3.2.1, **step 5**.

13. Scan the microarray slide with a microarray scanner. Before scanning, allow the lasers to warm up for at least 15 min. Place the microarray slide into the microarray scanner with the printed side facing down. Choose laser and filter suitable for the fluorescent dye of the anti-mouse IgG antibody. The GenePix Pro software supplied with the instrument allows a "Preview scan" to detect possible saturation signals. Start with a PMT gain of 400 and adjust it such that good binding signals are visible, but no saturated signals are seen. Once the appropriate PMT gain is found, use the "Data scan" function to acquire a high-resolution scan. Save the scan as a .tif file.

14. Compare the oligosaccharide-specific IgG responses with the GenePix Pro software. Binding signal intensities may be estimated visually or by determining the mean fluorescence intensity signals of the relevant spots. Select the mouse with the highest binding signal towards the oligosaccharide antigen to perform a second boosting immunization at day 28 and harvest of splenocytes and cell fusion at day 35 (*see* **step 1**).

3.2.3 Cell Fusion

1. Murine P3X63Ag8.635 myeloma cells should be thawed about 2 weeks in advance of cell fusion. Propagate cells in myeloma growth medium at $<2 \times 10^6$ cells per mL in cell culture flasks at 37 °C in 5 % CO_2 in a humidified cell culture incubator. You will need 2×10^8 myeloma cells for the fusion. Replace the medium with fresh myeloma growth medium 24 h prior to the fusion.

2. Before starting cell fusion, read the complete Cell Fusion protocol (Subheading 3.2.3) thoroughly, since timing is crucial and extended incubation times will lower the fusion efficiency. Before beginning the fusion, pre-warm (1) 50 % PEG 1500 (w/v), (2) RPMI 1640 without supplements, (3) heat-inactivated FCS, and (4) hybridoma selection medium in a 37 °C water bath. Place a 500 mL glass beaker filled with water at 37 °C into the cell culture hood.

3. Briefly before fusion at day 35 (*see* Subheading 3.2.2, **step 1**), count myeloma cells with a hemocytometer. Spin down 2×10^8 cells at $300 \times g$, carefully aspirate the supernatant with a Pasteur pipette, and resuspend cells in 35 mL pre-warmed (37 °C) RPMI 1640 medium without supplements. Keep this cell suspension in a conical 50 mL centrifuge tube at 37 °C.

4. Euthanize the selected mouse and remove the spleen under aseptic conditions. Place the euthanized mouse on its back, wet the fur with 70 % ethanol in ddH_2O (v/v) to sterilize the

area, and make a midline incision with sterile scissors. Carefully remove the spleen with sterile forceps and place it into a sterile Petri dish containing 5 mL pre-warmed (37 °C) RPMI 1640 medium without supplements. Carefully remove fat and connective tissue. Instructions for spleen removal can also be found in Reference [21].

5. Remove the plunger from a sterile 10 mL plastic syringe and use it to carefully mash the spleen to release the splenocytes into the Petri dish with grinding circular movements. Remove the remainings of the spleen, which should acquire a whitish color once all splenocytes are removed.

6. Transfer the 5 mL cell suspension from the Petri dish into a new conical 50 mL centrifuge tube. Rinse the Petri dish once with 10 mL pre-warmed (37 °C) RPMI 1640 medium without supplements and transfer to the same 50 mL tube. Pipet the cell suspension up and down several times to disintegrate cell clumps.

7. Pipet the 15 mL splenocyte cell suspension into the 35 mL myeloma cell suspension of **step 3**. Mix well by pipetting up and down several times.

8. Centrifuge at $300 \times g$ for 10 min at RT.

9. Carefully remove the supernatant with a Pasteur pipette.

10. Gently disrupt the cell pellet by tapping the bottom of the tube. Place the tube into the 37 °C water-filled glass beaker in the cell culture hood.

11. Add 1.5 mL pre-warmed 50 % PEG (w/v) dropwise over a period of 1 min to the cell suspension while gently swirling the cells with the pipette tip. Continue swirling for one additional minute.

12. While gently swirling, slowly add 20 mL pre-warmed RPMI 1640 without supplements as follows: 1 mL over 1 min, 3 mL over 1 min, and 16 mL over 2 min.

13. Immediately spin down cells at $300 \times g$ for 10 min at RT.

14. Incubate the tube for 5 min at 37 °C. Then carefully remove the supernatant with a Pasteur pipette.

15. Gently resuspend the cells in 10 mL pre-warmed heat-inactivated FCS.

16. Transfer 50 % of the cell suspension (5 mL) to a bottle containing 200 mL pre-warmed hybridoma selection medium. Using a multichannel micropipette and a disposable reagent reservoir, distribute this cell suspension in ten 96-well flat-bottom tissue culture plates (200 μL per well). Gently mix the cell suspension after completing each 96-well plate to ensure even distribution. Incubate the 96-well plates at 37 °C in 5 % CO_2 in a humidified cell culture incubator.

17. To the remaining 50 % cell suspension, add 10 % (0.5 mL) cell culture grade DMSO, mix gently, and dispense into cryogenic vials. Freeze overnight at –80 °C. Then, transfer vials to a liquid nitrogen tank. This serves as a backup for the cloning procedure described below.

3.2.4 Glycan Microarray-Assisted Cloning and Purification of Oligosaccharide-Specific Monoclonal Antibodies

1. Let the 96-well plates from Subheading 3.2.3, **step 16** incubate for 7–14 days before the first subcloning step. Check regularly for cell growth under a phase contrast microscope. Under the described conditions, about 25–50 % of the wells should exhibit clonal growth of hybridoma cells. Usually the cells must be fed 5–7 days after fusion. Wells that require feeding can be spotted when the culture medium turns from red to orange/yellow. Feed cells by carefully aspirating ~150 μL of the cell culture supernatant of the respective well (avoid contacting the bottom), then add ~150 μL fresh, pre-warmed (37 °C) hybridoma selection medium.

2. The first screening for oligosaccharide-specific IgGs should usually be done 7–14 days after the fusion. Screen wells when cell growth is clearly visible under the phase contrast microscope. Usually colonies of a few hundred cells produce sufficient amounts of antibodies that can be detected by glycan microarray. Test wells before proliferating cells have reached ~25 % confluence. Note that growth rates of individual clones can vary substantially.

3. Screen wells with growing cells with an appropriate number of glycan microarray slides (from Subheading 3.2.1) by blocking according to Subheading 3.2.2, **step 5** and applying 64-well incubation chambers (*see* Subheading 3.2.2, **step 6**). Follow the glycan microarray procedure described in Subheading 3.2.2, **steps 8–13**, but instead of serum samples, add 50 μL undiluted hybridoma cell culture supernatants.

4. Those wells in which oligosaccharide-specific IgGs are detected should be subjected to the first round of subcloning. We typically also detect clones that produce IgGs against the carrier protein or the spacer moiety (*see* **Note 13**).

5. Check the cell morphology of positive wells under the phase contrast microscope. In case of scattered cell distribution, vigorously mix the well content by pipetting up and down with a 200 μL micropipette and transfer about 100 μL (depending on the cell density) to the upper left well ("A1") of a new 96-well flat-bottom cell culture plate containing 100 μL pre-warmed (37 °C) hybridoma selection medium in each well. In case of positive wells with one or more spatially defined colonies, "pick" individual clones with a 200 μL micropipette by moving the pipette tip to the bottom of the well where the colony is located and pipetting up ~50 μL (depending on the size of the colony)

while gently moving the pipette tip within the constraints of the colony. Pipet into the "A1" well of new 96-well flat-bottom cell culture plate containing 100 µL pre-warmed (37 °C) hybridoma selection medium in each well. After picking, check if the majority of cells of the respective colony has been successfully removed under the phase contrast microscope. If not, repeat.

6. Each new 96-well plate is used for subcloning of one clone. Well A1 should contain a total volume of 200 µL. In each plate, prepare serial 1:2 dilutions in row "A" left-to-right by pipetting 100 µL from well A1 to A2, A2 to A3, and so on, with a micropipette. In each dilution step, mix the cell suspension by gently pipetting up and down twice. Transfer the 100 µL from well A12 to well A1 and add 100 µL of pre-warmed (37 °C) hybridoma selection medium to wells A2 to A12, so that each well in row "A" contains 200 µL. Next, using a multichannel micropipette, prepare serial 1:2 dilutions top-to-bottom by pipetting 100 µL of wells from row A to row B, row B to row C, and so on. In each dilution step, mix the cell suspensions by pipetting up and down twice. Transfer the 100 µL from row H to row A, so that each well of row "A" contains 200 µL. Fill the volume of the remaining wells up to 200 µL by adding 100 µL fresh, pre-warmed (37 °C) hybridoma selection medium. Incubate the plate at 37 °C in 5 % CO_2 in a humidified cell culture incubator.

7. Check the plates regularly for cell growth under the phase contrast microscope. Identify wells containing single, spatially defined colonies, which should be of appropriate size after 7–14 days. Check wells of interest for oligosaccharide-specific IgGs by glycan microarray as described above. For positive clones, continue with a second round of subcloning as described in **step 6**.

8. Every primary hybridoma clone requires at least three consecutive rounds of subcloning. Typically, we recover 3–10 stably growing hybridoma clones producing oligosaccharide-specific IgGs after three rounds of subcloning.

9. Do not discard 96-well plates from which cells have been removed for subcloning, but aspirate the complete medium of each well with a Pasteur pipette (try to avoid touching the bottom) and carefully add ~100 µL hybridoma selection medium with 10 % cell culture grade DMSO. Freeze these plates at –80 °C. They serve as backups in case clones are lost during the subcloning procedure.

10. During each round of subcloning, we recommend preparing cryostocks of selected clones. Expand selected clones in hybridoma selection medium consecutively in 48-well and 6-well flat-bottom cell culture plates and grow to confluence. Resuspend

cells in 1 mL hybridoma selection medium with 10 % cell culture grade DMSO, transfer to a cryogenic vial, freeze overnight at –80 °C, and transfer to a liquid nitrogen tank.

11. After three rounds of subcloning, prepare at least ten cryostocks of each positive clone as described in **step 10**.

12. Expand selected clones in serum-free hybridoma medium. Allow the cells to adapt slowly to the new medium. First, gradually replace the hybridoma selection medium with serum-free hybridoma medium, but always add heat-inactivated FCS to a final concentration of 10 % (v/v). Then, reduce the FCS concentration to 0 % by gradually replacing the medium with serum-free hybridoma medium without adding FCS. Split cells or transfer to larger cell culture plates or flasks just before confluence is reached. Note that hybridoma cells tend to grow more slowly under reduced serum conditions. Once serum-free conditions are reached, let the cells grow to confluence in large (150 or 300 cm²) cell culture flasks with a high volume of serum-free hybridoma medium (150 or 300 mL, respectively) until confluence is reached. Then, let the cells stand for one additional week. Typically, about 1 mg mAbs per 100 mL will be secreted into the medium, but this may vary substantially between individual clones.

13. Transfer the complete hybridoma culture supernatant to conical 50 mL centrifuge tubes and spin down cells at $1200 \times g$ for 5 min at RT. Pass the supernatant through a filter with 0.2 μm pore size into a sterile bottle. Store at 4 °C for up to 1 month or immediately continue with the purification of mAbs.

14. Concentrate the filtered supernatant with a centrifugal filter device with 50,000 Da exclusion size by successively loading the supernatant to the upper chamber and centrifugation similar to Subheading 3.1.2, **step 3**, to a final volume of 12–15 mL. Purify mAbs from the concentrated supernatant with an antibody purification kit according to the manufacturer's recommendations. Determine the protein concentration of the eluate with a protein determination kit or by UV–Vis spectroscopy at 280 nm. The eluates are typically stable for at least 1 month at 4 °C. You may add 0.02 % sodium azide (w/v) to increase the stability. For long-term storage, add 50 % glycerol (v/v) and store at –20 or –80 °C.

15. The purified mAbs are now ready to be tested for your application of choice. They can be characterized in terms of specificity, affinity, and epitope recognition using glycan microarray, surface plasmon resonance, and saturation transfer difference-NMR, respectively, as described [9, 10]. We have used such mAbs for the specific detection of bacteria by immunostaining [1–4].

4   Notes

1. We have used the nontoxic diphtheria toxin variant CRM_{197} (purchased from Pfénex, Inc.), bovine serum albumin (BSA), and chicken ovalbumin (OVA, purchased from Hyglos) to prepare glycoconjugates. Other frequently used carrier proteins are keyhole limpet hemocyanin (KLH), diphtheria toxoid (DT), and tetanus toxoid (TT).

2. Oligosaccharides can be chemically synthesized with amine-terminal linkers that allow their orientation-specific conjugation to proteins and immobilization on microarray surfaces [22].

3. The DSAP spacer used in this protocol quickly hydrolyzes in the presence of water. Therefore, DMSO used to dissolve DSAP spacer and the oligosaccharide should be anhydrous. For this purpose, sodium alumino-silicate molecular sieves of 4 Å (0.4 nm) can be added to the DMSO to remove residual water.

4. DSAP quickly hydrolyzes! It should be stored under anhydrous conditions, e.g., under an argon atmosphere in a desiccator. Di(N-succinimidyl) adipate can be purchased from Synchem, or can be synthesized as follows. To a solution of 1.453 mL adipoyl chloride in 90 mL THF, add 2.3 g N-hydroxysuccinimide (NHS) and 2.79 mL triethylamine. After 24 h at RT, evaporate the solvent and partition the residue between dilute aqueous HCl and chloroform. The organic layer is separated and washed successively with water and brine, dried over sodium sulfate, filtered and evaporated to give a white solid. The final product is recrystallized from isopropyl alcohol.

5. Nonpolymerized acrylamide is a neurotoxin! Handle with care and always wear a lab coat, gloves, and safety glasses. The 30 % acrylamide/bis-acrylamide solution can be readily purchased or be prepared as described in reference [23].

6. Fresh ammonium persulfate "crackles" when water is added. If it does not crackle anymore, fresh ammonium persulfate should be purchased. We recommend freshly preparing the ammonium persulfate solution prior to use. Alternatively, stocks may be prepared, which are stable for 1 month at −20 °C.

7. Chloroform is highly volatile, harmful, and irritant. Handle with care, perform all pipetting steps in a fume hood, and always wear work coat, gloves, and safety glasses. Chloroform may explode if it comes in contact with methanol, acetone, alkalis, aluminum, lithium, perchlorate pentoxide, bis-dimethylanstannane, potassium, or sodium. Chloroform should be protected from heat and light.

8. The glycoconjugates prepared with CRM_{197} (Pfénex, Inc.) are usually stable for about 1 month in ddH_2O at 4 °C. Stability

may vary for different carrier proteins. For long-term storage, the glycoconjugate can be lyophilized. To do so, first prepare 5× lyophilization buffer by dissolving 2.5 g sucrose, 11 mg anhydrous NaH_2PO_4, and 150 mg anhydrous Na_2HPO_4 in 10 mL ddH_2O. Add 3 μL (~2.75 mg) polysorbate 80, mix well, and pass through an 0.2 μm filter. Add the appropriate amount of 5× lyophilization buffer to the glycoconjugate solution to reach 1× concentration, e.g., add 0.2 mL 5× lyophilization buffer to 0.8 mL glycoconjugate solution. The sample can then be frozen and lyophilized and will be stable for several months when stored at −20 °C.

9. Direct determination of protein concentrations using UV/Vis spectroscopy relies on absorbance at 280 nm. We have noticed that the absorbance at this wavelength is unrealistically high for glycoconjugates prepared with the herein described method. This might be due to the presence of *N*-hydroxysuccinimide ions that are formed during the conjugation reaction, which show an absorbance peak ranging from ~230 to 300 nm [24].

10. Coomassie gel staining solution and destaining solution contain methanol. Dispose any waste in an appropriate flameproof hazardous waste container.

11. For instance, when using CRM_{197} as carrier protein with a molecular weight of 58.4 kDa, the lower m/z detection limit should be around 25,000 m/z to allow detection of its doubly charged ion (theoretical $m/z = 29,200$). We usually choose an upper detection limit of >200,000 m/z.

12. Broader peaks of glycoconjugates compared to the nonconjugated carrier proteins are the result of glycoconjugate species with different conjugation ratios, whose relative abundances should be roughly normally distributed, resulting in symmetric bell-shaped peaks. Therefore, the mean m/z value can be used to estimate the average conjugation ratio. For glycoconjugates prepared with CRM_{197} we usually achieve conjugation ratios ranging between 3 and 10.

13. We have noticed that in some cases, especially when using weakly immunogenic small oligosaccharide antigens, e.g., monosaccharides or disaccharides, some mice produce antibodies against the carrier protein only, which can be immunodominant. Also the generic spacer moiety usually is immunogenic. Thus, in order to detect mAbs that solely recognize the employed oligosaccharide antigen, we recommend printing of the nonconjugated carrier protein as well as a dummy conjugate.

14. Each spot has a volume of 0.5 nL. If printing a substance on 20 slides with 64 fields in duplicate, the required volume is

0.5 nL $\times 20 \times 64 \times 2 = 1.28$ µL. Always add at least one additional microliter to the uptake volume.

15. When handling microarray slides, always wear disposable laboratory gloves and be careful not to touch the surface.

16. Make sure that all procedures involving animals are according to your local animal protection regulations. Guidelines for the good practice to administration of substances and blood removal for various animals by the European Federation of Pharmaceutical Industries Associations (EFPIA) and the European Centre for the Validation of Alternative Methods (ECVAM) can be found in Reference [25].

Acknowledgements

We thank Dr. You Yang for providing the Kdo antigen and Pfénex, Inc. for providing CRM$_{197}$ at a reduced price for academic institutions. We acknowledge careful and critical reviewing of the manuscript by Andreas Geissner, Anika Reinhardt, Benjamin Schumann, and Stefan Matthies. We thank the Max Planck Society, the Körber Foundation (Körber Prize to PHS), and the German Federal Ministry of Education and Research (grant No. 0315447) for generous financial support.

References

1. Tamborrini M, Werz DB, Frey J et al (2006) Anti-carbohydrate antibodies for the detection of anthrax spores. Angew Chem Int Ed Engl 45:6581–6582

2. Tamborrini M, Oberli MA, Werz DB et al (2009) Immuno-detection of anthrose containing tetrasaccharide in the exosporium of *Bacillus anthracis* and *Bacillus cereus* strains. J Appl Microbiol 106:1618–1628

3. Tamborrini M, Holzer M, Seeberger PH et al (2010) Anthrax spore detection by a luminex assay based on monoclonal antibodies that recognize anthrose-containing oligosaccharides. Clin Vaccine Immunol 17:1446–1451

4. Anish C, Guo X, Wahlbrink A et al (2013) Plague detection by anti-carbohydrate antibodies. Angew Chem Int Ed Engl 52:9524–9528

5. Gao C, Liu Y, Zhang H et al (2014) Carbohydrate sequence of the prostate cancer-associated antigen F77 assigned by a mucin O-glycome designer array. J Biol Chem 289:16462–16477

6. Lee G, Cheung AP, Ge B et al (2012) CA215 and GnRH receptor as targets for cancer therapy. Cancer Immunol Immunother 61:1805–1817

7. Wang LX (2013) Synthetic carbohydrate antigens for HIV vaccine design. Curr Opin Chem Biol 17:997–1005

8. Lepenies B, Seeberger PH (2010) The promise of glycomics, glycan arrays and carbohydrate-based vaccines. Immunopharmacol Immunotoxicol 32:196–207

9. Oberli MA, Tamborrini M, Tsai YH et al (2010) Molecular analysis of carbohydrate-antibody interactions: case study using a *Bacillus anthracis* tetrasaccharide. J Am Chem Soc 132:10239–10241

10. Broecker F, Aretz J, Yang Y et al (2014) Epitope recognition of antibodies against a *Yersinia pestis* lipopolysaccharide trisaccharide component. ACS Chem Biol 9:867–873

11. Avery OT, Goebel WF (1931) Chemo-immunological studies on conjugated carbohydrate-proteins: V. The immunological specificity of an antigen prepared by combining the capsular polysaccharide of type III pneumococcus with foreign protein. J Exp Med 54:437–447

12. Jorge P, Abdul-Wajid A (1995) Sialyl-Tn-KLH, glycoconjugate analysis and stability by

high-pH anion-exchange chromatography with pulsed amperometric detection (HPAEC-PAD). Glycobiology 5:759–764

13. Borrow R, Dagan R, Zepp F et al (2011) Glycoconjugate vaccines and immune interactions, and implications for vaccination schedules. Expert Rev Vaccines 10:1621–1631

14. Oberli MA, Hecht ML, Bindschädler P et al (2011) A possible oligosaccharide-conjugate vaccine candidate for *Clostridium difficile* is antigenic and immunogenic. Chem Biol 18:580–588

15. Anish C, Martin CE, Wahlbrink A et al (2013) Immunogenicity and diagnostic potential of synthetic antigenic cell surface glycans of Leishmania. ACS Chem Biol 8:2412–2422

16. Martin CE, Broecker F, Oberli MA et al (2013) Immunological evaluation of a synthetic *Clostridium difficile* oligosaccharide conjugate vaccine candidate and identification of a minimal epitope. J Am Chem Soc 135:9713–9722

17. Eriksson M, Serna S, Maglinao M et al (2014) Biological evaluation of multivalent lewis X-MGL-1 interactions. Chembiochem 15:844–851

18. Wenzel T, Sparbier K, Mieruch T et al (2006) 2,5-Dihydroxyacetophenone: a matrix for highly sensitive matrix-assisted laser desorption/ionization time-of-flight mass spectrometric analysis of proteins using manual and automated preparation techniques. Rapid Commun Mass Spectrom 20:785–789

19. Yang Y, Oishi S, Martin CE et al (2013) Diversity-oriented synthesis of inner core oligosaccharides of the lipopolysaccharide of pathogenic Gram-negative bacteria. J Am Chem Soc 135:6262–6271

20. Mechetner E (2007) Development and characterization of mouse hybridomas. Methods Mol Biol 378:1–13

21. Reeves JP, Reeves PA (2001) Removal of lymphoid organs. Curr Protoc Immunol Chapter 1, Unit 1.9

22. Geissner A, Anish C, Seeberger PH (2014) Glycan arrays as tools for infectious disease research. Curr Opin Chem Biol 18:38–45

23. Moelling K, Broecker F, Kerrigan JE (2014) RNase H: specificity, mechanisms of action, and antiviral target. Methods Mol Biol 1087:71–84

24. Miron T, Wilchek M (1982) A spectrophotometric assay for soluble and immobilized N-hydroxysuccinimide esters. Anal Biochem 126:433–435

25. Diehl KH, Hull R, Morton D, European Federation of Pharmaceutical Industries Association and European Centre for the Validation of Alternative Methods et al (2001) A good practice guide to the administration of substances and removal of blood, including routes and volumes. J Appl Toxicol 21:15–23

Chapter 6

Murine Whole-Blood Opsonophagocytosis Assay to Evaluate Protection by Antibodies Raised Against Encapsulated Extracellular Bacteria

Guillaume Goyette-Desjardins, René Roy, and Mariela Segura

Abstract

In vaccine development, especially against pathogenic encapsulated extracellular bacteria, functional assays such as the opsonophagocytosis assay (OPA) are preferred to ELISA titers for evaluating protection against infection. Such assays are normally performed using phagocytic cell lines or purified cell types, which underestimate the complexity of blood bactericidal activity. Here, we describe an OPA using murine whole-blood as effector cells, in a small format (0.2 ml), which requires small quantities of sera (80 μl or less) from immunized individuals. Easy to develop and perform, this OPA can be readily adapted to various pathogens and could be used to evaluate sera from human or animal clinical trials of carbohydrate-based vaccines.

Key words Opsonophagocytosis assay, Whole-blood, Serum, Correlate of protection, *Streptococci*, Vaccine development

1 Introduction

In vaccine development, correlates of immunity correspond to measurable signs showing that an individual is protected against an infection, such as specific antibody titers or as functional antibody activity [1–3]. The use of an opsonophagocytosis assay (OPA) as correlate of immunity is preferred to ELISA titers for evaluating protection against invasive bacterial diseases, such as pneumonia, meningitis, and septicemia [4]. OPAs have been mainly described for several encapsulated gram-positive bacteria, including *Streptococcus pneumoniae*. OPA is based on the fact that opsonization by specific immunoglobulins (antibodies) at the bacterial surface will activate the classical pathway of complement, leading to complement deposition [5]. Together, deposited immunoglobulins and/or complement components will be recognized by Fc receptors and complement receptors, respectively, triggering an enhanced immune response by blood leukocytes which results in

Bernd Lepenies (ed.), *Carbohydrate-Based Vaccines: Methods and Protocols*, Methods in Molecular Biology, vol. 1331, DOI 10.1007/978-1-4939-2874-3_6, © Springer Science+Business Media New York 2015

bacterial phagocytosis and bactericidal activity [6–8]. Specific cell type activation depends on the immunoglobulin isotypes present in the immune serum, since each isotype possesses different binding preferences to Fc receptors, which differently influences the cell response [8]. As such, OPAs are performed by mixing and incubating effector cells (phagocytes), the target (bacteria) and specific antibodies (naive or immune sera). The presence of specific and functionally active antibodies in the test sera will result in target elimination by effector cells. For gram-negative bacteria, the generally accepted format is the serum bactericidal assay, which uses serum complement as the sole effector for bacterial killing [9–11], mostly due to the important role played by the complement membrane attack complex in the direct lysis of gram-negative bacteria [12]. However, OPAs, as correlate of immunity, have also been applied to some gram-negative encapsulated bacteria, such as *Neisseria meningitidis* [13].

While most OPAs are performed using phagocytic cell lines, here we describe a method for such a test using a more complete murine whole-blood model [4]. Instead of using a cell line or a single purified cell type, the OPA requires whole blood from naive mice. This model takes into account all leukocytes present in the blood and thus represents a more realistic model of the complex interactions between all immune cells, plasma proteins/components, and bacteria during a systemic infection. Other advantages of this assay are the use of small volumes of reagents (only 80 μl of diluted whole-blood and 80 μl or less of serum for a final volume of 200 μl) and the fact that the same assay microtube can be used for multiple time-points.

To develop this method, we used *Streptococcus pneumoniae* serotype 14 and *Streptococcus suis* serotype 2 as target bacterial models. *Streptococcus pneumoniae* serotype 14 is one of the most important serotypes causing invasive pneumococcal disease and is still included in all multivalent glycoconjugate vaccines [14]. *Streptococcus suis* serotype 2 is an important swine pathogen and zoonotic agent, being the most frequently isolated and associated with disease, for which no vaccine is currently available [15]. Both pathogens are extracellular, and antibodies against surface-exposed bacterial components (such as capsular polysaccharides and cell wall proteins) play a major role in host defense to fight these infections [5–8]. We also present results for the negative and positive controls used for the OPA with *S. pneumoniae* serotype 14 and *S. suis* serotype 2. This developed method can be applied to other extracellular bacterial species, especially when measuring the activity of antibodies generated by glycoconjugate vaccines and directed against bacterial surface carbohydrates.

2   Materials

Prepare all solutions using fresh ultrapure deionized water (such as Milli-Q purified) and analytical grade reagents. All materials and reagents that are to be in contact with the bacteria used in the assay must be sterile and endotoxin-free (*see* **Note 1**). All manipulations, except before sterilization, must be performed aseptically either using a flame or under a biological cabinet. Prepare or store all reagents at 4 °C (unless indicated otherwise).

2.1   Bacterial Growth, Preparation and Viable Counts

1. Columbia Agar with 5 % sheep blood (Oxoid, Nepean, ON, Canada).

2. Todd-Hewitt Broth (THB; Becton Dickinson, Mississauga, ON, Canada). Prepare according to the manufacturer's instructions. Autoclave to sterilize.

3. Sterile 15 ml polypropylene conical tubes.

4. Sterile 100 mm polystyrene petri dishes.

5. Todd-Hewitt Broth Agar (THA) plates. Dissolve 30 g of THB (Becton Dickinson) in 1000 ml of water. Add 15 g of agarose. Autoclave to sterilize and dissolve agar. Keep the solution warm in a water bath heated to 65 °C until plates are poured Pour 15 ml of THA per plate aseptically. Let the plates cool down for about 1 h until solidified, then incubate for 24 h at 37 °C without CO_2 to confirm the absence of plate contamination. Store plates in their plastic sleeves at 4 °C.

6. Sterile and endotoxin/pyrogen-free phosphate buffered saline (PBS): 0.01 M $H_2PO_4^-$/HPO_4^{2-}, 0.138 M NaCl, 0.0027 M KCl, pH 7.4 (Gibco/Invitrogen, Burlington, ON, Canada).

7. Sterile 1.5 ml microtubes (Eppendorf, Hamburg, Germany).

8. 1.5 ml microtubes containing either 180 or 900 µl of PBS for serial dilutions. Store at room temperature.

2.2   Blood Collection and OPA

1. Sodium heparin solution: 140 USP/ml sodium heparin in PBS. Consult the lot certification for the specific activity (in USP/mg). Weigh the required quantity and dissolve in 5 ml of sterile PBS. Aseptically filter through a 0.22 µm syringe filter and aliquot 50 µl of the sterile sodium heparin solution into 1.5 ml microtubes.

2. Five- to eight-week-old female C57BL/6 mice (*see* **Note 2**).

3. Sterile 1 ml syringes mounted with a sterile 25 G 5/8″ needle.

4. 5 ml polypropylene conical tubes (Eppendorf) (*see* **Note 3**).

5. 1.5 ml polypropylene microtubes (Eppendorf).

6. Complete RPMI medium: RPMI 1640 (no glutamine, no HEPES, with phenol red) supplemented with 5 % heat-inactivated fetal bovine serum, 10 mM HEPES, 2 mM L-glutamine and 50 µM 2-mercaptoethanol (Gibco/Invitrogen). To 500 ml of RPMI 1640, add 25 ml of heat-inactivated fetal bovine serum (*see* **Note 4**), 5 ml of 1 M HEPES, 5 ml of 200 mM L-glutamine and 0.45 ml of 55 mM 2-mercaptoethanol.

7. Red Cell Lysis Buffer (eBiosciences, San Diego, California, USA).

8. Trypan blue solution.

9. Hemacytometer.

10. Inverted microscope for cell counting.

11. Sterile 25 G 5/8″ needle.

3   Methods

3.1   Bacterial Culture and Preparation for OPA

Here we present a general protocol for the growth and preparation of *Streptococci*. Please adapt growth medium and optimal growth conditions to your pathogen accordingly. All microbiological manipulations should be performed aseptically using either a flame or under a biological cabinet (*see* **Note 1**). Follow diligently all governmental and institutional regulations regarding the manipulation of pathogens.

1. On day −3, prepare a fresh bacterial culture by plating on blood agar. Incubate for 16–24 h at 37 °C with 5 % CO_2 (*see* **Note 5**).

2. On day −2, check the purity of the culture. Plate three colonies using an inoculation loop onto a new blood agar plate and incubate for 16–24 h at 37 °C with 5 % CO_2.

3. On day −1, check the purity of the culture. Inoculate 5 ml of THB using an inoculation loop with three colonies. Mix the contents of the tube for a few seconds using a vortex. Incubate the tubes standing for 16 h at 37 °C with 5 % CO_2.

4. On the day of the OPA, inoculate 10 ml of fresh THB with 0.1 ml from the previous 16 h-culture. Mix the tube for a few seconds using a vortex. Incubate the tubes standing at 37 °C with 5 % CO_2 until the culture reaches the mid-logarithmic phase.

5. Wash the bacteria by centrifuging at $7400 \times g$ for 5 min at 4 °C and resuspending the pellet in 10 ml of PBS.

6. Repeat the wash a second time and resuspend the final pellet in 5 ml of PBS. Using a spectrophotometer, adjust the bacterial suspension by diluting with PBS in order to reach an $OD_{600} = 0.6$. Bacterial cultures must have been previously

Table 1
Required concentrations for the diluted bacterial and whole-blood suspensions in function of the desired multiplicity of infection (MOI)

		Concentrations for desired MOI (CFU/ml or leukocytes/ml)				
		1	0.5	0.1	0.05	0.01
Bacteria	Diluted (to add)	1.25×10^7	6.25×10^6	1.25×10^6	6.25×10^5	1.25×10^5
	Final (Assay)	2.5×10^6	1.25×10^6	2.5×10^5	1.25×10^5	2.5×10^4
Leukocytes (whole-blood)	Diluted (to add)			6.25×10^6		
	Final (Assay)[a]			2.5×10^6		

[a]For a total of 5×10^5 leukocytes in a final volume of 200 µl

standardized in order to accurately determine the bacterial concentration in colony-forming units (CFU)/ml at this specific turbidity. Once the culture conditions are standardized, always use the same protocol.

7. Perform serial dilutions in complete RPMI medium to obtain the desired bacterial concentration (in CFU/ml) for the assay (*see* Table 1). Keep the final suspension of bacteria in complete RPMI medium on ice until you are ready to perform the OPA (*see* **Notes 6** and **7**).

3.2 Blood Collection and Preparation for OPA

All experiments and manipulations involving animals must be conducted in accordance with the governmental and institutional guidelines and policies. Personnel must be qualified to handle laboratory animals. All manipulations should be performed aseptically and all blood suspensions must be kept at room temperature (*see* **Note 8**).

1. Humanely euthanize one mouse at a time and collect blood by intracardiac puncture using a 1 ml syringe mounted with a 25 G 5/8″ needle (*see* **Note 9**).

2. Remove the needle and distribute the blood by adding approximately 450 µl of blood in the microtubes containing the sodium heparin solution (*see* **Note 10**).

3. Quickly mix the microtubes by gentle hand-agitation, ten times (*see* **Note 11**).

4. Pool the blood recovered from all of the mice by transferring the blood from the microtubes to a 5 ml tube (*see* **Note 3**).

5. Count the leukocytes (white blood cells) from the pooled blood and dilute the pooled blood with complete RPMI medium to a final concentration of 6.25×10^6 leukocytes/ml (*see* **Note 12**).

3.3 Opsonophagocy-tosis Assay (OPA)

All microbiological manipulations should be performed aseptically using either a flame or under a biological cabinet. Follow diligently all governmental and institutional regulations regarding the manipulation of pathogens and biological samples.

Internal controls: to perform the OPA, it is important to have, in advance, a positive control sera or polyclonal or monoclonal antibodies directed against the target pathogen. In addition, respective matching negative control sera or polyclonal or monoclonal antibodies must be included as negative control. The positive and negative control sera/antibodies will be used as internal controls (*see* **Note 13**). The concentration of positive and negative control sera/antibodies to be added to the OPA must be standardized in advance. The goal is to obtain >90 % of killing with the positive internal control.

The test sera must include not only the sera from placebo and immunized animal groups but also sera from naive (strict control) animals from the same animal lot as that used for the vaccine trial (*see* **Note 14**).

1. Distribute the diluted whole-blood (effector cells) and test sera (including naive, placebo and immunized sera) to the 1.5 ml microtubes. The volumes to be added to obtain the different OPA conditions are given in Table 2. Also include microtubes containing the standardized concentration of internal control sera/antibodies. Mix well the blood and sera together by gently tapping with a finger on the side of the microtube (*see* **Note 15**).

2. Distribute the final suspension of bacteria to the assay microtubes. The concentrations for the 40 µl volumes to be added are given in Table 1. This step corresponds to the start of the assay ($t = 0$ min). Mix well the contents by gently tapping with a finger on the side of the microtube (*see* **Note 15**).

Table 2
Volumes of the components to be added to the OPA microtubes, in respective order

	Final test serum concentration			
	40 %	**20 %**	**10 %**	**5 %**
Diluted whole-blood 6.25×10^6 *leukocytes/ml*	80 µl	80 µl	80 µl	80 µl
Test serum (*naive, placebo, or immunized*)	80 µl	40 µl	20 µl	10 µl
Complete RPMI medium	0 µl	40 µl	60 µl	70 µl
Diluted bacteria[a]	40 µl	40 µl	40 µl	40 µl
Total volume	200 µl	200 µl	200 µl	200 µl

[a]Different multiplicity of infection ratios can be performed; *see* Table 1 for details on how to obtain the desired bacterial concentration

3. Pierce the top of every assay microtube with a sterile 25 G 5/8″ needle to allow cell respiration.

4. Incubate the assay microtubes at 37 °C with 5 % CO_2. Every 20 min, mix the microtube contents by gently tapping the side in order to increase contact between cells and bacteria (*see* **Note 15**). Optimal incubation times vary from one target pathogen to another and must be standardized.

3.4 Bacterial Viable Counts

All microbiological manipulations should be performed aseptically under a biological cabinet.

1. If several incubation times are being tested, you can use the same OPA microtube by retrieving a small volume at each time point. To this aim, at the desired incubation time points, mix well each assay tube (by gently tapping on the side), retrieve 20 μl of the content, and return the assay tube to the incubator. Dilute the 20 μl sample in a microtube containing 180 μl of PBS (10^{-1} dilution).

2. Mix the 10^{-1} dilution thoroughly by vortexing. Perform tenfold serial dilutions by retrieving 100 μl from the previous dilution and diluting it in a microtube containing 900 μl of PBS. Mix thoroughly each microtube by vortexing (*see* **Note 16**).

3. Spread aseptically either 50 or 100 μl from selected dilutions on THA plates (*see* **Note 16**).

4. Incubate all plates for 24–48 h at 37 °C with 5 % CO_2.

5. After incubation, count the CFU and determine the bacterial killing using the following formula:

$$\text{Killing \%} = \left(1 - \frac{\text{CFU from test serum}}{\text{CFU from naive serum}}\right) \times 100$$

Where "CFU from test serum" represents the CFU counts from one OPA test tube incubated with either placebo or immunized sera, and "CFU from naive serum" represents the average CFU counts from three OPA test tubes incubated with the pool of naive sera included each time the assay is performed (*see* **Note 17**).

The killing percentage represents the proportion of bacteria killed in a test tube compared to the tubes incubated with naive sera. In the case where CFU from test serum is higher than CFU from naive serum (resulting in a negative killing value), report killing percentage value as 0. The test sera (placebo and immunized) can be performed in duplicate or triplicate if the amount of serum sample is sufficient.

For internal controls, killing percentages are calculated using the same mathematical formula, except by using matched positive and negative serum/antibodies as "test serum" and "naive serum" values, respectively.

3.5 Examples of Results and Further Applications

Examples of results obtained with various positive and negative serum controls for *S. pneumoniae* and *S. suis* are shown in Figs. 1 and 2, respectively. These results represent the different bacterial behaviors that would be observed during initial standardization of

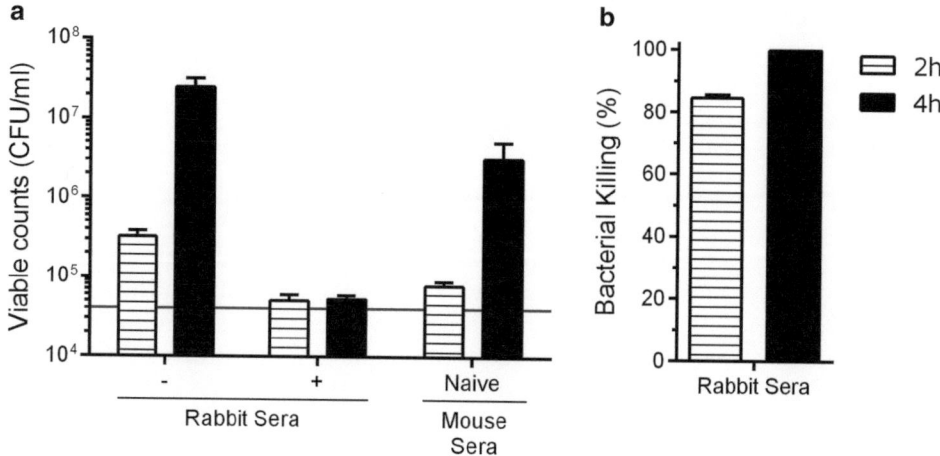

Fig. 1 Optimal assay conditions for *S. pneumoniae* serotype 14 are a multiplicity of infection (MOI) of 0.01, an incubation time of 4 h, and addition of 20 % of test sera. (**a**) Viable counts with an MOI of 0.01 obtained at 2 and 4 h for negative (–) and positive (+) rabbit sera (internal controls) and for a pool of naive mouse sera (at 20 %). The *grey line* represents the bacterial inoculum value (time = 0 min). (**b**) Bacterial killing values at 2 and 4 h were calculated for positive rabbit control serum (using matching negative rabbit serum for the formula). The OPA was performed at a serum concentration of 20 % and an MOI of 0.01. Results are expressed as mean ± SEM from at least three independent experiments

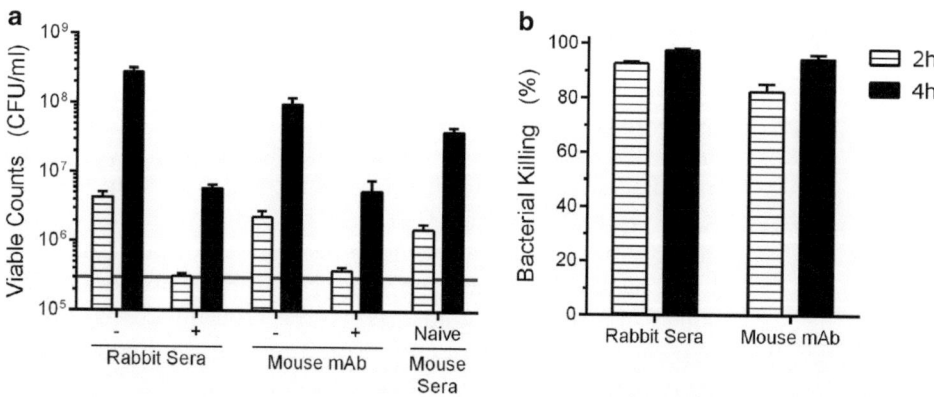

Fig. 2 Optimal assay conditions for *S. suis* serotype 2 are a multiplicity of infection (MOI) of 0.1, an incubation time of 2 h, and addition of 40 % of test sera. (**a**) Viable counts with an MOI of 0.1 obtained at 2 and 4 h for negative (–) and positive (+) rabbit sera or matched positive (anti-*S. suis*) and negative monoclonal antibodies (mAb), used as internal controls or for a pool of naive mouse sera (at 40 %). The *grey line* represents the bacterial inoculum value (time = 0 min). (**b**) Bacterial killing values at 2 and 4 h were calculated for positive rabbit control serum and for positive control mAb (using matching negative rabbit serum or negative mAb for the formula, respectively). The OPA was performed at a serum/mAb concentration of 40 % and an MOI of 0.1. Results are expressed as mean ± SEM from at least three independent experiments

the OPA and allow selection of optimal multiplicity of infection (MOI), incubation time, and/or serum concentrations to be used.

The OPA method described in this chapter has been developed for mouse test sera and rabbit sera (mainly used as internal control). Immunoglobulins share common structures due to their function in immunity, thus enabling cross-species functional antibodies. As such, this OPA method could be easily adapted for other test sera, such as those from human vaccine clinical trials.

4 Notes

1. Working in sterile and endotoxin-free conditions will ensure intra- and inter-assay consistency. Microbial or endotoxin contamination might activate leukocytes and lead to high variations in the results.

2. Using mice with the same genetic background will ensure inter-assay consistency. Inbred mice, such as C57BL/6 or BALB/c, are recommended. Avoid using aged mice as the functionality of immune cells might be compromised. Both male and female mice can be used. We preferred female mice as they are less aggressive than males for housing and handling.

3. A conical 15 ml tube is too long and too narrow in order to retrieve the diluted blood with a 1000 μl micropipette without risking contamination. Conical 5 ml tubes (Eppendorf) were found to work perfectly for this purpose.

4. To obtain the greatest reproducibility, when preparing cell culture medium always use certified serum grade, which is guaranteed to be exempt of endotoxin contamination.

5. When starting a culture from a frozen aliquot, bacterial growth is more uniform and consistent after a minimum of two passages on blood agar. Blood agar plates used to start broth cultures can be kept for up to 1 week at 4 °C with Parafilm to prevent dehydration. After 1 week of storage, a new blood agar plate should be inoculated.

6. When preparing the inoculum by diluting bacteria in complete RPMI medium, prepare at least an extra 0.5 ml for viable counts and purity verification.

7. If one person alone performs the assay, it is advisable to prepare, wash, and dilute in complete RPMI medium the bacterial culture first, and keep it on ice while collecting mouse blood. During this waiting time, bacterial growth and/or death is greatly reduced if kept on ice. Nevertheless, this waiting time should not be longer than 3–4 h. When the OPA assay is started, viable counts for the initial inoculum (at $t = 0$ min) can be performed. Control the purity of your final suspension by inoculating on blood agar.

8. Keeping blood suspensions at room temperature, instead of keeping on ice or at 4 °C, helps to prevent hemolysis/cell lysis.

9. Depending on the age of the mice and the experience of the person performing the intracardiac puncture, 0.4–0.8 ml of blood is normally obtained from a single mouse. In this case, depending on number of tubes required for the assay, several mice are necessary in order to provide sufficient blood volume. It is recommended to perform one mouse at a time to maximize the amount of blood recovered per mouse.

10. It is important to remove the needle before flushing the blood out of the syringe to prevent lysis of effector cells. Indeed, flushing the blood through a narrow needle will cause animal cells to lyse.

11. From the moment the mouse is sacrificed, blood clotting begins. Thus, blood collection by intracardiac puncture, transfer to a microtube and mixing with the heparin solution must be performed as quickly as possible, in less than 3 min per mouse.

12. To count the leukocytes, aliquot 25 µl of the pooled blood in a microtube, add 250 µl of 1× Red Cell Lysis Buffer and incubate for 5 min at room temperature. After incubation, add 500 µl of room temperature PBS and quickly spin at $13,000 \times g$ for 12 s at room temperature. Discard the supernatant, resuspend the cells with 25 µl of PBS, and dilute 1/20 with PBS followed by a 1/5 dilution with Trypan Blue (final dilution of 1/100). Determine the leukocyte concentration by counting the cells using a hemacytometer. We normally obtain approximately 7.5×10^6 leukocytes/ml of whole-blood.

13. Internal controls: when designing the assay, always include matching positive and negative sera, polyclonal antibodies or isotype-matched monoclonal antibodies as internal controls to ensure inter-assay consistency. These control sera/antibodies can be produced in your own laboratory or purchased when commercially available for your pathogen. The species origin of these sera/antibodies can be different from mouse. For example, rabbit serum works very well as an internal control. If inconsistencies are observed in the values for positive and negative internal controls between tests, results should be discarded and the assay repeated. If inconsistencies persist, perform troubleshooting to find the problem (i.e., bacterial culture, cell viability, mouse lot, etc.).

14. Naive mouse sera: when testing immunized/placebo mouse sera from the vaccine trial, sera from a number of naive mice of the same age, sex, and genetic background (same lot if possible) might be collected and pooled to create the reference negative (naive) control used in the mathematical formula to calculate the % of killing. Do not use placebo sera as "naive sera"

since animals might react to the injection with vehicle solution used as placebo (especially when containing adjuvants). Consequently, sera from placebo animals might give variable % of killing in the OPA test.

15. Whenever whole-blood is involved, mixing by using a vortex is highly discouraged. Such agitation will cause lysis of the leukocytes and result in a lower killing rates.

16. To perform viable bacterial counts, long vortex mixing times such as ≥30 s are recommended to ensure maximal recovery of viable bacteria (internalized or not). In our experience, there was no need for an additional cell lysis step. The optimal dilutions to plate will depend on the chosen plating method (i.e., manual plating or automatic plating). It is recommended to use a large number of dilutions during preliminary standardization and then, based on these results, to select ranges of dilutions that will allow an accurate CFU counting. In the case of both negative and positive internal controls, a range of two dilutions will be sufficient. For the test sera, select at least four dilutions: two high range dilutions if a negative/low killing value is expected and two low range dilutions if a high % of killing is expected.

17. Naive serum assay tubes must be performed in each assay in triplicate in order to eliminate inter-tube variations. The average for the viable counts is then used in the formula to determine killing percentages.

Acknowledgement

This work was mainly supported by the Natural Sciences and Engineering Research Council of Canada (NSERC) through a grant to MS (#342150). Partial contribution was also provided by Canadian Institutes of Health Research grant to MS and RR. The authors wish to thank Dr. Mario Feldman (University of Alberta) and Dr. Marcelo Gottschalk (University of Montreal) for their generous gifts of reference antisera and antibodies against *S. pneumoniae* and *S. suis*, respectively.

References

1. Plotkin SA (2001) Immunologic correlates of protection induced by vaccination. Pediatr Infect Dis J 20:63–75

2. Plotkin SA (2008) Vaccines: correlates of vaccine-induced immunity. Clin Infect Dis 47:401–409

3. Plotkin SA (2010) Correlates of protection induced by vaccination. Clin Vaccine Immunol 17:1055–1065

4. Song JY, Moseley MA, Burton RL et al (2013) Pneumococcal vaccine and opsonic pneumococcal antibody. J Infect Chemother 19:412–425

5. Abbas AK, Lichtman AH, Pillai S (2007) Effector mechanisms of humoral immunity. In: Schmidtt W (ed) Cellular and Molecular Immunology, 6th edn. Saunders, Philadelphia, PA, pp 321–348

6. Underhill DM, Ozinsky A (2002) Phagocytosis of microbes: complexity in action. Annu Rev Immunol 20:825–852

7. Ricklin D, Hajishengallis G, Yang K et al (2010) Complement: a key system for immune surveillance and homeostasis. Nat Immunol 11:785–797

8. Guilliams M, Bruhns P, Saeys Y et al (2014) The function of Fcg receptors in dendritic cells and macrophages. Nat Rev Immunol 14:94–108

9. Borrow R, Balmer P, Miller E (2005) Meningococcal surrogates of protection–serum bactericidal antibody activity. Vaccine 23:2222–2227

10. Martin D, McCallum L, Glennie A et al (2005) Validation of the serum bactericidal assay for measurement of functional antibodies against group B *meningococci* associated with vaccine trials. Vaccine 23:2218–2221

11. Balmer P, Borrow R (2004) Serologic correlates of protection for evaluating the response to meningococcal vaccines. Expert Rev Vaccines 3:77–87

12. Groves E, Dart AE, Covarelli V et al (2008) Molecular mechanisms of phagocytic uptake in mammalian cells. Cell Mol Life Sci 65:1957–1976

13. Ison CA, Anwar N, Cole MJ et al (1999) Assessment of immune response to meningococcal disease: comparison of a whole-blood assay and the serum bactericidal assay. Microb Pathog 27:207–214

14. Tan TQ (2012) Pediatric invasive pneumococcal disease in the United States in the era of pneumococcal conjugate vaccines. Clin Microbiol Rev 25:409–419

15. Goyette-Desjardins G, Auger JP, Xu J et al (2014) *Streptococcus suis*, an important pig pathogen and emerging zoonotic agent-an update on the worldwide distribution based on serotyping and sequence typing. Emerg Microbes Infect 3:e45

Chapter 7

Determination of *N*-linked Glycosylation in Viral Glycoproteins by Negative Ion Mass Spectrometry and Ion Mobility

David Bitto, David J. Harvey, Steinar Halldorsson, Katie J. Doores, Laura K. Pritchard, Juha T. Huiskonen, Thomas A. Bowden, and Max Crispin

Abstract

Glycan analysis of virion-derived glycoproteins is challenging due to the difficulties in glycoprotein isolation and low sample abundance. Here, we describe how ion mobility mass spectrometry can be used to obtain spectra from virion samples. We also describe how negative ion fragmentation of glycans can be used to probe structural features of virion glycans.

Key words Virus, Glycosylation, Structure, Mass spectrometry, Glycoprotein

Abbreviations

2-AA	2-Aminobenzoic acid (anthranilic acid)
2-AB	2-Aminobenzamide
BHK	Baby hamster kidney
CCD	Charge-coupled device
CID	Collision-induced dissociation
DC-SIGN	Dendritic cell-specific intercellular adhesion molecule-3-grabbing non-integrin
DHB	2,5-Dihydroxybenzoic acid
DMSO	Dimethylsulfoxide
DMT-MM	4-(4,6-Dimethoxy-1,3,5-triazin-2-yl)-4-methylmorpholinium chloride
EDTA	Ethylenediaminetetraacetate
ESI	Electrospray ionization
Fuc	Fucose
Gal	Galactose
GC/MS	Gas chromatography/mass spectrometry
GlcNAc	*N*-acetylglucosamine

Bernd Lepenies (ed.), *Carbohydrate-Based Vaccines: Methods and Protocols*, Methods in Molecular Biology, vol. 1331, DOI 10.1007/978-1-4939-2874-3_7, © Springer Science+Business Media New York 2015

GMEM	Glasgow's Minimum Essential Medium
H20N100E2	20 mM HEPES 100 mM NaCl, 2 mM EDTA
HEPES	4-(2-Hydroxyethyl)piperazine-1-ethanesulfonic acid
HIV	Human immunodeficiency virus
HPLC	High-performance liquid chromatography
MALDI	Matrix-assisted laser desorption/ionization
Man	Mannose
MoI	Multiplicity of infection
MS	Mass spectrometry
Neu5Ac	N-acetylneuraminic acid (sialic acid)
Neu5Gc	N-glycoylneuraminic acid
PCR	Polymerase chain reaction
PGC	Porous graphitized carbon
PNGase F	Protein-N-glycosidase F
PBS	Phosphate-buffered saline
Q	Quadrupole
SDS-PAGE	Sodium dodecylsulfate-polyacrylamide gel electrophoresis
THAP	2,4,6-Trihydroxyacetophenone
TOF	Time-of-flight
Tris-base	2-Amino-2-(hydroxymethyl)-1,3-propanediol
UUKV	Uukuniemi virus

1   Introduction

Membrane-embedded glycoproteins that extend from the virion surface are key determinants of host cell tropism and pathobiology [1]. An understanding of the structure of such viral glycoproteins is key for revealing mechanisms of host infection and molecular targets of the immune response.

Biophysical tools such as macromolecular crystallography and electron microscopy are important for studying viral glycoprotein functionality at the amino-acid level [2]. However, due to the inherent flexibility and heterogeneity of glycans, these techniques are frequently unsuitable for defining glycan structure and composition. As a result, glycans are often removed prior to structural analysis [3] and, as such, their structure and function remain poorly understood with respect to protein-counterparts.

Virion-presented N-linked glycosylation is known to play multiple roles in the virus-life cycle. Whilst often a basic requirement for host-directed protein biosynthesis and folding, more diverse roles have been identified in processes such as host cell infection and immune evasion. For example, the glycoproteins from a number of viruses, including flaviviruses, alphaviruses, and phleboviruses, have been observed to display high-mannose type glycans, which act as attachment receptors for host C-type lectins, such as DC-SIGN [4]. N-linked glycans are also known to play a role in viral immune evasion. For example, in the case of human immunodeficiency virus

(HIV)-1, where glycosylation content accounts for around 50 % of the molecular mass of the gp120 viral attachment protein, the high-mannose content of the "glycan shield" creates an immunologically "non-self" mannose patch [5–7]. The recent discovery of a library of monoclonal antibodies that target both these glycan patches and mixed protein-glycan epitopes [8–13] provides hope for the future development of an anti-glycan inducing vaccine [14–16].

The extent to which host-derived biosynthetic enzymes influence viral glycan composition is fundamentally dependent upon the extraneous protein environment during glycoprotein biosynthesis and folding [17]. As such, accurate glycan structure and compositional analysis necessitates a sample most closely resembling that produced in infected tissue. Here, using Uukuniemi phlebovirus (UUKV) as a model system, we present a robust ion mobility mass spectrometry-based methodology for compositional and structural analysis of *N*-linked glycans derived from intact virions produced in cell culture (Fig. 1), the results of which have recently been reported [18]. This methodology relies upon the production of viral particles to a purity sufficient for isolation by SDS-PAGE analysis. To confirm virion integrity, electron microscopy is utilized as a validation tool, prior to mass spectrometric analysis. We anticipate that the sensitivity of this approach will enable study of low-titre and difficult-to-isolate viruses, previously not tractable for glycan analysis.

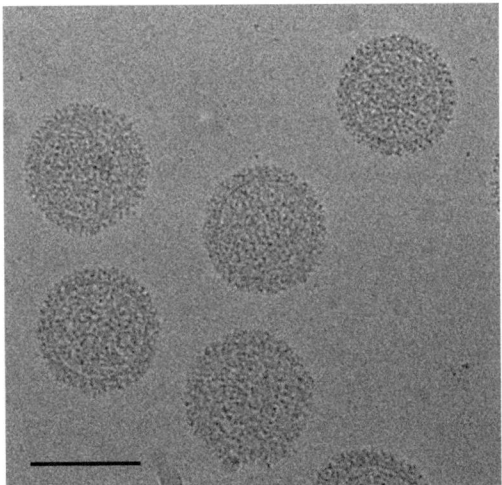

Fig. 1 Two-dimensional projection image of concentrated UUKV supernatant, plunge-frozen at liquid nitrogen temperature. Image was acquired on a FEI Tecnai Polara electron microscope, at −5 μm defocus, 300 kV, and a nominal magnification of 59,000× (75,000× calibrated magnification), leading to a calibrated pixel size of 2 Å/pixel. The images were recorded with a charge-coupled device (CCD) camera (Ultrascan 4000, Gatan, USA), in low dose mode, using 20 electrons/Å². Scale bar corresponds to 100 nm

1.1 Mass Spectrometry

1.1.1 Ionization

Mass spectra of the *N*-glycans can be obtained using either matrix-assisted laser desorption/ionization (MALDI) or electrospray ionization (ESI). MALDI in positive ion mode is most appropriate for obtaining profiles of neutral glycans because of the production of singly charged [M+Na]+ ions. Sialylated glycans, however, are relatively unstable and readily eliminate sialic acid as the result of the mobile nature of the carboxyl proton. This decomposition can be prevented by derivatization of the carboxyl group by esterification [19, 20] or amide formation [21]. Methyl esters can conveniently be prepared by reaction of the sodium salts of the sialic acids with methyl iodide [19]. A more useful reaction is that with methanol catalyzed by 4-(4,6-dimethoxy-1,3,5-triazin-2-yl)-4-methylmorpholinium chloride (DMT-MM). This reaction produces methyl esters from α2-6-linked sialic acids and lactones from α2-3-linked acids [20]. The mass difference between these products is 32 units, allowing the linkage of the sialic acids to be determined directly by mass measurement. The same catalyst can be used to synthesize amides.

ESI has the disadvantage of producing fragmentation in the ion source of the mass spectrometer and production of ions in several charge states, often leading to several ions from each glycan, and hence the advantage of MALDI for glycan profiling. In addition, several adducts are sometimes produced, particularly in negative ion mode, further complicating the spectra. This situation, however, can be improved by adding various salts to direct adduct formation in only one direction, as described below

1.1.2 Fragmentation

Carbohydrates fragment predominantly by two mechanisms: glycosidic cleavages between the sugar rings and cross-ring cleavages formed by cleavage of two bonds within the rings. Glycosidic cleavages give information on constituent monosaccharide sequences, whereas the cross-ring cleavages yield information on the positions of attachment of the various constituent sugars. Although fragmentation in positive ion mode has been the favored method in the past, negative ion fragmentation has been shown to produce simpler but much more structurally informative spectra [22–25], and is used here. Main disadvantages of positive ion spectra are the production of glycosidic fragments by loss of residues from several sites and the generally low abundance of cross-ring cleavage ions. Negative ion spectra, on the other hand, contain abundant cross-ring cleavage products as the result of proton loss from specific hydroxyl groups. They enable structural features, such as the branching pattern of the glycan, the location of fucose, the presence of bisects, and the number of isomers to be determined directly; features that are difficult to determine by positive ion fragmentation. Some typical spectra are shown in Figs. 2 and 3 and details of the diagnostic ions are given in Subheading 3.5.3.1.

In negative ion mode, carbohydrates naturally form [M–H]− ions or [M–Hn]n− ions if several acid groups are present. [M–H]−

ions are relatively unstable leading to extensive fragment ion production. However, neutral glycans can also be made to form stable $[M+X]^-$ ions where X is an anion such as a halogen, nitrate, phosphate, or sulfate. Nitrate, chloride, and phosphate adducts all fragment similarly by first eliminating the adduct together with a proton to leave what is essentially a $[M-H]^-$ ion. Although nitrate adducts give the cleanest glycan profiles, samples from biological sources invariably contain phosphate and some chloride. Thus, for the present work, formation of $[M+H_2PO_4]^-$ ions is maximized by addition of ammonium phosphate to the sample solution. Sulfate and iodide adducts give very strong spectra but do not produce fragment ions. Fragmentation spectra of sialylated glycans are not as informative as those of the $[M+H_2PO_4]^-$ ions from the neutral glycans because of formation of $[M-Hn]n^-$ ions. These ions are formed by loss of protons from the sialic acids rather than the OH groups, thus inhibiting the formation of the main diagnostic ions that were present in the spectra of the neutral glycans. Derivatization, as described above, removes this problem.

The scheme that is universally used to name the fragment ions is that devised by Domon and Costello in 1988 [26]. For ions with charge retention on the reducing end of the ion, glycosidic cleavage ions are labeled Y (cleavage on the non-reducing end of the linking oxygen) and Z, with subscript numbers starting with 1 for the reducing terminal glycan as shown in Fig. 4. Corresponding glycosidic cleavages with charge retention on the non-reducing end of the ion are labeled B and C, with subscript numbers starting from the non-reducing end. Cross ring cleavages are A and X, with preceding superscript numbers denoting which bonds are cleaved. Negative ion spectra tend to contain larger amounts of B, C and particularly A-type fragments. We have modified this system somewhat when discussing fragmentation of the reducing terminus. Here, under the Domon and Costello system, the subscript numbers change as the result of differing chain lengths. In order to avoid the subsequent confusion, A, B, and C ions are given the subscript R for cleavages at the reducing terminal GlcNAc, R-1 for the penultimate GlcNAc and R-2 for the branching mannose. Ions formed by a specific loss of the 3-antenna and chitobiose core, i.e., they contain the intact 6-antenna and branching mannose, are called D ions. A cross ring cleavage ion from the mannose residue in the 3-antenna containing carbons 1–4 and, consequently the chains linked to carbons 2 and 4 in some triantennary glycans is referred to as an E-type ion. The D and E nomenclature is not part of the Domon and Costello system.

1.1.3   Ion Mobility

Work reported in this paper uses the Waters Synapt G2 mass spectrometer, which has a Q-Tof-type configuration with a traveling-wave ion mobility cell positioned between the quadrupole and the TOF analyzer. A trap collision cell precedes the ion mobility cell

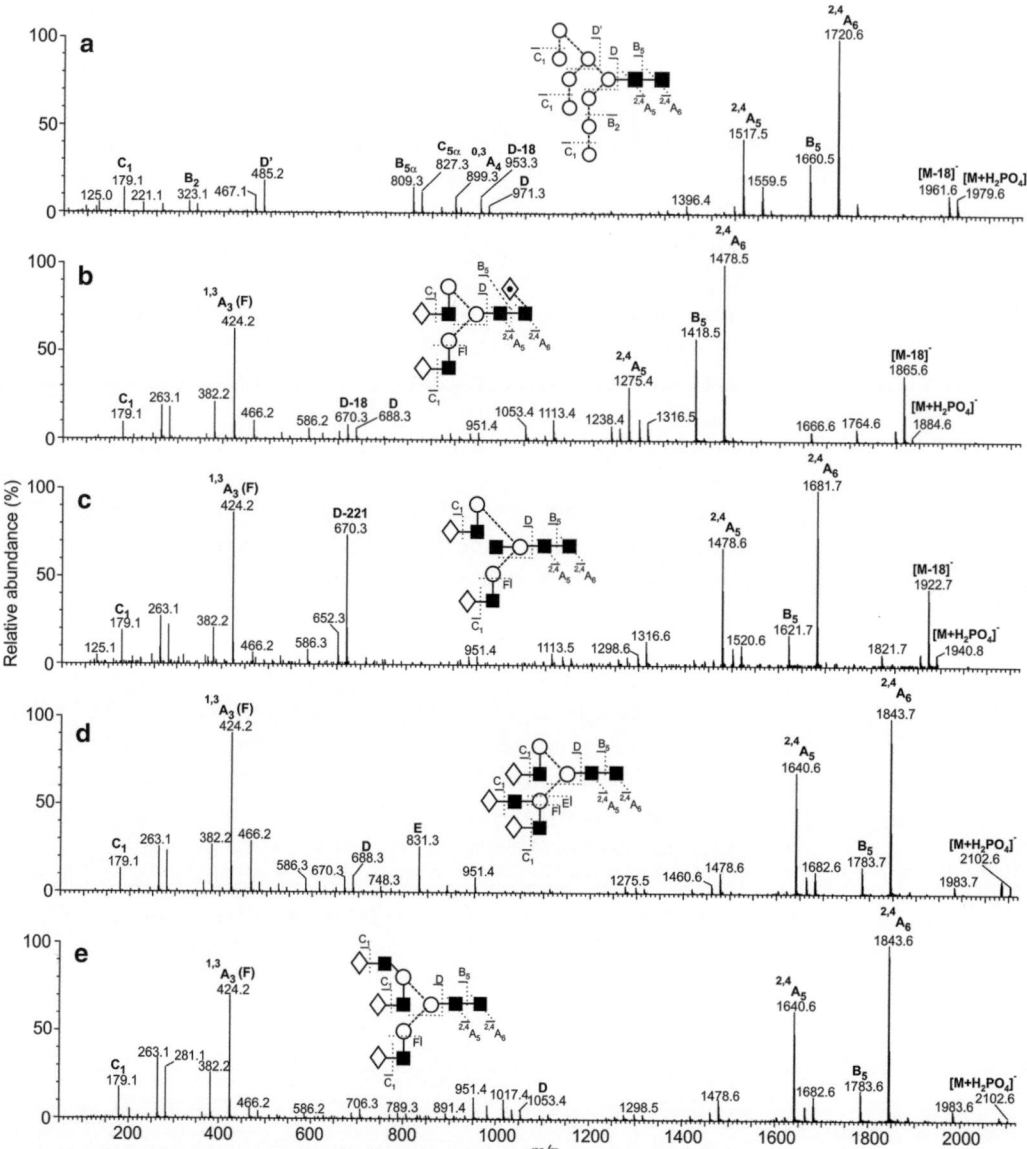

Fig. 2 Negative ion CID spectra of typical *N*-glycans (phosphate adducts). (**a**) High-mannose glycan Man$_9$GlcNAc$_2$. ^{2,4}A$_6$, B$_5$, and ^{2,4}A$_5$ ions defining the trimannosylchitobiose core are at *m/z* 1720, 1660, and 1517 respectively. The C$_1$ ion at *m/z* 179 shows hexose (mannose) at the non-reducing termini of the antennae. The D, D-18, ^{0,4}A$_4$, ^{0,3}A$_4$, and B$_{3\alpha}$ ions defining the composition of the 6-antenna are at *m/z* 971, 953, 899, 869, and 809, respectively, and the D′ ion at *m/z* 485 shows the composition of the 6-branch of the 6-antenna. (**b**) Fucosylated biantennary glycan Gal$_2$Man$_3$GlcNAc$_4$Fuc$_1$. The presence of the 6-linked core fucose is revealed by the 405 mass unit difference between the molecular ion and the ^{2,4}A$_6$ ion at *m/z* 1478.5. The C$_1$ ion at *m/z* 179 shows hexose (galactose) at the termini of the antennae, and the ^{1,3}A$_3$ cross-ring ion at *m/z* 424 (labeled *F*) confirms the Gal-GlcNAc composition of the antennae. The D and D-18 ions at *m/z* 688 and 670, respectively, show the composition of the 6-antenna. (**c**) Bisected biantennary glycan Gal$_2$Man$_3$GlcNAc$_5$. Most diagnostic ions are as for the glycans in spectra (**a**) and (**b**). The presence of the bisecting GlcNAc is revealed by the very prominent ion at *m/z* 670 corresponding to D-221 mass units. (**d**) Triantennary glycan Gal$_3$Man$_3$GlcNAc$_5$. Branching of the 3-antenna gives rise to the ion at *m/z* 831 (labeled *E*). The unbranched 6-antenna produces

and a second collision cell, known as the transfer cell follows it. For the work reported here, collision-induced decomposition (CID) is performed in the transfer cell. The ion mobility cell separates ions on the basis of charge and shape and is used in this work both to separate ions in different charge states and to remove contaminants [27, 28]. Figure 5a shows the ESI spectra of *N*-glycans from UUKV. The spectrum is weak, as would be expected from the small amount of material available. Figure 6 shows a plot (Waters Driftscope) of ion mobility drift time against m/z showing fractionation of the glycan ions according to charge. By selecting the regions highlighted in the figure, the profiles of glycans with one, two and three/four charges can be extracted and displayed as in panels c, d, and e of Fig. 5. Not only does this method allow ions in specific charge states to be extracted, but much of the chemical noise is rejected, giving clean spectra with good signal–noise ratios. Ion mobility can be used in the same way to clean fragment ions by removing those fragments produced by co-selected parent ions with equivalent m/z values. A good example of this is the singly charged ion at m/z 1007 corresponding to $Man_3GlcNAc_2$ and the triply charged ion from $Gal_3Man_3GlcNAc_5Fuc_1Neu5Ac_3$, both of which are commonly found in *N*-glycan mixtures.

Negative ion fragmentation combined with ion mobility can be used to detect the presence of isomeric glycans. Such compounds cannot be separated by mass spectrometry alone but, frequently, their negative ion fragmentation spectra contain ions of different mass. By plotting arrival time distributions of these fragments it can often be seen that there is a small separation between them, confirming the presence of isomers. Another parameter that can be derived is the collisional cross section of the ions. This parameter is independent of the instrument used to record the cross section and can, thus, be used in databases in a similar way to the use of glucose units in HPLC.

2   Materials

1. BHK-21 cells (in this work provided by Dr. Anna Överby, Umeå, Sweden), cultured in Glasgow's Minimum Essential Medium (GMEM, Life Technologies, UK), supplemented with

---

Fig. 2 (continued) D and D-18 ions at m/z 688 and 670 as in the spectrum of the biantennary glycan (spectrum **b**). (**e**) Triantennary glycan $Gal_3Man_3GlcNAc_5$. The branching pattern is shown by the absence of ion E and the shift in the D and D-18 ions to m/z 1053 and 1035 respectively. These ions are accompanied by a third ion at m/z 1017 (D-36). Fragments are named according to the scheme proposed by Domon and Costello (*see* Fig. 4) [26]. Key to symbols used for the structural diagrams: ◇ = Gal, ■ = GlcNAc, ○ = Man, ◈ = Fuc, ★ = sialic acid. The angle of the lines linking the symbols denotes the linkage position (I = 2-link, / = 3-link, - = 4-link, \ =6-link) with full lines for β-bonds and broken lines for α-bonds. For further details, *see* [47]

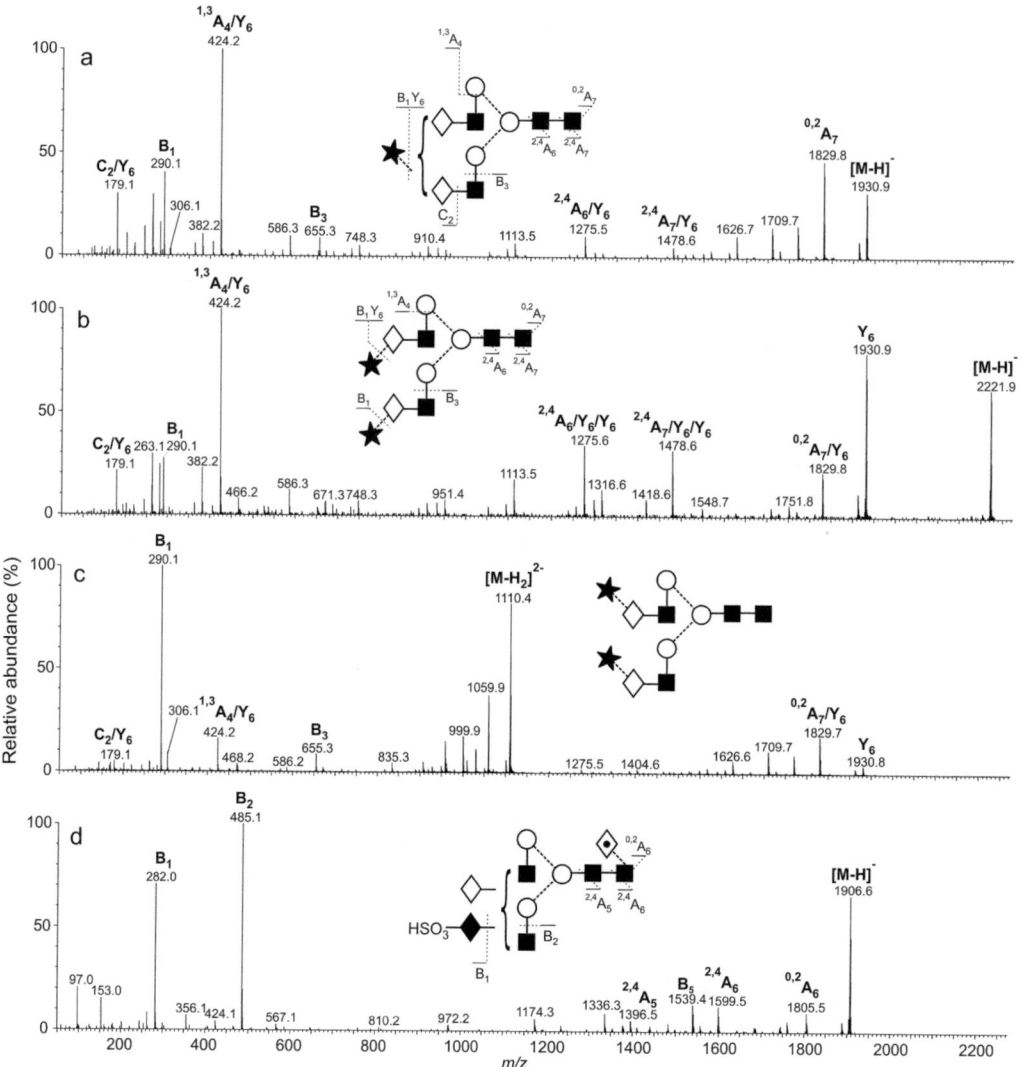

Fig. 3 Negative ion CID spectra of sialylated glycans. (**a**) Monosialylated biantennary glycan Gal$_2$Man$_3$GlcNAc$_4$Neu5Ac$_1$. The presence of the ^{0,2}A$_7$ (*m/z* 1829) ion is typical of these compounds. The sialic acid residue is seen as the B$_1$ ion at *m/z* 290 and this ion is accompanied by a relatively low abundance ion at *m/z* 306 showing the α2 → 6-linkage. ^{2,4}A$_6$, ^{2,4}A$_7$, and ^{1,3}A$_4$ ions are present (*m/z* 1275, 1478, and 424) but are formed with additional loss of sialic acid. (**b**) Di-sialylated biantennary glycan Gal$_2$Man$_3$GlcNAc$_4$Neu5Ac$_2$ (singly charged ion). Fragments in the spectrum of this compound are similar to those in the spectrum of the mono-sialylated glycan except that most of the main diagnostic fragments have lost both sialic acid moieties. The ^{2,4}A$_6$ and ^{2,4}A$_7$ ions (with losses of sialic acid) (*m/z* 1275 and 1478, respectively) are generally more abundant in the spectra of glycans containing α2 → 3-linked sialic acid (as shown here) than in the spectra of glycans bearing α2 → 6-linked sialic acids. (**c**) Di-sialylated biantennary glycan Gal$_2$Man$_3$GlcNAc$_4$Neu5Ac$_2$ (doubly charged ion). Some singly charged fragments (ions at higher mass that the [M−H$_2$]$^{2-}$ ion at *m/z* 1110) are formed by loss of one of the sialic acid groups. Most of the diagnostic ions seen in the spectra of the neutral compounds are suppressed, a trend that becomes more apparent as the number of sialic acid groups increases. (d) Sulfated biantennary glycan Gal$_1$GalNAc$_1$Man$_3$GlcNAc$_4$Fuc$_1$ (singly charged). The charge resides mainly on the sulfate group giving rise to the prominent B$_1$ and B$_2$ ions at *m/z* 282 and 485, respectively. Symbols for the structural diagrams are as defined in the footnote to Fig. 2 plus: ★ = Neu5Ac, ◈ = GalNAc

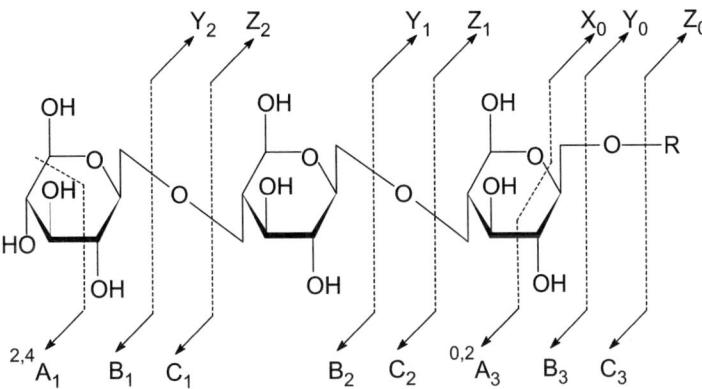

Fig. 4 Domon and Costello system for naming the fragment ions. R=H or attached lipid (as in gangliosides)

5 % v/v Fetal bovine serum, 20 mM 4-(2-Hydroxyethyl)piper-azine-1-ethanesulfonic acid (HEPES) buffer, and 10 % w/v Tryptose phosphate broth.

2. UUKV strain S23 (from Dr. Anna Överby, Umeå, Sweden) *see* [29].

3. Sterile phosphate buffered saline (PBS) and cell culture grade trypsin–ethylenediaminctctraacetate (EDTA) solution.

4. Three or more cell culture flasks, 175 cm²/flask surface area with filter cap (Greiner Bio-One, UK). One flask is used for cell counting and culturing. The other flasks are used for virus production.

5. Cell counting chamber (e.g., Neubauer counting chamber).

6. 50 ml Falcon tubes.

7. Sucrose.

8. Low-speed centrifuge (e.g., Beckman Coulter (UK) Ltd, Allegra X-12A).

9. Ultracentrifuge (e.g., Beckman Coulter (UK) Ltd, Optima L-80 XP) and ultracentrifuge rotor with compatible buckets (Beckman Coulter (UK) Ltd, SW28 or SW32).

10. Disposable ultracentrifuge tubes compatible with the buckets used (e.g., thin-wall polyclear, from Seton Scientific, USA).

11. HEPES buffering agent, EDTA, and sodium chloride powder.

12. 50 ml Plastic syringe and 0.2 μm Minisart plastic syringe filters (Sartorius Stedim, UK).

13. A plunge-freezing device (*see* **Note 1**).

14. Molybdenum grids with holey carbon film (Quantifoil, Microtools, Germany, *see* **Note 1** for possible alternatives) and grid boxes (e.g., Ted Pella, USA) for storing grids in liquid nitrogen.

15. Positive and negative action tweezers (Electron Microscopy Sciences, USA, *see* **Note 2**).

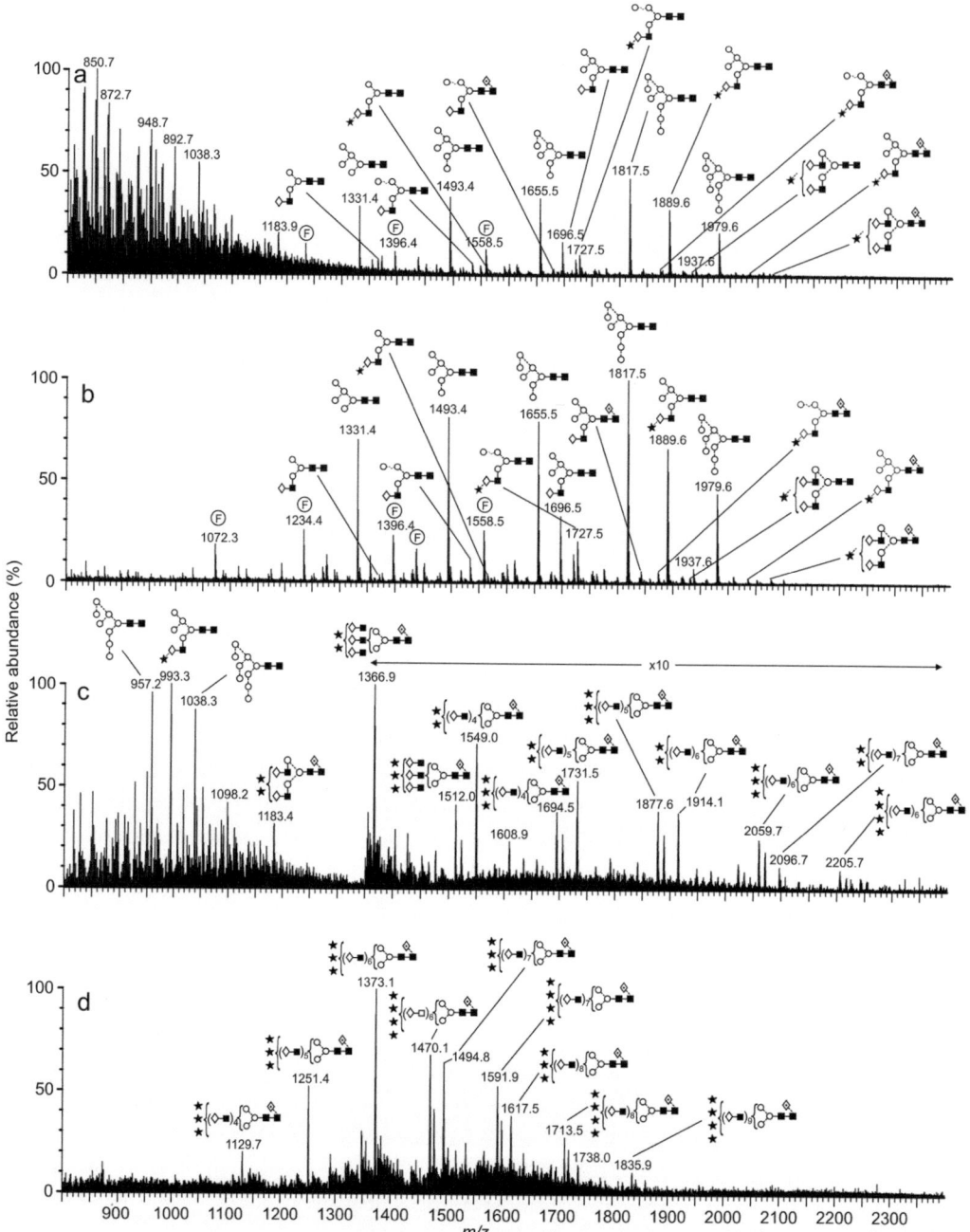

Fig. 5 (**a**) Electrospray spectrum of *N*-glycans released from the Gc virion glycoprotein from Uukuniemi virus. (**b**) Extracted singly charged ions from region 1 of Fig. 6. (**c**) Extracted doubly charged ions from region 2 of Fig. 6. (**d**) Extracted triply charged ions from region 3 of Fig. 6

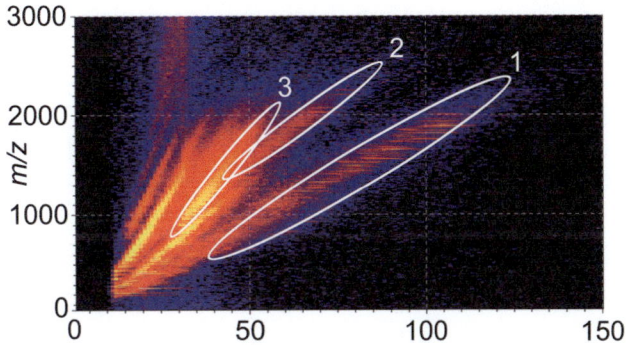

Fig. 6 Driftscope display (drift time:*m*/*z*) of *N*-glycans released from the Gc virion glycoprotein from Uukuniemi virus (UUKV). *Circled areas* are: 1 = singly charged ions, 2 = doubly charged ions, 3 = triply charged ions. Extracted spectra are shown in Fig. 5 (**b–d**). Further details are in ref. 18

16. Large tweezers for handling materials in liquid nitrogen.

17. A cryo grid box handling rod (e.g., Ted Pella, USA).

18. Liquid nitrogen storage dewars (e.g., Statebourne Cryogenics, UK).

19. Transfer dewars (e.g., Statebourne Cryogenics, UK).

20. For storage of grid boxes in liquid nitrogen, small holes (~0.5 cm diameter) must be drilled into the caps and the tubes prior to use to avoid pressure building inside of the tubes due to boiling of liquid nitrogen.

21. Liquid nitrogen and liquid ethane.

22. Plasma cleaner (Harrick Plasma, USA).

23. Plastic Petri dish.

24. Glass slides and Parafilm.

25. Blotting paper (Whatman, UK).

26. Preparation of concentrated UUKV, strain S23.

27. FEI F30 Polara electron microscope (FEI, Netherlands).

28. 4–12 % Bis–Tris gradient gel.

29. 2-Amino-2-(hydroxymethyl)-1,3-propanediol (Tris-base), sodium dodecyl sulfate, glycerol, glycine to mix the following components: Non-reducing SDS sample buffer (187.5 mM Tris pH 6.8, 2 % w/v SDS, 33 % v/v glycerol), SDS running buffer (25 mM Tris, 192 mM glycine, 0.1 % w/v SDS, pH 8.5).

30. Precision Plus Protein™ Kaleidoscope™ Standards (Bio-Rad Laboratories Ltd, Hemel Hempstead, Herts, UK).

31. Sterile disposable scalpel.

32. 1.5 ml Eppendorf tubes.

33. Safestain (Life technologies, UK) for staining of protein bands on SDS gels.

34. Cell culture incubator at 37 °C and 5 % CO_2.

35. PCR tubes or equivalent for handling µl quantities of solution.

36. Nafion membrane (Aldrich, Poole, UK).

37. Concentrated nitric acid.

38. 250 ml Beaker (for preparing Nafion membranes).

39. 0.5–2 µl and 2–20 µl Gilson pipettes and tips.

40. 1 M Sodium hydroxide solution.

41. Dowex AG50W resin (Bio-Rad Laboratories Ltd, Hemel Hempstead, Herts, UK).

42. Vacuum rotary evaporator.

43. Dimethylsulfoxide (DMSO).

44. Methanol.

45. Methyl iodide.

46. 4-(4,6-Dimethoxy-1,3,5-triazin-2-yl)-4-methylmorpholin-4-ium chloride (DMT-MM) (*see* **Note 3**).

47. Electric heating block or water bath.

48. Ammonium dihydrogen orthophosphate (referred to as ammonium phosphate).

49. Waters Synapt G2 mass spectrometer (Waters Corp. Manchester, UK).

50. Nanospray capillaries (*see* **Note 4**).

51. 2,5-Dihydroxybenzoic acid (DHB).

52. Acetonitrile, ethanol, acetone. About 1 ml each for preparation of MALDI matrices.

53. 2,4,6-Trihydroxyacetophenone (THAP) and ammonium nitrate, if negative ion MALDI is to be attempted.

54. Vacuum concentrator.

55. Bench top incubator at 37 °C.

56. Peptide-*N*-glycosidase F (PNGase F), glycerol-free (New England Biolabs, Ipswich, MA, USA).

57. α2-3,6,8,9 Neuraminidase (New England Biolabs, Ipswich, MA, USA).

58. 96-Well multiscreen IP filter plate with 0.45 µm pore size PVDF membrane (Merck Millipore, MA, USA).

59. 96-Well 2 ml collection plate.

60. 96-Well vacuum manifold assembly block and vacuum manifold pump.

3 Methods

3.1 Virus Production

The UUKV production and purification protocol is similar to previously established methods [30, 31].

1. Prepare the following solutions:

   20 mM HEPES, 100 mM NaCl, 2 mM EDTA (H20N100E2, *see* **Note 5**). 20 % sucrose in 20 mM HEPES, 100 mM NaCl, 2 mM EDTA. Phosphate buffered saline (PBS), to be autoclaved. Cool down the first two solutions to 4 °C and sterilize by filtration.

2. BHK-21 cells are propagated to ~80–90 % confluency in 175 cm² cell culture flasks, using PBS for cell wash steps and Trypsin–EDTA solution for cell detachment, as required for cell passaging of adherent cells. The total volume of complete growth media used is 20 ml per cell culture flask (*see* **Note 6**).

3. The amount of virus needed per flask is determined for a multiplicity of infection (MoI) of 0.1. To this end, determine the number of cells per flask by using a cell counting chamber. Scale the amount of virus according to the number of flasks used. Add a predetermined volume of concentrated UUKV S23 preparation corresponding to a MoI of 0.1 into a sterile container (e.g., an empty cell culture flask) that contains a multiple of 20 ml serum-free growth media (add some excess). Mix well by pipetting up and down.

4. Wash cells once with serum-free growth media. Add 20 ml of UUKV inoculum to each flask.

5. Incubate flasks for 1 h at 37 °C and 5 % CO_2.

6. Remove virus inoculum, add 20 ml of serum-free growth media, and incubate for 2 days, up to 48 h (*see* **Note 7**) at 37 °C and 5 % CO_2.

7. Collect virus-containing cell supernatant in 50 ml Falcon tubes, cool to 4 °C, and centrifuge for 30 min at $3000 \times g$, 4 °C.

8. After centrifugation, pool virus-containing supernatant (e.g., in a sterile T175 flask).

9. Add viral supernatant into plastic ultracentrifuge tubes compatible with Beckman-Coulter SW32/28 centrifuge buckets and carefully underlay the supernatant with 2 ml of the previously prepared 20 % sucrose solution, such that a sucrose cushion is formed at the bottom of the tube (*see* **Note 8**).

10. Place the plastic ultracentrifuge tubes containing both the infectious supernatant and sucrose cushion into the ultracentrifuge buckets and centrifuge for 2 h at 4 °C and $100,000 \times g$.

11. After centrifugation, discard the supernatant, blot the excess of liquid with paper, and immediately add 100–300 μl cold H20N100E2 buffer to the virus pellet. Dissolve the pellet overnight at 4 °C (*see* **Note 9**).

12. Gently but fully resuspend the virus pellet by pipetting on ice. At this point, aliquots can be prepared for long term storage at −80 °C (*see* **Note 10**).

3.2 Electron Cryo-Microscopy of UUKV Preparations

This section describes virus sample preparation for the qualitative assessment of native UUKV particles by electron cryo-microscopy. Alternatively, negative staining at room temperature microscopy can be used, although this method may be less suitable for assessing structural integrity of membraneous viruses in the unfixed state, because the stain (e.g., uranylacetate) would compromise the structural integrity of the virions. Electron cryo-microscopy acquisition parameters used for UUKV are given below, but for a detailed description of sample mounting and image acquisition using an FEI Polara electron microscope (as used here), please refer to [32]. A lower end electron microscope is suitable as long as the sample holder of the microscope can be used at liquid nitrogen temperature and the viral glycoprotein spikes and lipid bilayers are visible in the acquired projection images.

1. Cool the plunge-freezer (*see* **Note 1**) with liquid nitrogen (allow 20 min for cooling to liquid nitrogen temperature).

2. Wrap a glass slide in Parafilm and place electron microscopy grids onto the slide, with the carbon film facing upwards.

3. Place the slide into the plasma cleaner chamber under vacuum for at least 1 min. Perform glow discharge, which renders the carbon film of the copper grids hydrophilic, according to the manufacturer's instructions (*see* **Note 11**). With the Harrick plasma cleaner, the carbon-coated copper grids are glow-discharged for 20 s at the highest intensity. After glow discharge, store the slide with the grids in a plastic Petri dish, to prevent contact with dust.

4. Add liquid ethane to the dedicated liquid ethane cup.

5. Add a few hundred microliters of neutral pH buffer (e.g., H20N100) on top of a strip of Parafilm.

6. Select a glow-discharged grid by using inverted tweezers. Add virus solution (3 μl of virus preparation with a glycoprotein concentration in the lower micromolar range) onto the carbon coated side of the grid. Wash the carbon coated side facing the drop for 10–20 s against wash buffer (e.g., H20N100) to remove residual sucrose. Remove the grid sideways from the buffer to keep the liquid on the grid [33].

7. Mount tweezers onto the plunge freezing device and blot the liquid away with filter paper from the uncoated side of the grid,

until the liquid stops spreading on the paper (typically for 2–4 s). As soon as the liquid stops spreading, release the tweezers from the plunge-freezing rod to allow rapid immersion of the grid in liquid ethane for vitrification. Once the grid is frozen, it has to be kept at liquid nitrogen temperature. Unmount the tweezers from the rod, transfer to a grid box in liquid nitrogen, close the grid box lid using the grid box tool, and store in liquid nitrogen until required for further use.

8. Assess the UUKV preparation in any electron microscope equipped for analysis of vitrified specimens at liquid nitrogen temperatures. For simple projection image acquisition on the FEI Tecnai Polara, we acquired images at −5 μm defocus, 300 kV, and a nominal magnification of 59,000× (75,000× calibrated magnification), leading to a calibrated pixel size of 2 Å/pixel. The images were recorded with a charge-coupled device (CCD) camera (Ultrascan 4000, Gatan, USA), in low dose mode, using 20 electrons/Å^2 (Fig. 1).

3.3 SDS-PAGE of UUKV Preparation and Gel Band Excision

For a reproducible MS analysis of glycoprotein-derived *N*-glycans, glycoprotein bands on SDS gels need to contain protein in the range of several μg. In a previous study, as much as ~4 μg per viral glycoprotein were sufficient to generate reproducible MS data [34].

1. For UUKV, a second pelleting step over a sucrose cushion that increases the virus concentration is needed. To this end, dilute appropriate amounts of concentrated UUKV preparations from the first pelleting step in serum-free GMEM, and re-pellet over a sucrose cushion in a single ultracentrifuge tube.

2. Mix the virus sample with non-reducing SDS sample buffer at room temperature and load onto an SDS gel. Additionally, load 10 μl of Precision Plus Protein™ Kaleidoscope™ Standards.

3. Run the SDS gel for 1 h at 180 V.

4. Stain gels with Safestain and destain with deionized water.

5. Cut out bands corresponding to viral glycoproteins with a sterile scalpel on a clean surface (to avoid cross-contamination). Cut bands into pieces approximately 1 × 1 mm^2 and freeze at −20 °C until required for further use.

3.4 Release of Glycans from Within the SDS Gel Bands

1. Wash the gel pieces alternately with 1 ml of acetonitrile and deionized water. Perform five washing steps in total, starting and finishing with acetonitrile. On the final wash leave the gel pieces in acetonitrile for 10 min to allow dehydration of the gel. Discard the acetonitrile wash, and dry gel bands fully in the vacuum concentrator.

2. Dilute PNGase F 1:100 in deionized water and add 50 μl to each sample. Allow the gel pieces to rehydrate fully, adding

more PNGase F solution as necessary so the gel pieces are just covered.

3. Incubate samples for approximately 16 h (e.g. overnight) at 37 °C.

4. Transfer reaction supernatant to clean 1.5 ml Eppendorf tube. Wash gel pieces with 100 μl deionized water, each time vortexing well, and transfer washes to the stored supernatant.

5. Dry eluted glycans in the vacuum concentrator.

3.5 Analysis by Mass Spectrometry/Sample Preparation

Sialylated glycans can be profiled directly by negative ion electrospray mass spectrometry, where they will produce singly or multiply charged ions depending on the number of sialic acids, but fragmentation will provide little detailed structural information. Alternatively, glycan profiles can be obtained by MALDI after preparation of methyl esters or amides because underivatized sialylated glycans will eliminate much sialic acid under these conditions. Permethylation is not advised because such compounds do not give negative ion fragmentation spectra. A good compromise is to record the spectra by electrospray to obtain molecular weight and composition information together with information on the nature and linkage of the sialic acid, then desialylate and repeat the acquisition. The spectra of the resulting desialylated glycans will provide the necessary structural information, enabling the structures of the sialylated glycans to be deduced. No useful structural information is gained by reducing-terminal derivatization; in most cases, such derivatization has a detrimental effect on fragmentation. Glycans derivatized with 2-aminobenzoic acid, in particular, produce little useful information because the negative charge is localized on the derivative rather than on the glycan.

1. Glycan samples should be dissolved in water at approximately 1 μg/μl.

2. Prepare Nafion 117 membrane, as described by Bornsen et al. [35] (*see* **Notes 12** and **13**).

3. Cut the membrane into approximately 2×2 cm squares and float one of these in a small container of water. A Petri dish is a suitable container.

4. Place 1 μl samples of the glycan solutions on the surface of the membrane, cover the dish to limit evaporation, and leave for about an hour (*see* **Note 14**).

5. Analyze the sample by MALDI or ESI, as described below (Subheading 3.5.3).

If the glycans are found to be sialylated, the next step is either to remove the sialic acids and rerun the sample as neutral glycans, as above, or to derivatize the sialic acid moiety by forming methyl esters or amides to convert them to neutral compounds.

3.5.1   Desialylation

1. Clean 1 μl of the original glycan solution with the Nafion membrane, as above.

2. In an Eppendorf tube, add 1 μl of a non-linkage-specific neuraminidase to the glycans and make up to a total of 10 μl with deionized water.

3. Incubate for about 16 h at 37 °C.

4. Purify the glycans by applying them to a PVDF protein-binding membrane attached to a vacuum manifold: first activate the membrane with ethanol and wash with water; apply the sample and leave for 30 min; elute with water.

5. Dry glycans using a vacuum concentrator.

3.5.2   Methyl Ester Formation

Use of Methyl Iodide [19]

1. Prepare a small column of about 1 μl of Dowex AG50W resin (Bio-Rad Laboratories Ltd, Hemel Hempstead, Herts, UK), wash with 1 M sodium hydroxide solution (1 ml) followed by an equivalent amount of water.

2. Clean a 1 μl sample of the original glycan solution with Nafion, as above.

3. Dilute the sample with about 10 μl of water and apply to the column.

4. Elute with water (500 μl) into a small Eppendorf tube.

5. Evaporate to dryness with a rotary evaporator.

6. Wash the sides of the tube with a further 100 μl of water to concentrate the sample at the bottom of the tube and again evaporate to dryness.

7. Dissolve the sample in *dry* dimethylsulfoxide (DMSO, 1 μl) (*see* **Note 15**).

8. Add methyl iodide (1 ml, *see* **Note 16**) and allow the mixture to stand for 2 h at room temperature.

9. Add a further 5 μl of DMSO and evaporate the methyl iodide with a stream of nitrogen.

10. Evaporate to dryness with a rotary evaporator. The sample can now be redissolved in water and analyzed, as above.

11. If MALDI is used for analysis, 1 μl aliquots of the DMSO solution can be spotted directly onto the MALDI target and allowed to dry (*see* **Note 17**).

Use of Methanol and DMT-MM [20] (*See* Note 3)

1. Evaporate the cleaned glycan sample to dryness.

2. Add methanol (10 μl) and 1 μl of a methanolic solution (0.1 M) of DMT-MM (*see* **Note 18**).

3. Incubate at about 80 °C for 1 h.

4. Evaporate to dryness and analyze by MALDI or ESI.

3.5.3 Acquisition of Mass Spectra Data

Samples can be ionized either by ESI or by MALDI using a mass spectrometer such as the Waters Synapt G2 fitted with a MALDI ion source, or with a MALDI-TOF instrument. Although it would be possible to analyze the samples by directly coupled HPLC-MS, the procedure reported here is specifically designed for cleanup of the samples by ion mobility.

ESI

1. Remove the sample from the Nafion membrane (from Subheading 3.5.4) with a 2 μl pipette and place into a PCR tube (or small equivalent).

2. Add 3 μl water, 4 μl methanol and 1 μl of 0.05 M ammonium phosphate (*see* **Note 19**).

3. Spin at $10,000 \times g$ for 1 min to sediment any particulates.

4. Transfer the sample to a nanospray capillary (*see* **Note 4**).

5. Spray into the mass spectrometer (a Waters Synapt G2 instrument is used in our work and settings refer to this instrument). Suitable parameters are: Capillary potential, 1.1 kV; ion source temperature, 80 °C; cone voltage 100 V. These parameters may be altered depending on the sample.

6. Record the mass spectrum with the mobility cell switched on.

7. Record CID spectra for target ions, as described below.

MALDI Ionization

1. Remove the sample from the Nafion membrane with a 2 μl pipette and place on the MALDI target plate.

2. For positive ion spectra, add 1 μl of a solution of 2,5-dihydroxybenzoic acid (DHB, 10 mg DHB in 1 ml acetonitrile, *see* **Note 20**) and allow to dry under ambient conditions.

3. Redissolve in a minimum amount of ethanol (about 0.2 μl) and allow to dry (*see* **Note 21**).

4. For production of negative ion MALDI spectra, use an equivalent amount of 2,4,6-trihydroxyacetophenone (THAP, in acetone) as the matrix and add 0.5 μl of 0.5 M ammonium nitrate [36]. Allow to dry. Ethanol recrystallization is not needed because of the smaller crystals produced by THAP.

5. Introduce the target into the mass spectrometer and ionize with the minimum laser power that is consistent with sufficient signal strength.

Acquisition of Ion Mobility Data

1. Set up the Synapt instrument for either ESI or MALDI ionization, as above.

2. In ion mobility mode, set the ion mobility wave velocity to 450 m/s and the wave height to 40 V.

3. Record the spectrum over the mass range m/z 50–3500.

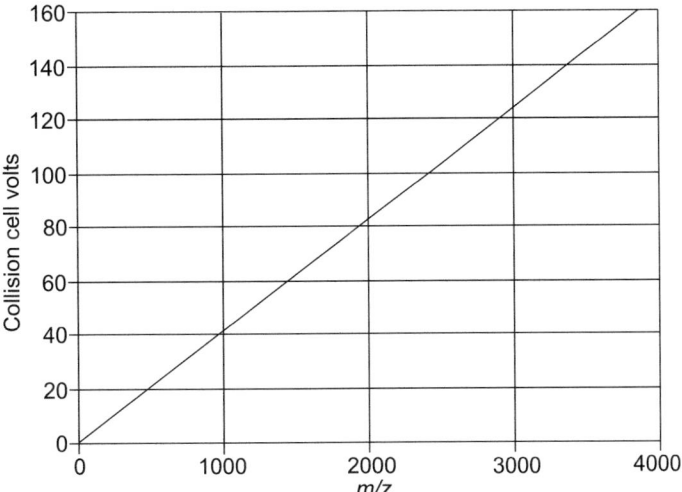

Fig. 7 Guide for setting the collision cell voltage with respect to mass for acquisition of CID spectra. The optimum voltage will depend on the glycan structure and could vary by about ±10 V

Acquisition of Fragmentation Data by Collision-Induced Dissociation (CID)

1. Set up the Synapt instrument in IMS mode for either ESI or MALDI ionization, as above.

2. Record CID spectra for target ions in the transfer cell (*see* **Note 22**), accumulating the signal until a satisfactory signal–noise ratio is obtained (*see* **Note 23**). The voltage on the collision cell should be set to be appropriate to the *m/z* value of the ions being fragmented using the information in Fig. 7 (*see* **Note 24**). Spectra are interpreted as described in Subheading 3.5.5.

3.5.4 Extraction of Glycan Ions Using Ion Mobility

1. Open the mass spectral file with DriftScope.

2. Select the ions of interest (click the "+" button followed by one of the blue symbols) (*see* **Note 25**).

3. Export the selected region to MassLynx (Use "File", "Export to MassLynx", "Retain drift time"). DriftScope appends consecutive numbers to the file name to identify the selected region.

4. Open the selected file in MassLynx, chromatogram window and drag with the right mouse button across the display to show the mass spectrum in the spectrum window.

5. Dragging across selected ions or groups of ions (usually isotope peaks) with the right mouse button in the spectrum window displays the ion mobility data for the selected ions in the chromatogram window. These profiles should be roughly Gaussian for single components but may be asymmetric if isomers or conformers are present (*see* **Note 26**).

6. CID spectra are best opened in DriftScope and the ions of interest selected, as above (*see* **Note 27**).

3.5.5 Interpretation of Negative Ion CID Data of Neutral Ions (Phosphate Adducts, Singly Charged Ions, with Reference to Fig. 2) (See Note 28)

1. Locate the $^{2,4}A_R$, B_{R-1}, and $^{2,4}A_{R-1}$ fragment ions (*see* **Note 29**).

2. Use Table 1 to determine which adduct is present and, hence, the molecular weight of the glycan.

3. From the molecular weight, determine the isobaric monosaccharide content (*see* **Notes 30** and **31**).

4. The mass difference between the molecular ion and the $^{2,4}A_R$ ion (*see* Table 1, phosphate adducts) shows the presence or absence of fucose at the 6-position of the reducing terminal GlcNAc. Absence of the $^{2,4}A_R$ ion indicates substitution (fucose) at the 3-position of this GlcNAc.

5. Residues at the non-reducing terminus are identified by the masses of the C_1 fragments listed in Table 2.

6. The composition of the 6-antenna is determined by the masses of the D, D-18, $^{0,3}A_{R-2}$, and $^{0,4}A_{R-2}$ ions (Table 3).

7. The branching pattern of the triantennary glycans can be determined by the presence of the ions listed in Table 4 [37] (*see* Fig. 2).

8. The antenna composition is specified by the prominent cross-ring cleavage ion, (monosaccharides) + CH=CH-O⁻ ions listed in Table 5.

9. The presence of bisecting GlcNAc is revealed by the absence of a D ion and the presence of a prominent D-221 ion (loss of the bisecting GlcNAc) with masses as for the D-18 ions in Table 3 (Fig. 2c).

10. Sialylated glycans (Neu5Ac) produce a prominent B_1 ion at m/z 290 (Fig. 3a–c). An additional ion of 10–15 % relative abundance at m/z 306 indicates that the sialic acid is α2-6-

Table 1
Determination of adduct

[M+X]⁻—$^{2,4}A_R$ᵃ	Fuc on R	Adduct (X)
259	0	$[H_2PO_4]^-$
405	1	$[H_2PO_4]^-$
197	0	$[Cl]^-$
343	1	$[Cl]^-$
224	0	$[NO_3]^-$
370	1	$[NO_3]^-$

ᵃReducing terminal GlcNAc.

Table 2
Masses of the C_1 fragments

m/z	Monosaccharide
179	Hexose
220	HexNAc
325	Hex-Fuc

Table 3
Ions that specify the composition of the 6-antenna

Ion					
D	D-18	D-36	0,3A	0,4A	Composition
647	629	–	575	545	Hex$_3$
809	791	–	737	707	Hex$_4$
971	953	–	899	677	Hex$_5$
688	670	–	616	586	Hex-HexNAc
850	832	–	778	748	Hex-Hex-HexNAc
834	826	–	762	732	Hex-HexNAc-Fuc
1053	1035	1017	981	951	(Hex-HexNAc)$_2$
1199	1181	1163	1127	1097	(Hex-HexNAc)$_2$Fuc
526	508	–	454	424	HexNAc

Table 4
Ions specifying the branching pattern of triantennary glycans

D	D-18	D-36	D-221	E[1]	Structure
688	670	–	–	831	3-Branched
1053	1035	1017	–	–	6-Branched
–	–	–	670	–	Bisected (Gal on bisecting GlcNAc)
688	670	–	–	977	3-Branched with fucose

linked [38]. Neu5Gc-substituted glycans produce corresponding ions at *m/z* 306 and 322.

11. Glycans containing sulfated GalNAcGlcNAc moieties fragment to give two very prominent ions at *m/z* 282 and 485 (Fig. 3c).

Table 5
Masses of the (monosaccharides) + CH-CH$_2$-O$^-$ ion specifying the antenna structure

m/z	Composition
262	HexNAc-CH=CH$_2$-O$^-$
424	Hex-HexNAc-CH=CH$_2$-O$^-$
570	Hex-HexNAc-Fuc-CH=CH$_2$-O$^-$
465	HexNAc-HexNAc-CH=CH$_2$-O$^-$
716	Hex-HexNAc-Fuc$_2$-CH=CH$_2$-O$^-$

Table 6
Residue masses of common monosaccharides[a]

Monosaccharide	Residue formula	Residue mass
Pentose	C$_5$H$_8$O$_4$	132.042 132.116
Deoxy hexose	C$_6$H$_{10}$O$_4$	146.078 146.143
Hexose	C$_6$H$_{10}$O$_5$	162.053 162.142
Hexosamine	C$_6$H$_{11}$NO$_4$	161.069 161.157
HexNAc	C$_8$H$_{13}$NO$_5$	203.079 203.179
Methyl hexose	C$_7$H$_{12}$O$_5$	194.079 194.185
Hexuronic acid	C$_6$H$_8$O$_6$	176.032 176.126
N-acetyl-neuraminic acid	C$_{11}$H$_{17}$NO$_8$	291.095 291.258
N-glycoyl-neuraminic acid	C$_{11}$H$_{17}$NO$_9$	307.090 307.257

[a]Top figure = monoisotopic mass (based on C = 12.000000, H = 1.007825, N = 14.003074, O = 15.994915), Lower figure = average mass (based on C = 12.011, H = 1.00794, N = 14.0067, O = 15.9994)

The masses of the intact glycans can be obtained by addition of the residue masses given above, plus the mass of the terminal group (H$_2$O for an unmodified glycan) 18.011 (monoisotopic) and 18.152 (average), the mass of the adduct and the mass of any reducing-terminal or other derivative

4   Notes

1. The manually operated plunge freezer device used in this protocol was custom-built in the Max Planck Institute for Biochemistry, Martinsried, Germany. Semiautomatic alternatives include, for example, those manufactured by FEI (Vitrobot; FEI, Netherlands) and Gatan (Cryo plunge 3; Gatan, USA). If one of the latter devices is used, the blotting protocols may need to be adjusted, and it is of advantage to keep the grids, once loaded with virus sample, in a humid atmosphere. Additionally, other types of EM grids (molybdenum or copper) may be purchased, such as C-flat grids (Protochips, USA), as long as they have a holey carbon film.

2. Negative action tweezers can be modified to fit a custom-built plunge-freezer. If a commercial plunge-freezer is used, follow the manufacturer's guidance regarding which tweezers are most appropriate.

3. DMT-MM can be purchased from Sigma-Aldrich (Product code 74104) or synthesized from *N*-methylmorpholine and 4-chloro-2,6-dimethoxytriazine by the method described by Kunishima et al. [39].

4. Waters long thin wall capillaries (part No. M956232AD1–S) are satisfactory or capillaries can be made "in house".

5. Alternative buffers may be equally suitable as long as they contain 100 mM NaCl and are held at neutral pH (e.g., pH 7.2–7.5). For example, as an alternative one may use 20 mM Tris-buffer, 100 mM NaCl pH 7.4. The use of EDTA in resuspension buffer and sucrose solution is not essential, but may limit aggregation. If no EDTA is used in the virus resuspension buffer, it needs to be present in the SDS sample buffer (~2 mM EDTA), otherwise there will be aggregation in the wells of the SDS-gel.

6. All the procedures that involve living cells and virus should be carried out in a standard class 2 biosafety cabinet.

7. Infectious supernatant can also be harvested before 48 h incubation, usually once the cells start detaching, which may happen after ~43 h of incubation.

8. Sucrose may be added using 1 ml filter tips or serological pipettes. Care needs to be taken not to overfill the tubes before adding the cushion from the bottom. Ideally, about 30 ml of infectious supernatant can be added first and then 2 ml of sucrose solution. Remaining supernatant may then be added carefully from the top to fill the tube to the maximal recommended volume.

9. For the overnight incubation step, the ultracentrifuge tubes may be transferred to 50 ml plastic Falcon tubes, sealed and

stored in a fridge or cold room, to avoid incubation of open containers on ice. Virus pellets should be dissolved for at least 2 h.

10. The amount of viral glycoprotein present in the virus preparation may be estimated by SDS-PAGE analysis using different volumes (e.g., 5–20 μl of virus preparation), using a reference protein with known concentration. Viral glycoprotein concentrations of Uukuniemi virus pellets are usually in the lower μM range.

11. Depending on the exact device available, the glow discharging procedure may be different. In principle, after a vacuum has been pumped, a small amount of air is allowed to flow back into the chamber, to allow plasma formation, which in turn will ionize the carbon film, rendering it more hydrophilic.

12. Cleaning the sample with a Nafion membrane is only one of several methods that could be adopted. Others include the use of porous graphitized carbon (PGC) [40] or cellulose [41].

13. Nafion acts as a dialysis membrane allowing ions from salts etc. to diffuse into the water. It also has the ability to adsorb hydrophobic compounds; thus it is not appropriate for cleaning glycans that have been derivatized with, for example, 2-aminobenzamide (2-AB).

14. The length of time that the sample should remain on the membrane will depend on the amount of contaminants and should be determined by trial and error. Heavily contaminated samples benefit from leaving for several hours, e.g. overnight. If the sample contains glycerol (found in some PNGase F preparations), water will diffuse into the sample, causing swelling and the possibility of samples merging. Such PNGase F preparations should be avoided.

15. DMSO can be dried with a (4 Å, 14 × 30 mesh) molecular sieve.

16. Methyl iodide is toxic; consequently these steps should be performed in a fume hood.

17. DMSO has a low volatility and drying may take several hours.

18. DMT-MM solutions in methanol are relatively unstable and must be made up immediately before use. Alternatively, a small crystal of DMT-MM can be added to the methanolic glycan solution.

19. Other ammonium salts could be used at this stage to form the corresponding [M+X]$^-$ ions.

20. Many other matrices have been reported for analysis of glycans [42] but DHB is the most widely used.

21. DHB typically dries to give a target containing large crystals of DHB that point from the periphery of the target towards the

center. The purpose of the ethanol is to produce a more uniform target with smaller crystals.

22. Fragment ions from the transfer cell all have the same mobility and are displayed vertically in a drift time (x axis):m/z plot and are easily separated from contaminating ions. Fragmentation can also be performed in the trap cell (before fragmentation) allowing mobility data to be collected from each fragment.

23. For complicated samples, there is a balance between the signal-noise ratio and the number of spectra that can be recorded before the sample is exhausted. Selected ions can be targeted if known or, if not, a good strategy is to target the abundant ions first, as these will require less time for spectral acquisition allowing the maximum number of glycans to be fragmented before the sample is exhausted.

24. The voltage on the collision cell should not be allowed to rise more than about 30 V above the recommended value in order to prevent the spectrum changing to one in which most diagnostic ions are missing [43].

25. Ion selection buttons in DriftScope: The square highlights a rectangular region by dragging the mouse. The triangle extracts a triangular section. The circle allows the user to draw around the region of interest. The half-filled square is used to highlight a region by clicking at various points around the region of interest (double-click to close the selected region).

26. Isomers and conformers can usually be differentiated by plotting corresponding mobilograms of fragment ions. If these have the same profile as that of the molecular ion, asymmetry is probably due to conformers. If not, isomers are present.

27. Any ion that falls within the window used by the quadrupole will produce fragments. With mixtures, it is often the case that several ions can be selected and each will produce fragments. See the example of m/z 1007 discussed above.

28. With ammonium phosphate added to the ESI solvent, most adducts should be phosphate (addition of 96.969 mass units). However, there is the possibility that sufficient chloride is present to produce additional chloride adducts (addition of 34.969 units (^{35}Cl isotope)). Thus, it is important to determine which adduct is present. Chloride adducts can often be identified by the presence of the ^{35}Cl and ^{37}Cl chloride isotopes. Negative ion MALDI spectra with ammonium nitrate added to the matrix will, of course, produce nitrate adducts (addition of 61.989 units). Positive ion MALDI adducts are invariably sodium (22.989 units).

29. The B_{R-1} ion is 60 mass units less in mass than the $^{2,4}AR$ ion and the $^{2,4}A_{R-1}$ ion is 203 mass units lower if the penultimate

GlcNAc contains no fucose, or 349 units lower if it contains a single fucose.

30. The isobaric monosaccharide content is the number of hexose, HexNAc residues etc. The mass of the resulting ion is given by adding the residue masses of the monosaccharides (listed in Table 6), the mass of the adduct and the mass of one molecule of water (18.01). A spread sheet, a home-constructed software program or the use of an internet-available program such as GlycoMod [44] or that in GlycoWorkbench (http://glycomics.ccrc.uga.edu/eurocarb/gwb/home.action) can be used to obtain the composition from the measured mass. Unfortunately, the GlycoWorkbench algorithm does not work in negative ion mode but compositions can be obtained by using a mass corresponding to the experimentally measured mass, by subtraction of the mass of the adduct minus 1 (i.e., the mass of the $[M+1]^+$ ion) from that of the measured ion. It must be remembered that this composition does not specify individual monosaccharide types such as galactose, mannose etc. that have the same mass. This information must be obtained by other methods such as GC/MS following hydrolysis or exoglycosidase digestions. Also, attempts to identify a structure completely by simply matching the composition to structures in a database is not rigorous enough because of a potentially large number of glycans that could have the same composition. More information (e.g., fragmentation, exoglycosidase digestion, methylation analysis) must be obtained. Experimental details for performing GC/MS analyses can be found in ref. 40, and suitable exoglycosidases and incubation protocols are described in the same reference. However, for these virus samples, there probably will not be enough material for GC/MS linkage analysis.

31. A potential problem occurs when preparations of PNGase F contain dithiothreitol (added to denature proteins). This compound can liberate H_2S which competes with water when the glycosylamine that is released by the enzyme is hydrolysed to the glycan. Instead of an OH group at the anomeric position, H_2S introduces an SH group. The mass difference between this group and OH is 16, which is the mass of oxygen. Thus, the reducing terminal GlcNAc and its attached substituents, such as fucose would appear to contain an extra oxygen. If fucose is present the result can be a mis-diagnosis as hexose. Because the anomeric position and the fucose are lost in formation of the $^{2,4}A_R$ ion, the fragmentation spectrum is fully consistent with this incorrect deduction [45]. Other groups have also been reported to attach to the reducing terminus in this way, an example being urea [46].

Acknowledgments

M.C. is supported by the Center for HIV/AIDS Vaccine Immunology and Immunogen Discovery Grant (UM1AI100663) and the International AIDS Vaccine Initiative through the Neutralizing Antibody Consortium and Bill and Melinda Gates Center for Vaccine Discovery. M.C. is a Fellow of Oriel College, Oxford. We also thank the Wellcome Trust (grant number 090532/Z/09/Z), the Academy of Finland (grant numbers 130750 and 218080 to J.T.H.), and the MRC (MR/L009528/1 to T.A.B. and MR/K024426/1 to K.J.D) for funding. The Wellcome Trust Centre for Human Genetics is supported by Wellcome Trust Centre grant 090532/Z/09/Z

References

1. Bowden TA, Jones EY, Stuart DI (2011) Cells under siege: viral glycoprotein interactions at the cell surface. J Struct Biol 175:120–126

2. Rossmann MG (2013) Structure of viruses: a short history. Q Rev Biophys 46:133–180

3. Chang VT, Crispin M, Aricescu AR et al (2007) Glycoprotein structural genomics: solving the glycosylation problem. Structure 15:267–273

4. Lozach PY, Kuhbacher A, Meier R et al (2011) DC-SIGN as a receptor for phleboviruses. Cell Host Microbe 10:75–88

5. Alexandre KB, Gray ES, Lambson BE et al (2010) Mannose-rich glycosylation patterns on HIV-1 subtype C gp120 and sensitivity to the lectins, Griffithsin, Cyanovirin-N and Scytovirin. Virology 402:187–196

6. Bonomelli C, Doores KJ, Dunlop DC et al (2011) The glycan shield of HIV is predominantly oligomannose independently of production system or viral clade. PLoS One 6:e23521

7. Doores KJ, Bonomelli C, Harvey DJ et al (2010) Envelope glycans of immunodeficiency virions are almost entirely oligomannose antigens. Proc Natl Acad Sci U S A 107: 13800–13805

8. Walker LM, Huber M, Doores KJ et al (2011) Broad neutralization coverage of HIV by multiple highly potent antibodies. Nature 477:466–470

9. Scanlan CN, Pantophlet R, Wormald MR et al (2002) The broadly neutralizing anti-human immunodeficiency virus type 1 antibody 2G12 recognizes a cluster of alpha1→2 mannose residues on the outer face of gp120. J Virol 76:7306–7321

10. Kornfeld R, Kornfeld S (1985) Assembly of asparagine-linked oligosaccharides. Annu Rev Biochem 54:631–664

11. Walker LM, Phogat SK, Chan-Hui PY et al (2009) Broad and potent neutralizing antibodies from an African donor reveal a new HIV-1 vaccine target. Science 326:285–289

12. Bonsignori M, Hwang KK, Chen X et al (2011) Analysis of a clonal lineage of HIV-1 envelope V2/V3 conformational epitope-specific broadly neutralizing antibodies and their inferred unmutated common ancestors. J Virol 85:9998–10009

13. Falkowska E, Le KM, Ramos A et al (2014) Broadly neutralizing HIV antibodies define a glycan-dependent epitope on the prefusion conformation of gp41 on cleaved envelope trimers. Immunity 40:657–668

14. Crispin M, Bowden TA (2013) Antibodies expose multiple weaknesses in the glycan shield of HIV. Nat Struct Mol Biol 20:771–772

15. Dalziel M, Crispin M, Scanlan CN et al (2014) Emerging principles for the therapeutic exploitation of glycosylation. Science 343:1235681

16. Burton DR, Ahmed R, Barouch DH et al (2012) A blueprint for HIV vaccine discovery. Cell Host Microbe 12:396–407

17. Bowden TA, Crispin M, Harvey DJ et al (2010) Dimeric architecture of the Hendra virus attachment glycoprotein: evidence for a conserved mode of assembly. J Virol 84:6208–6217

18. Crispin M, Harvey DJ, Bitto D et al (2014) Uukuniemi phlebovirus assembly and secretion leave a functional imprint on the virion glycome. J Virol 88:10244–10251

19. Powell AK, Harvey DJ (1996) Stabilisation of sialic acids in *N*-linked oligosaccharides and gangliosides for analysis by positive ion matrix-assisted laser desorption-ionization mass spectrometry. Rapid Commun Mass Spectrom 10: 1027–1032

20. Wheeler SF, Domann P, Harvey DJ (2009) Derivatization of sialic acids for stabilization in matrix-assisted laser desorption/ionization mass spectrometry and concomitant differentiation of α(2-3) and α(2-6) isomers. Rapid Commun Mass Spectrom 23:303–312

21. Liu X, Qiu H, Lee RK et al (2010) Methylamidation for sialoglycomics by MALDI-MS: a facile derivatization strategy for both α2,3- and α2,6-linked sialic acids. Anal Chem 82:8300–8306

22. Harvey DJ (2005) Fragmentation of negative ions from carbohydrates: Part 2, Fragmentation of high-mannose N-linked glycans. J Am Soc Mass Spectrom 16:631–646

23. Harvey DJ (2005) Fragmentation of negative ions from carbohydrates: Part 1; Use of nitrate and other anionic adducts for the production of negative ion electrospray spectra from N-linked carbohydrates. J Am Soc Mass Spectrom 16:622–630

24. Harvey DJ (2005) Fragmentation of negative ions from carbohydrates: Part 3, Fragmentation of hybrid and complex N-linked glycans. J Am Soc Mass Spectrom 16:647–659

25. Harvey DJ, Royle L, Radcliffe CM et al (2008) Structural and quantitative analysis of N-linked glycans by MALDI and negative ion nanospray mass spectrometry. Anal Biochem 376:44–60

26. Domon B, Costello CE (1988) A systematic nomenclature for carbohydrate fragmentations in FAB-MS/MS spectra of glycoconjugates. Glycoconj J 5:397–409

27. Harvey DJ, Scarff CA, Crispin M et al (2012) MALDI-MS/MS with traveling wave ion mobility for the structural analysis of N-linked glycans. J Am Soc Mass Spectrom 23:1955–1966

28. Harvey DJ, Scarff CA, Edgeworth M et al (2013) Travelling wave ion mobility and negative ion fragmentation for the structural determination of N-linked glycans. Electrophoresis 34:2368–2378

29. Pettersson R, Kääriäinen L (1973) The ribonucleic acids of Uukuniemi virus, a noncubical tick-borne arbovirus. Virol J 56:608–619

30. Lozach PY, Mancini R, Bitto D et al (2010) Entry of bunyaviruses into mammalian cells. Cell Host Microbe 7:488–499

31. Pettersson R, Kääriäinen L, von Bonsdorff CH et al (1971) Structural components of Uukuniemi virus, a noncubical tick-borne arbovirus. Virology 46:721–729

32. Grassucci RA, Taylor D, Frank J (2008) Visualization of macromolecular complexes using cryo-electron microscopy with FEI Tecnai transmission electron microscopes. Nat Protoc 3:330–339

33. Cyrklaff M, Roos N, Gross H et al (1994) Particle surface interaction in thin vitrified films for cryo-electron microscopy. J Microsc 175: 135–142

34. Crispin M, Harvey DJ, Bitto D et al (2014) Structural plasticity of the Semliki Forest virus glycome upon interspecies transmission. J Proteome Res 13:1702–1712

35. Börnsen KO, Mohr MD, Widmer HM (1995) Ion exchange and purification of carbohydrates on a Nafion(R) membrane as a new sample pre-treatment for matrix-assisted laser desorption-ionization mass spectrometry. Rapid Commun Mass Spectrom 9:1031–1034

36. Domann P, Spencer DIR, Harvey DJ (2012) Production and fragmentation of negative ions from neutral N-linked carbohydrates ionized by matrix-assisted laser desorption/ionization. Rapid Commun Mass Spectrom 26:469–479

37. Harvey DJ, Crispin M, Scanlan C et al (2008) Differentiation between isomeric triantennary N-linked glycans by negative ion tandem mass spectrometry and confirmation of glycans containing galactose attached to the bisecting (β1-4-GlcNAc) residue in N-glycans from IgG. Rapid Commun Mass Spectrom 22:1047–1052

38. Wheeler SF, Harvey DJ (2000) Negative ion mass spectrometry of sialylated carbohydrates: discrimination of N-acetylneuraminic acid linkages by matrix-assisted laser desorption/ionization-time-of-flight and electrospray-time-of-flight mass spectrometry. Anal Chem 72:5027–5039

39. Kunishima M, Kawachi C, Morita J et al (1999) 4-(4,6-Dimethoxy-1,3,5-triazin-2-yl)-4-methyl-morpholinium chloride: an efficient condensing agent leading to the formation of amides and esters. Tetrahedron 55:13159–13170

40. Morelle W, Michalski JC (2007) Analysis of protein glycosylation by mass spectrometry. Nat Protoc 2:1585–1602

41. Selman MHJ, Hemayatkar M, Deelder AM et al (2011) Cotton HILIC SPE microtips for microscale purification and enrichment of glycans and glycopeptides. Anal Chem 83: 2492–2499

42. Hossain M, Limbach PA (2010) A comparison of MALDI matrices. In: Cole RB (ed) Electrospray and MALDI mass spectrometry: fundamentals, instrumentation, practicalities, and biological applications, 2nd edn. John Wiley and Sons Inc., Hoboken, NJ, pp 215–244

43. Harvey DJ, Edgeworth M, Krishna BA et al (2014) Fragmentation of negative ions from *N*-linked carbohydrates: Part 6: Glycans containing one *N*-acetylglucosamine in the core. Rapid Commun Mass Spectrom 28:2008–2018

44. Cooper CA, Gasteiger E, Packer NH (2001) GlycoMod - a software tool for determining glycosylation compositions from mass spectrometric data. Proteomics 1:340–349

45. Harvey DJ, Rudd PM (2010) Identification of by-products formed during the release of *N*-glycans with protein *N*-glycosidase F in the presence of dithiothreitol. J Mass Spectrom 45:815–819

46. Omtvedt LA, Royle L, Husby G et al (2004) Artefacts formed by addition of urea to *N*-linked glycans released with peptide-*N*-glycosidase F for analysis by mass spectrometry. Rapid Commun Mass Spectrom 18:2357–2359

47. Harvey DJ, Merry AH, Royle L et al (2009) Proposal for a standard system for drawing structural diagrams of *N*- and *O*-linked carbohydrates and related compounds. Proteomics 9:3796–3801

Chapter 8

N-Glycosylation Fingerprinting of Viral Glycoproteins by xCGE-LIF

René Hennig*, Erdmann Rapp*, Robert Kottler, Samanta Cajic, Matthias Borowiak, and Udo Reichl

Abstract

The ongoing threat of pathogens, increasing resistance against antibiotics, and the risk of fast spreading of infectious diseases in a global community resulted in an intensified development of vaccines. Antigens used for vaccination comprise a wide variety of macromolecules including glycoproteins, lipopolysaccharides, and complex carbohydrates. For all of these antigens the sugar composition plays a crucial role for immunogenicity and protective efficacy of the vaccine. Here, we provide a protocol for *N*-glycosylation fingerprinting utilizing high performance multiplexed capillary gel electrophoresis with laser-induced fluorescence detection (xCGE-LIF) technology. The method described, enables to analyze the *N*-glycosylation of specific proteins out of a complex sample or even the total of all *N*-glycans contained in such a sample. The protocol is exemplarily demonstrated for *N*-glycosylation fingerprinting of cell culture-derived influenza A and B viruses and their major antigens, the membrane glycoproteins hemagglutinin and neuraminidase.

Key words Influenza virus, Vaccine, Hemagglutinin, Neuraminidase, Glycosylation, *N*-glycan, Glycoanalysis, Fingerprinting, Capillary gel electrophoresis, xCGE-LIF, APTS

1 Introduction

Carbohydrates are one of the most abundant and structurally diverse classes of organic molecules in nature. In the form of glycoproteins, glycolipids, and polysaccharides, carbohydrates build the so-called glycocalyx of eukaryota and prokaryotic cells. These glycoconjugates are important virulence factors of many bacteria, which make them potential vaccine candidates [1, 2]. Several carbohydrate-based bacterial vaccines, e.g., against *Haemophilus influenzae b* (Hib), *Neisseria meningitides* or *Streptococcus pneumoniae*, are already licensed and even more are under development [3]. Furthermore, the membrane of enveloped viruses typically carries glycoproteins that are highly immunogenic. During the early

---

*Both authors contributed equally

Bernd Lepenies (ed.), *Carbohydrate-Based Vaccines: Methods and Protocols*, Methods in Molecular Biology, vol. 1331, DOI 10.1007/978-1-4939-2874-3_8, © Springer Science+Business Media New York 2015

years of vaccine development the knowledge regarding carbohydrate structures and composition and their impact on immunogenicity has been very limited. With the development of new glycoanalytical techniques [4, 5] and advances in glycochemistry [6, 7], detailed studies regarding the impact of glycosylation on immunogenicity can be performed, and targeted design of new classes of vaccines seems feasible. In addition, recent developments in glycobiotechnology will allow optimization of existing vaccine production platforms using a variety of host cell systems including glycoengineered mammalian [8, 9], yeast [10–12], and insect cells [13, 14].

Over the last two decades it could be shown that viral protein glycosylation is crucial for their survival and virulence, as well as for the recognition of pathogens by the host. On the one hand, proper protein glycosylation facilitates correct protein folding, virus entry into the host cell and an efficient virus particle release [15–17]. In addition, using the glycosylation machinery of the host cell may mask antigenic protein sites to support immune evasion [17, 18]. On the other hand, protein glycosylation can greatly influence uptake and proteolytic processing of antigens by the immune system and thus improves their presentation as glycopeptides by MHC-I and MHC-II receptors [19, 20]. Consequently, as exemplified for influenza viruses by Abdel-Motal et al. [15], De Vries et al. [21] and Hütter et al. [22], appropriate glycosylation of antigens might be exploited for the development of more efficacious viral vaccines in future.

Advances in antigen characterization were driven by the establishment of powerful glycoanalytical techniques. The most common methods used for glycoanalysis are based on mass spectrometry (MS) [4, 23–25], liquid chromatography (LC) [4, 26, 27], capillary electrophoresis (CE) [4, 5, 28], and nuclear magnetic resonance spectroscopy (NMR) [29]. As each of these glycoanalytical techniques has its benefits and limitations, they are often applied in combination. Recently, a comparative study of the most frequently used glycoprofiling techniques was published about human IgG N-glycosylation [30] illustrating advantages and disadvantages of each technique.

Regarding viruses, a common analytical technique is still the enzymatic treatment of the antigenic glycoproteins with endo-β-N-acetylglucosaminidase H (Endo H, cleaving high-mannose type N-glycans) or peptide N-glycosidase F (PNGase F, cutting of all N-glycan types). This results in a shift of the protein band to lower masses, which can be monitored by one-dimensional sodium dodecylsulfate polyacrylamide gel electrophoresis (1D-SDS-PAGE) [15, 21, 31, 32]. More detailed information about the N-glycosylation of viral glycoproteins can be obtained by using chromatographic methods. For example PNGase F released N-glycans can be fluorescently labeled and analyzed by size exclusion chromatography (SEC) [33, 34], ion exchange chromatography (IEC) [35], or hydrophilic interaction chromatography (HILIC) [32].

Recently, the toolbox for viral glycoanalysis has been extended by matrix-assisted laser desorption/ionization time of flight MS (MALDI-TOF-MS) [21, 36, 37], and multiplexed capillary gel electrophoresis with laser induced fluorescence detection (xCGE-LIF) [22, 38–41].

In the following, we describe (*see* Figs. 1, 2, 3, and 4) a universal protocol for the fingerprinting of viral protein *N*-glycosylation using xCGE-LIF. An overview of the workflow is shown in Fig. 1. Here, fluorescently labeled glycans are separated inside a polymer-

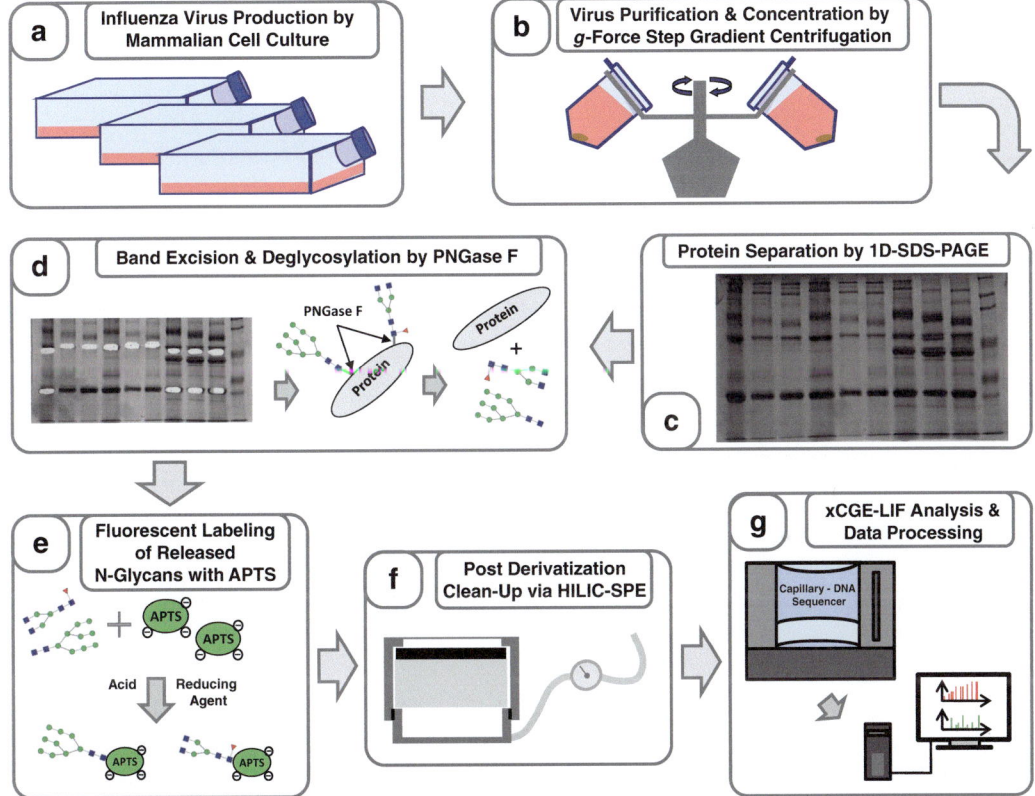

Fig. 1 Workflow to generate *N*-glycosylation fingerprints of influenza A and influenza B virus proteins, using multiplexed capillary gel electrophoresis with laser-induced fluorescence detection (xCGE-LIF). (**a**) Cell culture-derived Influenza A virus harvest [42] is used as starting material for the xCGE-LIF analysis. (**b**) The virus particle containing cell culture supernatant is purified by a centrifugation procedure with stepwise increasing *g*-forces to remove dead cells, cell debris, and other molecular aggregates. In a final step, the virus particles are pelletized via ultracentrifugation (for details *see* Subheading 3.1). (**c**) Virus proteins are separated by one-dimensional sodium dodecylsulfate polyacrylamide gel electrophoresis (1D-SDS-PAGE) (for details *see* Subheading 3.2). (**d**) Protein bands are excised from gel and disulfide bonds of proteins are reduced and alkylated before *N*-glycans are released by Peptide-*N*-Glycosidase F (PNGase F) treatment (for details *see* Subheading 3.3). (**e**) Released and extracted *N*-glycans are labeled via reductive amination with the fluorescent dye 8-aminopyrene-1,3,6-trisulfonic acid trisodium salt (APTS) (for details *see* Subheading 3.4). (**f**) APTS-labeled *N*-glycans are depleted from excess APTS, salt, and other impurities by hydrophilic interaction chromatography (HILIC) based solid phase extraction (SPE) (for details *see* Subheading 3.5). (**g**) The concentrated HILIC-SPE purified *N*-glycan pools are analyzed utilizing xCGE-LIF (for details *see* Subheading 3.6)

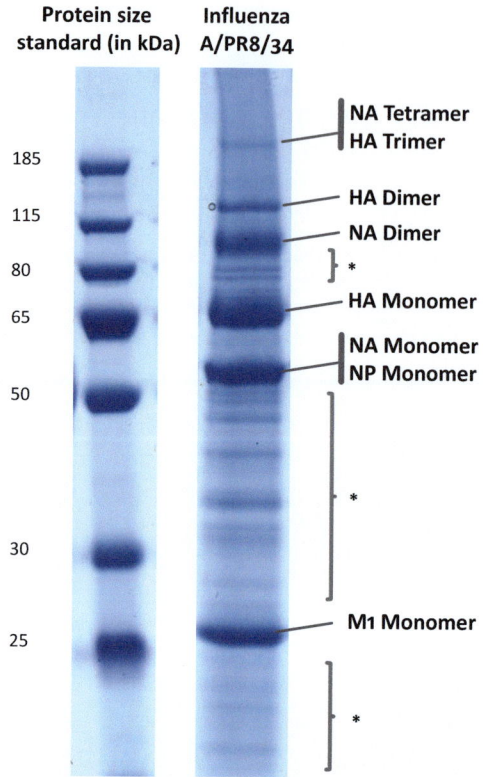

Fig. 2 A typical 1D-SDS-PAGE separation of influenza A virus proteins. 5 μL protein size standard PageRuler Plus (Thermo Scientific) (*left lane*) and 10 μg of influenza A/Puerto Rico/8/34 (H1N1) proteins (*right lane*) were applied on a 10 % Bis-Tris SDS-gel. Proteins were separated for 50 min at constant 200 V using a Life Technologies NuPAGE® system. Colloidal Coomassie colored bands are marked with the respective viral protein. Bands marked with * belong to host cell proteins attached to or encapsulated into virus particles

filled capillary by their charge, size and shape at very high resolution. Using laser excitation, emission signals of the fluorescently labeled glycans are recorded over their migration time as an electropherogram (*see* Fig. 4). Based on the highly sensitive LIF detection, analysis of the labeled *N*-glycans from viral glycoproteins is achieved in the "sub-μg range." An internal standard is used to normalize the migration time of the *N*-glycans resulting in a so called *N*-glycosylation "fingerprint" (*see* Figs. 5 and 6). Based on this internal normalization very high reproducibility of migration time of all glycans can be obtained.

In the following, the use of the method is examplified for cell culture-derived influenza A/Puerto Rico/8/34 (H1N1) and influenza B/Malaysia/2506/2004 virus. However, the method described can also be used for the characterization of

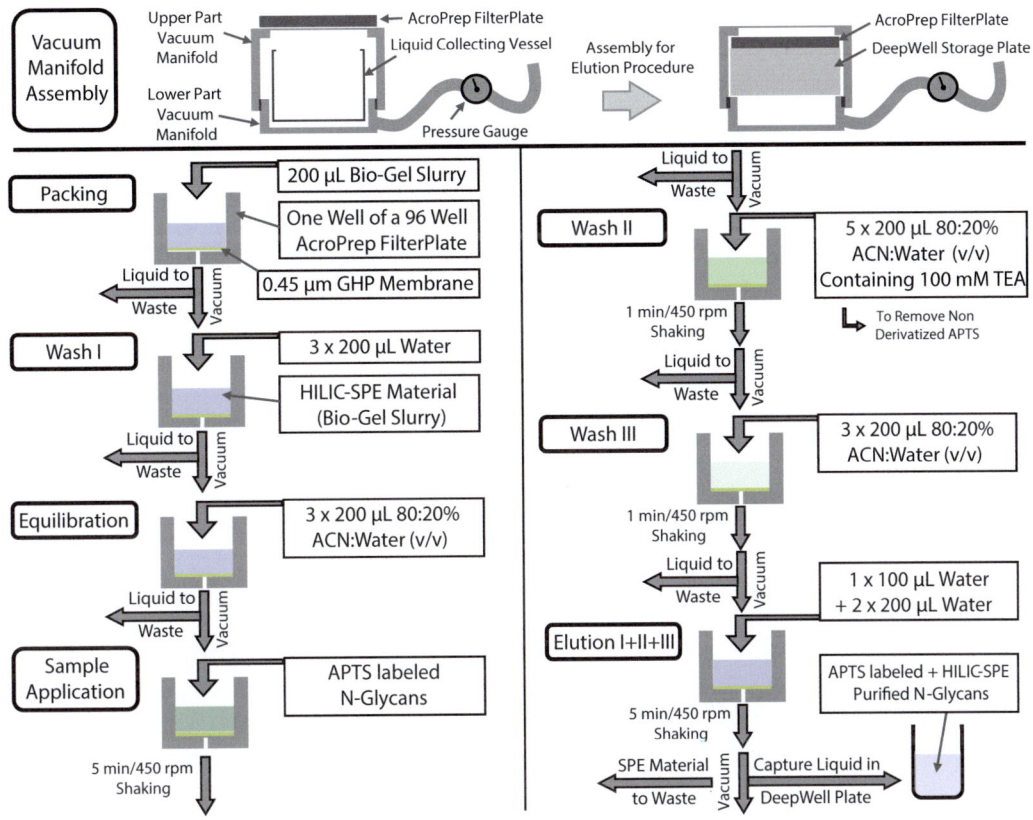

Fig. 3 Schematic overview of the HILIC-SPE post derivatization sample cleanup procedure of APTS-labeled *N*-glycans. For details regarding the post derivatization sample cleanup procedure, *see* Subheading 3.5

N-glycosylation of other enveloped virus particles, as well as for specific glycoproteins or distinct subunits thereof. Influenza virus is an enveloped negative-sense, single-stranded, segmented RNA virus containing the two membrane glycoproteins hemagglutinin (HA) and neuraminidase (NA). Within the viral membrane, HA assembles as a homo-trimer and NA as a homo-tetramer. Each HA monomer (HA_0) is composed of two subunits, a globular head (HA_1) and a stalk (HA_2), which anchors this protein in the viral envelope. Typical *N*-glycosylation fingerprints of the influenza A virus proteins HA and NA produced in cell culture are shown in Fig. 5a (HA, black; NA, blue). Summing up the individual fingerprints of these two glycoproteins in silico (*see* Fig. 5b, blue) results in an almost identical plot as the fingerprint from the *N*-glycosylation of the entire virus particle (*see* Fig. 5b, black). Only a few early peaks occur additionally, originating from other glycoproteins attached to (or encapsulated into) the virus particles. A similar result is obtained, comparing the fingerprint of the influenza B virus HA_0 with the summed up fingerprints of its subunits HA_1 and HA_2 (*see* Fig. 6).

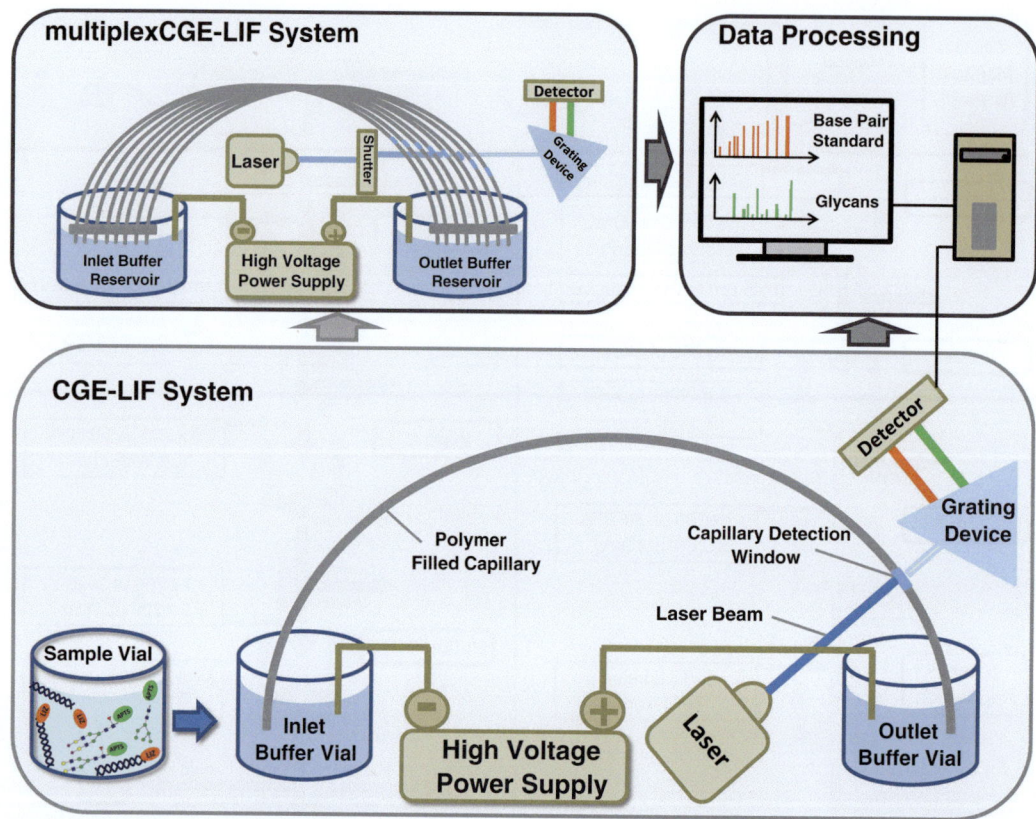

Fig. 4 Basic configuration of an xCGE-LIF system with 4–96 capillaries in parallel (multiplexed). Sample (containing the APTS-labeled *N*-glycans) and LIZ standard are co-injected and separated in a polymer filled capillary by applying an electric field. Passing the capillary detection window the fluorescent tags of both sample and standard are excited by an argon laser at 488 nm. The emitted light of LIZ standard and sample is wavelength resolved and recorded in two separate traces (APTS labeled *N*-glycans at 522 nm and LIZ standard at 650 nm). The internal LIZ standard is used for migration time normalization of the sample electropherogram, resulting in a so-called "fingerprint." For more detailed explanation, *see* Subheading 3.6

Following the given protocol and employing xCGE-LIF enables a straightforward and highly sensitive glycoprofiling of a large number of samples at high resolution within 2–4 days. This is in contrast to the currently prevailing LC and LC-MS methods, where multiplexing to achieve a similar throughput is extremely expensive, lab-space intensive, and ties up a lot of manpower.

2   Materials

All buffers and solutions were prepared using ultrapure water (18 MΩ cm at 25 °C; total organic carbon (TOC) < 5 ppb). All used chemicals and solvents were purchased in A.C.S. reagent grade or higher quality.

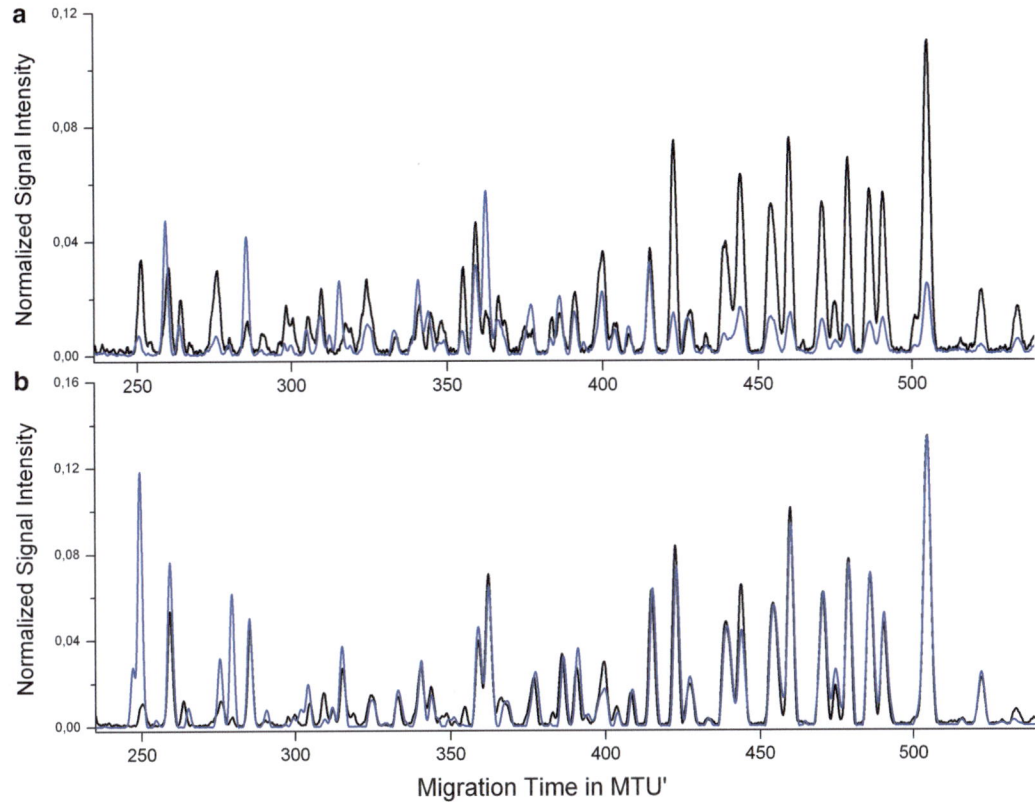

Fig. 5 *N*-glycosylation fingerprints derived from influenza A/Puerto Rico/8/34 (H1N1) virus, produced in adherent MDCK cells. Normalized signal intensity [-] is plotted over the normalized migration time [MTU′]. (**a**) Overlay of the *N*-glycosylation fingerprints of the influenza A virus proteins hemagglutinin (HA, *black*) and neuraminidase (NA, *blue*). (**b**) Overlay of the *N*-glycosylation fingerprint of the entire influenza virus (*black*) and an artificial fingerprint (*blue*), generated by addition of the individual *N*-glycosylation fingerprints of HA and NA from (**a**)

Storage conditions: Store all prepared buffers and solutions at 4 °C (unless indicated otherwise).

Shelf life: Buffers and solutions are stable for at least 2 weeks at recommended storage conditions (unless indicated otherwise).

All listed instruments, devices, and consumables are used in our laboratory. Equipment and materials from other manufacturers might be used, if they meet the requirements described in the respective paragraphs.

Larger Laboratory Equipment Required for Sample Preparation

1. Centrifuge: Centrifuges, equipped with a swing-out rotor, fixed angle rotor, and microplate rotor.

2. Ultracentrifuge: Optima LE-80K equipped with the fixed angle rotor Type 70Ti (both Beckman Coulter).

3. Rotational vacuum concentrator: Ice condenser combined with a rotational vacuum concentration unit and a vacuum pump.

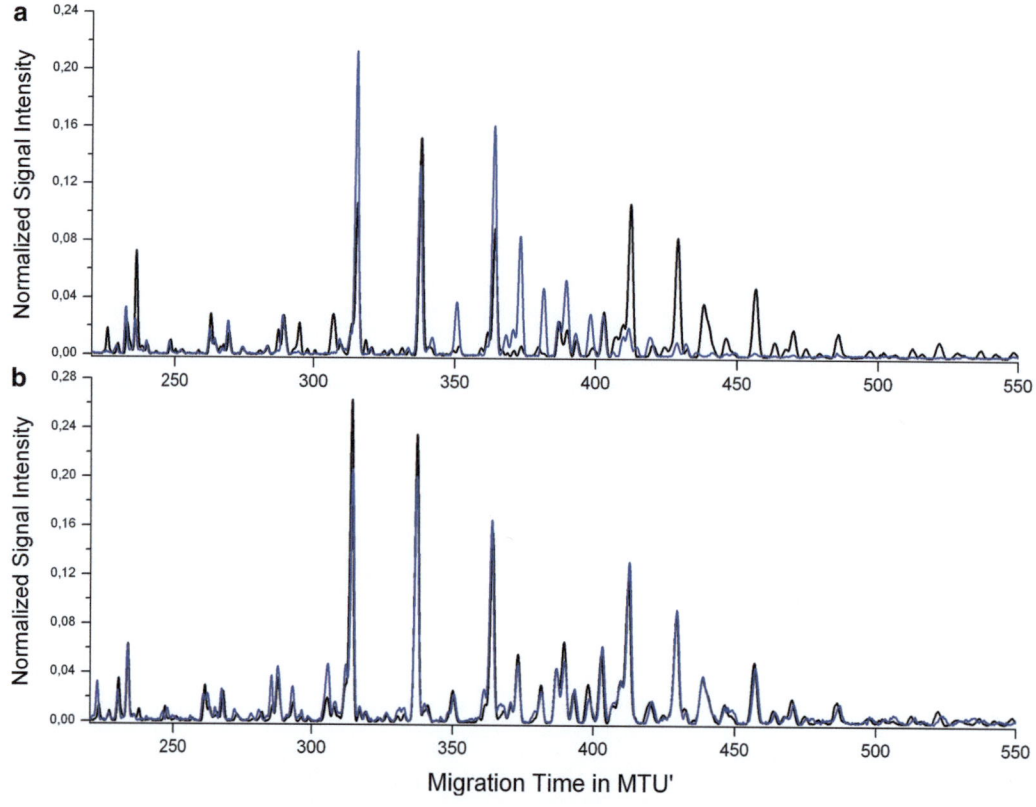

Fig. 6 *N*-glycosylation fingerprints of influenza B/Malaysia/2506/2004 virus produced in adherent MDCK cells. Normalized signal intensity [-] is plotted over the normalized migration time [MTU′]. (**a**) Overlay of the *N*-glycosylation fingerprints of the influenza A virus hemagglutinin (HA$_0$) subunits HA$_1$ (*black*) and HA$_2$ (*blue*). (**b**) Overlay of the *N*-glycosylation fingerprint of the intact HA$_0$ (*black*) and an artificial fingerprint (*blue*), generated by addition of the individual *N*-glycosylation fingerprints of the subunits HA$_1$ and HA$_2$ from (**a**)

4. Thermomixer: Thermoblock enabling simultaneously tempering and shaking, equipped with framework for 0.5 mL tubes or for microtiter plates.

5. Vacuum manifold, e.g., Millipore Multiscreen$_{HTS}$ Vacuum Manifold system.

6. xCGE-LIF system: 3130 Genetic Analyzer (Life Technologies).

2.1 Influenza A and B Virus Purification and Concentration by g-Force Step Gradient Centrifugation

1. Ultracentrifugation tube (UC-tube): 32.4 mL OptiSeal tubes (Beckman Coulter).

2. Dissolving buffer: 100 mM tris(hydroxymethyl)-aminomethan (Tris) pH 7.4. Fill 50 mL water in a 100 mL graduated cylinder. Weigh 1211.4 mg Tris and transfer to the graduated cylinder. Add water to a volume of 90 mL. Mix until everything has dissolved and adjust pH using 37 % hydrochloric acid (HCl). Fill up to 100 mL with water, mix, and transfer buffer in a 100 mL glass bottle.

3. Virus lysis buffer: 2 % (w/v) sodium dodecyl sulfate (SDS) containing 150 mM sodium chloride (NaCl), 5 mM ethylene-diaminetetraacetic acid disodium salt solution (Na$_2$EDTA) and 1 mM magnesium chloride hexahydrate (MgCl$_2$·6H$_2$O). Fill 50 mL water in a 100 mL graduated cylinder. Weigh 2 g of SDS, 876.6 mg NaCl, 186.1 mg Na$_2$EDTA, and 20.3 mg MgCl$_2$·6H$_2$O. Transfer weighed chemicals into the graduated cylinder and mix until everything has dissolved. Fill up to 100 mL with water, mix, and transfer virus lysis buffer in a 100 mL glass bottle.

2.2 Virus Protein Separation by 1D-SDS-PAGE

Virus protein separation by 1D-SDS-PAGE is examplified using the Life Technologies NuPAGE® SDS-PAGE system (*see* **Note 1**).

1. Sample buffer: 4× concentrated non-reducing sample buffer (*see* **Note 2**).

2. Running buffer: 1× SDS-PAGE running buffer (e.g., NuPAGE® MOPS SDS Running Buffer).

3. 10 % SDS-gel: 10 % Bis-Tris Protein Gel (*see* **Note 3**).

4. Protein size standard.

5. Fixing solution: 52.5:40:7.5 % (v/v) water–ethanol–acetic acid solution.

6. Coomassie solution A: 50 g/L Coomassie solution in water. Fill 80 mL water in a 100 mL glass bottle. Weigh 5 g Coomassie Brilliant blue. Mix until everything has dissolved. Fill up to 100 mL with water.

7. Coomassie solution B: Fill 300 mL water into a 500 mL graduated cylinder. Weigh 50 g ammonium sulfate ((NH$_4$)$_2$SO$_4$) and transfer into the graduated cylinder. Add 6 mL phosphoric acid (85 %) and mix until everything has dissolved. Fill up to 490 mL with water, mix, and transfer solution into a 500 mL glass bottle. Add 10 mL Coomassie solution A and mix.

8. Coomassie staining solution: Fill 200 mL Coomassie solution B into a glass bottle. Before use add 50 mL methanol and mix solution.

2.3 Protein Deglycosylation by PNGase F

1. Washing solution: 50:45:5 % (v/v) methanol–water–acetic acid solution. Prepare 1 mL washing solution per sample always fresh. For analysis of about 20 samples: Fill 10 mL methanol into a 25 mL graduated cylinder. Add 1 mL acetic acid. Fill up to 20 mL with water, mix, and transfer buffer into a 50 mL Falcon tube.

2. ABC buffer: 50 mM ammonium bicarbonate (ABC, NH$_4$HCO$_3$) buffer solution. Always freshly prepare 0.5 mL ABC buffer per sample. For analysis of about 20 samples: Fill 5 mL water into a 10 mL graduated cylinder. Weigh 39.5 mg

NH_4HCO_3 and transfer into the graduated cylinder. Mix until everything has dissolved. Fill up to 10 mL with water, mix, and transfer buffer into a 15 mL Falcon tube.

3. DTT solution: 45 mM DL-dithiothreitol (DTT) solution. Prepare 0.5 mL DTT solution per sample always fresh. For analysis of about 20 samples: Fill 5 mL water into a 10 mL graduated cylinder. Weigh 69.4 mg DTT and transfer into the graduated cylinder. Mix until everything has dissolved. Fill up to 10 mL with water, mix, and transfer buffer into a 15 mL Falcon tube.

4. IAA solution: 100 mM iodoacetamide (IAA) solution. Always freshly prepare 0.5 mL IAA solution per sample. For analysis of about 20 samples: Fill 5 mL water into a 10 mL graduated cylinder. Weigh 185 mg IAA and transfer into the graduated cylinder. Mix until everything has dissolved. Fill up to 10 mL with water, mix, and transfer buffer into a 15 mL Falcon tube.

5. PBS buffer: Phosphate buffered saline buffer (PBS) at pH 7.4. Fill about 500 mL water into a 1000 mL graduated cylinder. Weigh 8 g NaCl, 2 g potassium chloride (KCl), 0.2 g potassium dihydrogen phosphate (KH_2PO_4), and 1.15 g disodium hydrogen phosphate (Na_2HPO_4). Transfer all weighed chemicals into the graduated cylinder. Mix until everything has dissolved. Fill up to 1000 mL with water, mix, and transfer buffer into a 1000 mL glass bottle.

6. PNGase F solution: 1 U/µL PNGase F (P7367, Sigma-Aldrich) in PBS buffer.

7. IGEPAL solution: 8 % (v/v) octylphenoxypolyethoxyethanol (IGEPAL CA-630) in PBS buffer. Fill 8 mL PBS buffer into a 15 mL graduated Falcon tube. Add 800 µL IGEPAL CA-630. Fill up to 10 mL with PBS buffer and mix until everything has dissolved.

2.4 Fluorescent Labeling of Released N-Glycans with APTS

1. CA solution: 3.6 M citric acid (CA) solution. Fill 18 mL water into a 25 mL graduated cylinder. Weigh 18.91 g citric acid monohydrate and transfer into the graduated cylinder. Mix until everything has dissolved (*see* **Note 4**). Fill up to 25 mL with water, mix, and transfer buffer into a 50 mL Falcon tube.

2. APTS solution: 20 mM 8-aminopyrene-1,3,6-trisulfonic acid trisodium salt (APTS) in 3.6 M CA solution. Add the 5 mg APTS into a 1.5 mL tube and fill 478 µL CA solution into the tube. Mix until everything has dissolved and store at –20 °C (stable for 6 months).

3. PB solution: 0.2 M 2-picoline borane complex (PB) solution in dimethyl sulfoxide (DMSO). Fill 5 mL DMSO into a 10 mL

graduated cylinder. Weigh 214 mg PB and transfer into the graduated cylinder. Mix until everything has dissolved. Fill up to 10 mL with DMSO and mix. Aliquot PB solution into 1.5 mL tubes and store at –20 °C (stable for 6 months).

2.5 HILIC-SPE Post derivatization Sample Cleanup

1. Storage solution: 70:20:10 % (v/v) water–ethanol–acetonitrile (ACN) solution. Fill 140 mL water into a 200 mL graduated cylinder. Add 40 mL ethanol. Fill up to 200 mL with ACN, mix, and transfer solution into a 50 mL Falcon tube.

2. Bio-Gel slurry: 100 mg/mL Bio-Gel P-10 Slurry in 70:20:10 % (v/v) water–ethanol–ACN. Fill 20 mL storage solution into a 25 mL glass bottle. Weigh 2 g Bio-Gel P-10 particles (Bio-Rad) and transfer into the glass bottle.

3. ACN solution: 80:20 % (v/v) ACN–water solution. Fill 160 mL ACN into a 200 mL graduated cylinder. Fill up to 200 mL with water mix and transfer solution into a glass bottle.

4. ACN solution containing TEA: 80:20 % (v/v) ACN-water solution containing 100 mM triethylamine (TEA) at pH 8.5. Fill 160 mL ACN into a 200 mL graduated cylinder. Add 2.77 mL TEA and 0.48 mL acetic acid. Fill up to 200 mL with water, mix, and transfer solution into a 250 mL glass bottle.

5. FilterPlate: AcroPrep™ 96-well filter plate, 350 μL, 0.45 μm GHP membrane (Pall Corporation).

6. DeepWell plate: 0.8 mL deep-well storage plate (Thermo Scientific).

2.6 N-Glycosylation Fingerprinting via xCGE-LIF

All consumables are from Life Technologies.

1. HiDi: Hi-Di™ Formamide.

2. LIZ standard: Dilute 1 μL GeneScan™ 500 LIZ™ dye size standard in 49 μL HiDi.

3. 1× running buffer: xCGE-LIF system compatible 10× running buffer. Dilute 10× running buffer 1:10 with water. Fill 50 mL 10× running buffer into a 500 mL graduated cylinder. Fill up to 500 mL with water, mix, and transfer solution into a 500 mL glass bottle.

4. Capillary array: e.g., a 3130/3100-Avant Genetic Analyzer 4-Capillary Array, 50 cm.

5. POP-7 polymer: POP-7™ Polymer for 3130/3130xl Genetic Analyzers.

6. 384 well plate: xCGE-LIF system compatible clear 384 well PCR microplate.

7. Septa: 384 well plate compatible silicon sealing mat.

3   Methods

Carry out all procedures at room temperature (RT, 21 °C), unless specified otherwise. A brief overview of the presented workflow is shown in Fig. 1.

To demonstrate the applicability of the protocol described in this chapter, N-glycosylation fingerprinting is shown for viral glycoproteins of influenza A and influenza B virus. Virions produced in animal cell culture were used as starting material as described by Genzel et al. [42]. Briefly, human influenza virus A/Puerto Rico/8/34 (H1N1) (Robert Koch Institute, Germany) and B/Malaysia/2506/2004 were produced in adherent MDCK cells (No. 841211903, ECACC), cultivated at 37 °C in T75-flasks with 50 mL serum-containing cell growth medium until a confluent monolayer of cells was obtained. For virus production, cell growth medium was replaced by serum-free virus maintenance media, and MDCK cells were infected by influenza virus seed at a low multiplicity of infection. Several hours after infection, MDCK cells start to release virus particles into the cell culture supernatant. After about 2–3 days virus production is ceased, due to virus-induced apoptosis and cell lysis. Virus was harvested by transferring virus containing cell culture supernatant into a 50 mL Falcon tube without previous inactivation (*see* **Note 5**).

3.1   Influenza Virus Purification and Concentration by g-Force Step Gradient Centrifugation

Virus harvest is clarified by a g-force step gradient centrifugation procedure, to remove dead cells, cell debris, and other molecular aggregates from the supernatant as described below. In a final step, influenza virus particles are concentrated via ultracentrifugation, forming a virus pellet.

1. Removal of cells from cell culture supernatant. Transfer 50 mL influenza virus containing supernatant (after cell culture based influenza virus production in T75-flasks) into a 50 mL Falcon tube (*see* **Note 6**). Centrifuge at $100 \times g$ for 20 min at 4 °C in a centrifuge equipped with a swing-out rotor. Transfer virus containing supernatant into a new 50 mL Falcon tube. Discard cell pellet.

2. Removal of cell debris from cell culture supernatant. Use a centrifuge equipped with a fixed angle rotor. Centrifuge at $4000 \times g$ for 35 min at 4 °C. Transfer virus containing supernatant to a new 50 mL Falcon tube. Discard cell debris pellet. Centrifuge at $10,000 \times g$ for 35 min at 4 °C. Transfer virus containing supernatant into an UC-tube. Discard cell debris pellet.

3. Virus concentration. Use the Beckman ultracentrifuge equipped with a 70Ti rotor. Centrifuge at $100,000 \times g$ for 90 min at 4 °C with maximum acceleration and deceleration. Use a marker to indicate the small virus pellet at the bottom of the UC-tube after centrifugation (*see* **Note 7**).

4. Resuspending the virus pellet. Discard the supernatant of the UC-tube and resuspend the virus pellet in dissolving buffer (*see* **Note 8**). Store the resuspended virus pellet until use at −80 °C.

5. Estimate the protein concentration of the virus pellet using the Quant-iT™ Protein Assay Kit (Q33210, Life Technologies, Germany) (*see* **Note 9**). Mix resuspended virus pellet and transfer 5 μL to a new 0.5 mL tube. Add 5 μL of virus lysis buffer and homogenize by pipetting up and down. Incubate the sample for 5 min at 95 °C (*see* **Notes 5** and **10**). To estimate protein concentration, take 2 μL of virus lysate and carefully follow the instructions provided in the kit. The remaining 8 μL will be used for protein deglycosylation (*see* Subheading 3.3.1).

3.2 Virus Protein Separation by 1D-SDS-PAGE

To analyze the *N*-glycosylation of a specific protein out of a complex sample, the protein first needs to be purified. If no specific antibody against the protein is available, a separation by 1D-SDS-PAGE is the method of choice (*see* Subheading 3.1). Virus concentration and purification by *g*-force step gradient centrifugation results in a sufficiently purified influenza virus pellet for further analysis. Performing 1D-SDS-PAGE, viral proteins can be separated as shown in Fig. 2. By excising the respective protein band from the gel, glycoanalysis of a single viral glycoprotein can be performed.

The virus protein separation by 1D-SDS-PAGE is exemplarily described using the Life Technologies NuPAGE® SDS-PAGE system (*see* **Note 1**).

1. Load protein size standard into the first well of the 10 % SDS-gel.

2. Mix an aliquot of the resuspended virus pellet, containing 10 μg viral proteins, 3:1 with sample buffer and heat samples at 70 °C for 10 min (*see* **Note 5**).

3. Load all sample on the 10 % SDS-gel. Fill empty wells with 10 μL sample buffer.

4. Start gel electrophoretic protein separation (run), using running buffer.

5. Stop run before the blue dye front reaches the end of the gel.

6. To immobilize all proteins inside the gel, incubate the gel for 1 h with fixing solution.

7. Wash the gel twice for 10 min with water.

8. Stain the proteins inside of the gel with Coomassie staining solution overnight.

9. Wash the gel twice for 10 min with water, to eliminate remaining Coomassie dye.

10. Scan the gel for documentation purpose. A scanned 1D-SDS-PAGE gel lane of an influenza virus is shown in Fig. 2.

3.3 Protein Deglycosylation by PNGase F

The *N*-glycosylation fingerprinting can be performed for the entire influenza virus (all glycoproteins) or for a specific virus protein thereof (after 1D-SDS-PAGE separation). For *N*-glycosylation fingerprinting of a specific protein, first, the gel-band(s) containing the respective protein are excised from gel. During multiple incubation steps the excised gel band(s) are washed and the disulfide bonds of the incorporated protein are reduced and alkylated. To release the *N*-glycans from the glycoprotein an enzymatic reaction, using PNGase F is performed (so called "in-gel deglycosylation"). For *N*-glycosylation fingerprinting of the entire influenza virus, PNGase F is added directly to the lysed and homogenized virus pellet (so called "in-solution deglycosylation"). Both procedures are described in detail in the following. Typical fingerprints of the influenza A virus membrane glycoproteins HA and NA are depicted in Fig. 5a.

3.3.1 In-Gel Deglycosylation

All incubation steps are done in a temperature controlled Thermomixer at RT, unless stated otherwise.

1. Excise the HA band from 1D-SDS-PAGE gel with a scalpel (*see* **Note 11**) and transfer the band to a 1.5 mL tube. Cut the band inside of the 1.5 mL tube into smaller pieces (about 2 × 2 mm).

2. Add 200 μL washing solution and incubate for 60 min at 900 rpm. Replace the used washing solution by new 200 μL washing solution and incubate for additional 30 min at 900 rpm.

3. Discard washing solution. Add 200 μL ACN and incubate for 5 min at 900 rpm.

4. Discard ACN. Dry gel pieces via rotational vacuum concentrator for 5 min (*see* **Note 12**).

5. Add 200 μL DTT solution and incubate samples at 56 °C for 30 min at 900 rpm.

6. Discard DTT solution. Add 200 μL ACN and incubate for 5 min at 900 rpm. Discard ACN.

7. Add 200 μL IAA solution and incubate samples in darkness for 30 min at 900 rpm (*see* **Note 13**).

8. Discard IAA solution (*see* **Note 14**). Add 200 μL ACN and incubate for 5 min at 900 rpm. Discard ACN.

9. Add 200 μL ABC buffer and incubate for 10 min at 900 rpm. Discard ABC buffer.

10. Add 200 μL ACN and incubate for 5 min at 900 rpm. Discard ACN. Dry gel pieces via rotational vacuum concentrator for 5 min (*see* **Note 12**).

11. Dilute PNGase F solution 1:10 with ABC buffer to a final concentration of 0.1 U/μL (prepare always fresh). Add 10 μL of the PNGase F dilution on top of the dried gel pieces.

12. Allow the swelling of the gel pieces for 5 min. Add 50 µL ABC buffer and incubate samples at 37 °C for at least 3 h at 450 rpm (*see* **Note 15**).

13. After incubation with PNGase F transfer the ABC buffer around the gel pieces to a new 1.5 mL tube, further referred to as "*N*-glycan pool" (*see* **Note 16**).

14. To extract remaining *N*-glycans from gel pieces, add 200 µL of water and incubate for 10 min at 1000 rpm. Transfer the 200 µL water into the 1.5 mL *N*-glycan pool tube.

15. Repeat **step 14** twice. Dry *N*-glycan pool using a rotational vacuum concentrator (*see* **Note 17**).

3.3.2 In-Solution Deglycosylation

1. Take the 0.5 mL tube, containing the remaining 8 µL of lysed and homogenized virus from sample preparation **step 5** (*see* Subheading 3.1, protein concentration estimation by Quant-iT™ Protein Assay Kit).

2. Add 2 µL IGEPAL solution and mix carefully.

3. Dilute PNGase F solution 1:10 with ABC buffer to a final concentration of 0.1 U/µL (prepare always fresh). Add 1 µL of the PNGase F dilution to the sample and incubate at 37 °C for at least 3 h at 450 rpm (*see* **Note 15**).

4. Dry the solution containing released *N*-glycans using a rotational vacuum concentrator.

3.4 Fluorescent Labeling of Released N-Glycans with APTS

Glycans are only poorly detectable by light spectroscopic methods. To overcome this problem, glycans are often chemically modified with fluorescent dyes to enable use of highly sensitive laser induced fluorescence (LIF) detectors. Usually, an acid-catalyzed reductive amination is used to covalently link the fluorescence tag to released *N*-glycans. Due to its arylamine group, its three negative charges and its particular excitation and emission wavelengths, the fluorescent dye APTS is used for labeling the released *N*-glycans, and for xCGE-LIF based analysis. For effective labeling a large excess of APTS is necessary.

1. Dispense 2 µL water on top of the dried samples containing released *N*-glycans.

2. Add 2 µL APTS solution and 2 µL PB solution.

3. Mix sample, centrifuge (with benchtop centrifuge) and incubate at 37 °C for 16 h in darkness.

4. Stop the labeling reaction after 16 h by adding 100 µL 80:20 % ACN-water solution. Store samples until purification at 4 °C (for max. 60 min!).

3.5 HILIC-SPE Post-Derivatization Sample Cleanup

To improve sensitivity and to avoid possible complications during xCGE-LIF measurement, APTS-labeled *N*-glycans need to be purified from excess APTS, salts, acids and other impurities.

Hydrophilic interaction chromatography based solid phase extraction (HILIC-SPE), using the polyacrylamide-based stationary phase Bio-Gel P-10, has been shown to be an effective purification method [43]. The complete workflow of the HILIC-SPE cleanup procedure is shown in Fig. 3 and explained in detail in the following.

1. Assemble the vacuum manifold as shown in Fig. 3. Place a FilterPlate on top of the vacuum manifold. Place a liquid collecting vessel below the FilterPlate, inside of the vacuum manifold.

2. Dispense 200 µL Bio-Gel slurry into all used wells of the FilterPlate (*see* **Note 18**).

3. Seal unused wells of the FilterPlate with adhesive tape for proper formation of vacuum.

4. Remove solvent of the Bio-Gel slurry by applying vacuum to the FilterPlate. The vacuum should never exceed 5 in. Hg (*see* **Note 19**).

5. Wash the Bio-Gel particles three times with 200 µL water.

6. Equilibrate the Bio-Gel phase by washing three times with ACN solution (*see* **Note 20**).

7. Make sure that no liquid remains in the FilterPlate wells. Take the FilterPlate off the vacuum manifold and dry its lower part with a lint-free tissue.

8. Apply the samples on top of the Bio-Gel phase.

9. Shake the sample containing FilterPlate for 5 min at 450 rpm on a thermomixer (*see* **Note 21**). Remove solvent using the vacuum manifold (*see* **Note 22**).

10. Wash the samples five times with 200 µL ACN solution containing TEA (*see* **Note 23**).

11. Wash the samples three times with 200 µL ACN solution (*see* **Notes 23** and **24**).

12. Make sure that no liquid remains in the FilterPlate wells. Take the FilterPlate off the vacuum manifold and dry its lower part with a lint-free tissue.

13. To elute the sample put the FilterPlate on top of a DeepWell plate and fix both with adhesive tape.

14. Dispense 100 µL water to each sample containing FilterPlate well. Shake the FilterPlate/DeepWell plate assembly for 5 min at 450 rpm on a thermomixer (*see* **Note 25**).

15. For the elution procedure replace the liquid collecting vessel from the inside of the vacuum manifold by the FilterPlate/DeepWell plate assembly as shown in Fig. 3.

16. Elute the samples into the DeepWell plate by applying vacuum.

17. Dispense 200 µL water inside each sample containing FilterPlate well. Shake the FilterPlate/DeepWell plate assembly for 5 min at 450 rpm on a thermomixer and elute the samples into the DeepWell plate by applying vacuum.

18. Repeat **step 15** (*see* **Note 26**).

19. Transfer all samples into 0.5 mL tubes and dry the samples using a rotational vacuum concentrator.

20. Redissolve the dried samples in 30 µL water.

3.6 N-Glycosylation Fingerprinting via xCGE-LIF

1. Mix 2 µL HILIC-SPE purified sample with 7 µL HiDi and 1 µL LIZ standard to a final volume of 10 µL (*see* **Note 27**).

2. Fill all 10 µL into a 384 well plate and seal the plate with a septum.

3. Centrifuge sealed 384 well plate for 1 min at 150×*g*. Transfer the 384 well plate into the xCGE-LIF system.

4. The sample measurement is performed using an xCGE-LIF system (*see* Fig. 4 and **Note 28**), equipped for example with a four capillary array of 50 cm in length. The capillaries are filled with POP-7 polymer and the system is operated with 1× running buffer. Samples are injected electrokinetic and separated by applying a voltage of 15 kV for 45 min.

5. Convert xCGE-LIF generated data files to *.xml format using the DataFileConverter (provided free of charge by Life Technologies).

6. To analyze the converted *.xml files, e.g., the MATLAB based software glyXdata [44] or the Java based software glyXtool can be used (for both, see www.glyXera.com) (*see* **Note 29**).

7. Typical influenza virus derived *N*-glycosylation fingerprints can be seen in Figs. 5 and 6.

4   Notes

1. For the one-dimensional separation of virus proteins, other 1D-SDS-PAGE systems can be used. Good results were also obtained using the SERVA Electrophoresis systems or the Bio-Rad Mini-PROTEAN system.

2. Use a non-reducing sample buffer to analyze the *N*-glycosylation of the entire influenza HA (HA$_0$) protein. For the analysis of the influenza HA$_0$ subunits HA$_1$ and HA$_2$ use a reducing sample buffer (converts HA$_0$ into its subunits HA$_1$ and HA$_2$).

3. Use commercially available precasted gels to prevent sample contamination with oligohexose ladders and to ensure reproducible separation quality.

4. Heat the solution (maximal 50 °C) to facilitate dissolution of citric acid.

5. Handling of infectious influenza virus material has to be performed under biosafety level 2 conditions! After virus inactivation by treatment with virus lysis buffer (Subheading 3.1, **step 5**) or SDS-PAGE sample buffer (Subheading 3.2, **step 2**) at elevated temperature, biosafety level 1 is sufficient.

6. A minimum HA titers of 1.7 $\log_{10}$ HA units per 100 μL is required for subsequent analyses.

7. Marking the location of the virus pellet at the UC-tube is crucial for subsequent location of the pellet for resuspension.

8. For maximum virus recovery, remove the supernatant carefully. Remove about 10 % of the supernatant by squeezing the UC-tube, with the tube opening facing down (make sure that during this step the virus pellet is covered with supernatant). Turn the UC-tube until the virus pellet is positioned inside of the generated air bubble (this prevents the loss of the virus pellet during further supernatant removal). Squeeze the remaining supernatant out of the UC-tube. Cut the upper half of the UC-tube. Add 20 μL of dissolving buffer on top of the marked virus pellet and resuspend it by destroying the pellet with a pipet tip. Transfer all virus-containing liquid to a 0.5 mL tube. Add additional 20 μL dissolving buffer on top of the marked virus pellet and wash the virus pellet position. Transfer the additional 20 μL into the 0.5 mL tube.

9. Residuals of phenol red pH indicator from cell culture media can be present in the resuspended virus pellet. Its interference with common colorimetric protein estimation assays (like BCA assay and Bradford assay) can result in measurement of incorrect protein concentrations. The fluorescence-based protein quantitation kit Quant-iT™ is not influenced by phenol red.

10. For accurate protein estimation, all virus particles need to be lysed and homogenized. Otherwise, the protein concentration is underestimated.

11. HA monomer, dimer or trimer will result in the same N-glycan fingerprint. For best results cut the band with the strongest color (highest protein concentration). Also NA bands can be cut for N-glycan analysis. The lower detection limit of N-glycan analysis strongly depends on the glycoprotein itself. Approximately, this range is between 0.5 μg for more glycosylated proteins (e.g., influenza HA, bovine fetuin) and 4 μg for less glycosylated proteins (e.g., immunoglobulin G).

12. ACN induces shrinking of the gel pieces. Shrunk and dried gel pieces will more effectively absorb added chemical solutions or enzyme containing solutions.

13. Cover the sample rack of the thermomixer with aluminum foil for incubation in darkness.

14. DTT and IAA treatment will reduce and alkylate the disulfide bonds of proteins irreversibly, improving the accessibility of *N*-glycans by PNGase F.

15. Incubation overnight is also possible.

16. Do not discard the ABC buffer, which was used for PNGase F incubation. A large amount of released *N*-glycans (after PNGase F treatment) is contained in this liquid.

17. To confirm the presence of the desired protein in the excised gel band, a proteomics approach can be performed. Briefly, add 200 μL ACN on the remaining gel pieces (after *N*-glycan extraction). Incubate for 5 min at 900 rpm. Discard ACN. Dry gel pieces using a rotational vacuum concentrator for 5 min. Add trypsin on top of the dried gel pieces and incubate at 55 °C for 2 h. Extract peptides with 50:49:1 % ACN-water-trifluoroacetic acid.

18. The Bio-Gel particles will settle fast in the slurry suspension. Before dispensing the slurry into the FilterPlate wells, homogenize the Bio-Gel slurry by mixing. Cut the top of a pipet tip to facilitate uptake of the homogenized Bio-Gel slurry. For purification one Bio-Gel filled well is needed per sample.

19. Check the pressure at pressure gauge. The vacuum should never exceed 5 in. Hg during the whole cleanup procedure.

20. The Bio-Gel particles will shrink during the equilibration step.

21. Interaction time is important during HILIC-SPE purification.

22. Sticking of APTS labeled *N*-glycans and non-derivatized APTS to the Bio-Gel particles will be indicated by a green color.

23. Switch off the vacuum pump. Add the solvent, shake the FilterPlate for 1 min at 450 rpm on a thermomixer. Transfer the FilterPlate back on vacuum manifold and switch on vacuum pump to remove the solvent.

24. To elute non-derivatized APTS, TEA is introduced to the washing solution to disrupt possible interactions of the strong negative charges of APTS with the stationary phase. Bio-Gel particles are discolored due to the loss of non-derivatized APTS after all washing steps.

25. Bio-Gel particles will swell in water and consume a large portion of the 100 μL water.

26. After three elution steps the final sample volume is about 400 μL.

27. The LIZ standard is used for internal migration time normalization allowing reproducible measurements.

28. Alternatively the Genetic Analyzer models 3130xl, 3730, and 3730xl can also be used.

29. Both software tools support the automated migration time normalization to internal LIZ standard, data smoothing, and background adjustment, as well as automated peak picking and sample comparison.

Acknowledgements

This work was supported by the Max Planck Society and by the European Union's Seventh Framework Programme (FP7-Health-F5-2011) under grant agreement no. 278535 "HighGlycan."

References

1. Aich U, Yarema KJ (2008) Glycobiology and immunology. In: Carbohydrate-based vaccines and immunotherapies. Wiley, Hoboken, NJ, pp 1–53

2. Pon RA, Jennings HJ (2008) Carbohydrate-based antibacterial vaccines. In: Carbohydrate-based vaccines and immunotherapies. Wiley, Hoboken, NJ, pp 117–166

3. Astronomo RD, Burton DR (2010) Carbohydrate vaccines: developing sweet solutions to sticky situations? Nat Rev Drug Discov 9(4):308–324

4. Ruhaak LR, Zauner G, Huhn C, Bruggink C, Deelder AM, Wuhrer M (2010) Glycan labeling strategies and their use in identification and quantification. Anal Bioanal Chem 397(8): 3457–3481

5. Vanderschaeghe D, Festjens N, Delanghe J et al (2010) Glycome profiling using modern glycomics technology: technical aspects and applications. Biol Chem 391(2-3):149–161

6. Seeberger PH, Werz DB (2007) Synthesis and medical applications of oligosaccharides. Nature 446(7139):1046–1051

7. Wang P, Dong S, Shieh JH et al (2013) Erythropoietin derived by chemical synthesis. Science 342(6164):1357–1360

8. Meuris L, Santens F, Elson G et al (2014) GlycoDelete engineering of mammalian cells simplifies N-glycosylation of recombinant proteins. Nat Biotechnol 32(5):485–489

9. Kanda Y, Yamane-Ohnuki N, Sakai N et al (2006) Comparison of cell lines for stable production of fucose-negative antibodies with enhanced ADCC. Biotechnol Bioeng 94(4): 680–688

10. Callewaert N, Laroy W, Cadirgi H et al (2001) Use of HDEL-tagged Trichoderma reesei mannosyl oligosaccharide 1,2-α-D-mannosidase for N-glycan engineering in Pichia pastoris. FEBS Lett 503(2–3):173–178

11. Hamilton SR, Davidson RC, Sethuraman N et al (2006) Humanization of yeast to produce complex terminally sialylated glycoproteins. Science 313(5792):1441–1443

12. Jacobs PP, Geysens S, Vervecken W et al (2008) Engineering complex-type N-glycosylation in Pichia pastoris using GlycoSwitch technology. Nat Protoc 4(1):58–70

13. Aumiller JJ, Mabashi-Asazuma H, Hillar A et al (2012) A new glycoengineered insect cell line with an inducibly mammalianized protein N-glycosylation pathway. Glycobiology 22(3): 417–428

14. Cox MMJ (2012) Recombinant protein vaccines produced in insect cells. Vaccine 30(10): 1759–1766

15. Abe Y, Takashita E, Sugawara K et al (2004) Effect of the addition of oligosaccharides on the biological activities and antigenicity of influenza A/H3N2 virus hemagglutinin. J Virol 78(18):9605–9611

16. Klenk HD, Wagner R, Heuer D et al (2002) Importance of hemagglutinin glycosylation for the biological functions of influenza virus. Virus Res 82(1-2):73–75

17. Vigerust DJ, Shepherd VL (2007) Virus glycosylation: role in virulence and immune interactions. Trends Microbiol 15(5):211–218

18. Swarts BM, Guo Z (2008) Carbohydrate-based antiviral vaccines. In: Carbohydrate-based vaccines and immunotherapies. Wiley, Hoboken, NJ, pp 167–193

19. Wolfert MA, Boons G-J (2013) Adaptive immune activation: glycosylation does matter. Nat Chem Biol 9(12):776–784

20. Avci FY, Li X, Tsuji M et al (2011) A mechanism for glycoconjugate vaccine activation of the adaptive immune system and its implications for vaccine design. Nat Med 17(12):1602–1609

21. de Vries RP, Smit CH, de Bruin E et al (2012) Glycan-dependent immunogenicity of recombinant soluble trimeric hemagglutinin. J Virol 86(21):11735–11744

22. Hütter J, Rödig J, Höper D et al (2013) Toward animal cell culture-based influenza vaccine design: viral hemagglutinin N-glycosylation markedly impacts immunogenicity. J Immunol 190(1):220–230

23. Kailemia MJ, Ruhaak LR, Lebrilla CB et al (2014) Oligosaccharide analysis by mass spectrometry: a review of recent developments. Anal Chem 86(1):196–212

24. Wuhrer M, de Boer AR, Deelder AM (2009) Structural glycomics using hydrophilic interaction chromatography (HILIC) with mass spectrometry. Mass Spectrom Rev 28(2):192–206

25. Harvey DJ (2005) Proteomic analysis of glycosylation: structural determination of N- and O-linked glycans by mass spectrometry. Expert Rev Proteomics 2(1):87–101

26. Roth Z, Yehezkel G, Khalaila I (2012) Identification and quantification of protein glycosylation. Int J Carbohydr Chem 2012(2012):10

27. Anumula KR (2006) Advances in fluorescence derivatization methods for high-performance liquid chromatographic analysis of glycoprotein carbohydrates. Anal Biochem 350(1):1–23

28. Laroy W, Contreras R, Callewaert N (2006) Glycome mapping on DNA sequencing equipment. Nat Protoc 1(1):397–405

29. Raman R, Raguram S, Venkataraman G et al (2005) Glycomics: an integrated systems approach to structure-function relationships of glycans. Nat Methods 2(11):817–824

30. Huffman JE, Pučić-Baković M, Klarić L et al (2014) Comparative performance of four methods for high-throughput glycosylation analysis of immunoglobulin g in genetic and epidemiological research. Mol Cell Proteomics 13(6):1598–1610

31. Lin G, Simmons G, Pohlmann S et al (2003) Differential N-linked glycosylation of human immunodeficiency virus and Ebola virus envelope glycoproteins modulates interactions with DC-SIGN and DC-SIGNR. J Virol 77(2):1337–1346

32. Lin S-C, Jan J-T, Dionne B et al (2013) Different immunity elicited by recombinant H5N1 hemagglutinin proteins containing pauci-mannose, high-mannose, or complex type N-glycans. PLoS One 8(6):e66719

33. Geyer R, Diabaté S, Geyer H et al (1987) Carbohydrates of influenza virus. Structure of the oligosaccharides linked to asparagines 406 and 478 in the hemagglutinin of fowl plague virus, strain Dutch. Glycoconj J 4(1):17–32

34. Mir-Shekari SY, Ashford DA, Harvey DJ et al (1997) The glycosylation of the influenza A virus hemagglutinin by mammalian cells. A site-specific study. J Biol Chem 272(7):4027–4036

35. Yagi H, Watanabe S, Suzuki T et al (2012) Comparative analyses of N-glycosylation profiles of influenza A viruses grown in different host cells. Open Glycosci 5(1):2–12

36. Bateman AC, Karamanska R, Busch MG et al (2010) Glycan analysis and influenza A virus infection of primary swine respiratory epithelial cells: the importance of NeuAc 2-6 glycans. J Biol Chem 285(44):34016–34026

37. Doores KJ, Bonomelli C, Harvey DJ et al (2010) Envelope glycans of immunodeficiency virions are almost entirely oligomannose antigens. Proc Natl Acad Sci U S A 107(31):13800–13805

38. Schwarzer J, Rapp E, Reichl U (2008) N-glycan analysis by CGE–LIF: profiling influenza A virus hemagglutinin N-glycosylation during vaccine production. Electrophoresis 29(20):4203–4214

39. Schwarzer J, Rapp E, Hennig R et al (2009) Glycan analysis in cell culture-based influenza vaccine production: influence of host cell line and virus strain on the glycosylation pattern of viral hemagglutinin. Vaccine 27(32):4325–4336

40. Rödig J, Rapp E, Höper D et al (2011) Impact of host cell line adaptation on quasispecies composition and glycosylation of influenza A virus hemagglutinin. PLoS One 6(12):e27989

41. Rödig J, Rapp E, Bohne J et al (2013) Impact of cultivation conditions on N-glycosylation of influenza virus a hemagglutinin produced in MDCK cell culture. Biotechnol Bioeng 110(6):1691–1703

42. Genzel Y, Rödig J, Rapp E et al (2014) Vaccine production: upstream processing with adherent or suspension cell lines. Methods Mol Biol 1104:371–393

43. Ruhaak LR, Hennig R, Huhn C et al (2010) Optimized workflow for preparation of APTS-labeled N-glycans allowing high-throughput analysis of human plasma glycomes using 48-channel multiplexed CGE-LIF. J Proteome Res 9(12):6655–6664

44. Hennig R, Reichl U, Rapp E (2011) A software tool for automated high-throughput processing of CGE-LIF based glycoanalysis data, generated by a multiplexing capillary DNA sequencer. Glycoconj J 28:331

Chapter 9

Temporary Conversion of Protein Amino Groups to Azides: A Synthetic Strategy for Glycoconjugate Vaccines

Tomasz Lipinski and David R. Bundle

Abstract

Conjugation of synthetic oligosaccharides and native polysaccharides to proteins is an important tool in glycobiology to create vaccines and antigens to screen lectins, toxins, and antibodies. A novel approach to potentiate and profile the immune response to vaccines involves targeting antigens directly to dendritic cells (DCs), the key cells engaged in the immunization process. Inclusion of a carbohydrate ligand recognized by C-type lectins expressed on their cell surface ensures targeting of vaccines to DCs and improved immunological responses. Here we describe a strategy that permits three sequential orthogonal conjugation reactions to prepare glycoconjugates and apply them to the synthesis of a conjugate vaccine that is targeted for uptake by DCs. The carrier protein is treated with an azo-transfer reagent to convert accessible amino groups to azide and then amide bond formation via reaction with carboxylic acid side chains is used to attach amino tether groups of a ligand to the protein. Azide-alkyne Huisgen cycloaddition conjugation, "click chemistry" is used to attach a second ligand equipped with a propargyl group or an analogous terminal alkyne, and following reduction of protein azide groups back to amine, these amino acid side chains can be subjected to amide formation such as reaction with succinimide esters or homobifunctional coupling reagents such as dialkyl square.

Key words Azo transfer, Azidination of proteins, Protein carrier, Oligosaccharide–protein conjugation, Click chemistry

1 Introduction

A ß-mannan–tetanus toxoid conjugate vaccine against *Candida albicans* was highly immunogenic and effective in rabbits [1] but showed poor immunogenicity in mice [2]. Increasing awareness of the importance of DCs, a pivotal component of the adaptive immune system, in processing of antigens and presentation of antigenic fragments to T lymphocytes has suggested ways to design improved vaccines [3–5]. Since DCs possess C-type lectins on their cell surface, attachment of ligands that are bound by these lectins is one way to specifically deliver vaccines to DCs [6–10]. To improve the specific immune response to the cell mannan of the fungal pathogen *Candida albicans*, we aimed to target Dectin-1,

Bernd Lepenies (ed.), *Carbohydrate-Based Vaccines: Methods and Protocols*, Methods in Molecular Biology, vol. 1331, DOI 10.1007/978-1-4939-2874-3_9, © Springer Science+Business Media New York 2015

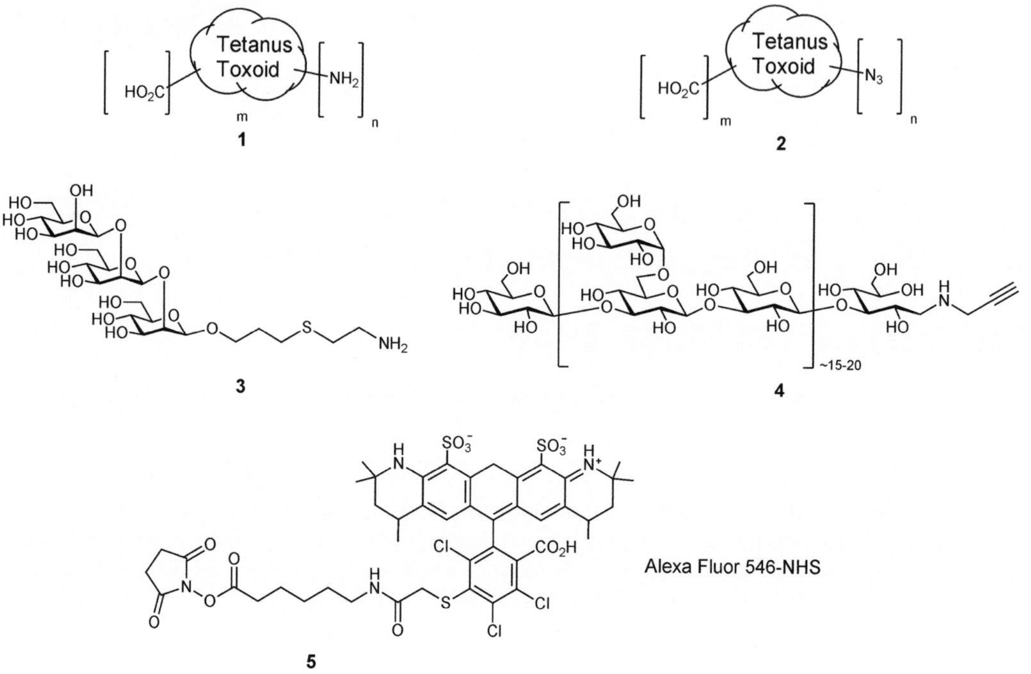

Fig. 1 Proteins, oligosaccharides and fluorescent dye derivatives used to synthesize a labelled tricomponent vaccine

a pattern-recognition receptor expressed on monocytes, macrophages and DCs [6–8]. We elected to create a conjugate vaccine bearing two carbohydrate epitopes, a ß-mannotriose and laminarin, a ß-glucan ligand of Dectin-1 [11]. In order to follow the uptake of conjugate vaccines by DCs during in vitro cell culture experiments we wished to attach a third molecule, a fluorescent tag to tetanus toxoid, the carrier protein (*see* Fig. 1).

To achieve covalent attachment of two ligands in a controlled manner we required two independent conjugation methods. Amide bond formation with either amino groups of lysine or carboxylic acid side chains of aspartic and glutamic acids has always had appeal but competing side reactions involving either cross-linking of protein molecules or internal amide bond formation within a protein molecule limit application of this approach [12, 13]. However, if accessible amines of tetanus toxoid **1** are first converted to azide by the azo transfer reagent, imidazole-1-sulfonyl azide hydrochloride [14, 15] it is then possible to engage aspartic and glutamic acids of **2** in efficient conjugation reactions with the mannotriose amine **3** without cross linking of protein amino acids. The azide groups of **2** are readily available for a second conjugation by click chemistry [16, 17] to propargylated laminarin **4**. After reduction of azide back to amino groups by trimethylphosphine, lysine residues can be acylated by the succinimide ester **5** of the fluorescent dye Alexa 546. This sequence; azo transfer conducted on accessible lysine residues of the protein, followed by amide bond formation with carboxylic acid sides chains, click chemistry with azide groups, reduction of azide group back to amines and reaction of these with activated esters is illustrated here for carbohydrate ligands but may be applied to any haptenic groups bearing suitably reactive functional groups.

The practical application of this conjugation strategy permitted a previously poorly immunogenic ß-mannan–tetanus toxoid conjugate vaccine against *Candida albicans* to be transformed in to an effective immunogen [11]. Our studies showed that the tricomponent vaccine was able to target cells bearing the Dectin-1 receptor via a ß-glucan ligand, laminarin. Antigen binding to Dectin-1 overexpressed on a macrophage cell line resulted in: (1) activation of SYK and Src-family-kinases revealed by their recruitment and phosphorylation in the vicinity of bound conjugate and (2) translocation of NF-κB to the nucleus. Treatment of immature bone marrow derived DCs with fluorescently labelled tricomponent vaccine confirmed that the ß-glucan enabled DC targeting and antigen uptake. Bone marrow derived DCs secreted increased levels of several cytokines including TGF-β and IL 6, which are known activators of Th17 helper T-cells. Immunization of mice with the novel vaccine resulted in an improved immune response manifested by high titers of antibody recognizing *C. albicans* β-mannan antigen. Vaccine containing laminarin also affected distribution of IgG subclasses showing that vaccine targeting to Dectin-1 receptor can benefit from augmentation and modulation of the immune response [11].

2 Materials

2.1 Synthesis of Propargylated Laminarin

1. 4 mL Kimball glass vial.

2. Laminarin is the soluble β-glucan from *Laminaria digitata*, (Sigma-Aldrich, St. Louis, MO).

3. Reductive amination buffer, 0.2 M phosphate buffer pH 6.0: Stock solutions: A, 0.2 M monobasic sodium phosphate (27.8 g of sodium dihydrogen phosphate is dissolved and made up to 1 L with water); B, dibasic sodium phosphate (53.65 g of disodium hydrogen phosphate heptahydrate is dissolved and made up to 1 L with water). 87.7 mL of A and 12.3 mL of B are mixed to provide 0.2 M buffer pH 6.0.

4. Magnetic stirrer.

5. 37 °C incubator.

6. Propargylamine hydrochloride.

7. Sodium cyanoborohydride.

8. Centrifuge.

9. Sephadex G-25 column (2.5×25 cm).

10. Fraction collector with refractive index monitor.

11. Silica gel TLC plate.

12. Spray for TLC plate: 5 % sulfuric acid in ethanol.

2.2 Azidination of Tetanus Toxoid

1. 4 mL Kimball glass vial.

2. Tetanus toxoid **1** (State Serum Institute, Copenhagen).

3. Azidination buffer, 0.5 M carbonate/bicarbonate buffer pH 9.8: Solution A 0.5 M sodium carbonate (53 g of anhydrous sodium carbonate is dissolved and made up to 1 L with water). Solution B 0.5 M sodium bicarbonate (42 g of sodium bicarbonate is dissolved and made up to 1 L with water). Buffer is prepared by mixing equal volumes of solution A and B, then pH is adjusted to 9.8 by titration with solution B controlled with pH meter.

4. UV spectrophotometer.

5. Imidazole-1-sulfonyl azide hydrochloride (CAUTION *see* **Note 1**).

6. 0.5 g $CuSO_4 \cdot 5H_2O$.

7. Magnetic stirrer.

8. Water containing 0.1 % Tween 20 (v/v).

9. Pellicon XL 50 ultradialysis unit with a 10 kDa cutoff TF membrane.

10. 10 mM EDTA in water (make up 3.72 g of disodium ethylenediamine tetraacetate dehydrate to 1 L with water).

11. 0.15 M NaCl in water (make up 8.76 g of sodium chloride to 1 L with water).

12. 100 mM 4-methylmorpholine in water (make up 11 mL (10.1 g) to 1 L with water).

13. Millipore centrifugal filter unit (10 kDa cutoff).

2.3 Conjugation of the Amino Tether with Protein Carboxylic Acid Residues

1. A solution of azidinated tetanus toxoid **2**.

2. Mannotriose **3** synthesized according to published methods [18].

3. 4-(4,6-Dimethoxy-1,3,5-triazin-2-yl)-4-methylmorpholinium chloride (DMTMM) [19].

4. Silica gel TLC plate.

5. TLC solvent dichloromethane methanol 20:1 (v/v).

6. 4-methylmorpholine.

7. Binding buffer, 20 mM Tris–HCl pH 8.5: Solution A, 0.2 M Tris (24.2 g of tris(hydroxymethyl)aminomethane dissolved and made up to 1 L with water). Solution B, conc. hydrochloric acid ~80 mL of A is titrated to pH 8.5 with hydrochloric acid using a pH meter and the final volume adjusted to 1 L with water.

8. DEAE Sepharose CL column (3 mL of gel volume).

9. Phosphate buffer saline (PBS): 0.2 M phosphate buffer pH 6.0: Stock solutions: A, 0.2 M monobasic sodium phosphate (27.8 g of sodium dihydrogen phosphate is dissolved and made up to 1 L with water); B, dibasic sodium phosphate (53.65 g of disodium hydrogen phosphate heptahydrate is dissolved and made up to 1 L with water). 72 mL of B is mixed with ~28 mL of A until the buffer is pH 7.2. Sodium chloride 8.18 g is dissolved in 50 mL of this solution which is made up to 1 L with water.

10. Superdex S-200 column (2 × 100 cm) equilibrated with PBS, the chromatography buffer.

11. Chromatography buffer, 0.1 M Tris–HCl pH 7.2, 1 M NaCl: Solution A, 0.2 M Tris (24.2 g of tris(hydroxymethyl)amino-methane dissolved and made up to 1 L with water). 100 mL of solution A is titrated with conc. hydrochloric acid to pH 7.2, then 11.7 g of sodium chloride is added and after dissolution the final volume adjusted to 200 mL.

12. Dialysis buffer, 0.2 M Tris–HCl pH 8.5: Solution A, 0.2 M Tris (24.2 g of tris(hydroxymethyl)aminomethane dissolved and made up to 1 L with water). Solution B, conc. hydrochloric acid ~80 mL of A is titrated to pH 8.5 with hydrochloric acid using a pH meter and the final volume adjusted to 100 mL.

13. Millipore centrifugal filter unit (10 kDa cutoff).

2.4 Conjugation of Propargylated Laminarin with Mannotriose–Azidinated Tetanus Toxoid Conjugate 6

1. 4 mL Kimball glass vial.

2. Trisaccharide–tetanus toxoid conjugate **6**.

3. Cycloaddition buffer 0.2 M Tris–HCl buffer pH 8.0: Solution A, 0.2 M Tris (24.2 g of tris(hydroxymethyl)aminomethane dissolved and made up to 1 L with water). Solution B, conc. hydrochloric acid. ~80 mL of A is titrated to pH 8.5 with hydrochloric acid using a pH meter and the final volume adjusted to 100 mL.

4. Propargylated laminarin.

5. Copper powder.

6. Isobutanol.

7. Bathophenanthroline/Cu^{+1} catalyst (*see* **Note 2**).

8. PBS.

9. Superdex S-200 column (2 × 100 cm).

2.5 Reduction of Protein Azide Groups to Amine

1. Conjugate **7**.

2. 1 M solution of trimethylphosphine in THF.

3. 0.5 M sodium carbonate: make up 53 g of anhydrous sodium carbonate to 1 L with water.

4. Thermo Scientific Slide-A-Lyzer Dialysis Cassette (10K molecular weight cutoff).

5. Sterile 0.22 μm filter.

2.6 Labeling Conjugates with Alexa Fluor 546

1. 4 mL Kimball glass vial.

2. Tetanus toxoid conjugate vaccine **8**.

3. PBS.

4. 1 M sodium bicarbonate: make up 8.4 g of sodium bicarbonate to 1 L with water.

5. Alexa Fluor 546 NHS ester (Molecular Probes, Life Technologies).

6. Dimethyl sulfoxide (DMSO).

7. PD-10 desalting column (GE Healthcare).

8. Amicon Ultra-4 Centrifugal Filter Unit (10 kDa).

3 Methods

3.1 Synthesis of Propargylated Laminarin [4] (See Scheme 1)

100 mg of laminarin is dissolved in 1 mL of the reductive amination buffer in a 4 mL glass vial with a magnetic stirring bar. Propargylamine hydrochloride (120 mg) is added followed by sodium cyanoborohydride (5 mg) and stirred in an incubator at 37 °C for 7 days. Additional portions of sodium cyanoborohydride (5 mg each) are added on days 2 and 4. After 7 days the reaction mixture is diluted with water to 10 mL and the polysaccharide is precipitated by the addition of ethanol (40 mL). The precipitate is collected by centrifugation ($7000 \times g \times 20$ min) and dissolved in a minimal amount of water (1 mL). This material is purified on a Sephadex G-25 column by elution with water. Fractions containing sugar are collected (see **Note 3**) and lyophilized (see **Note 4**).

3.2 Azidination of Tetanus Toxoid (See Scheme 2)

Tetanus toxoid **1** (36 mg, see **Note 5**) is dissolved in 1.8 mL of azidination buffer followed by the addition of 100 μL of the solution of imidazole-1- sulfonyl azide (stock solution 20 mg/mL in water) and 100 μL of a solution of CuSO4 (stock solution 10 mg/mL in water) as catalyst. This mixture is stirred on a magnetic stirrer for 9 h. The reaction mixture is diluted with 50 mL of water containing 0.1 % Tween 20 (to prevent protein aggregation) and then washed by ultradialysis using a TF membrane (Pellicon XL 50, 10 kDa cutoff) first with EDTA to remove copper (20 mL), then with NaCl (50 mL) and 4-methylmorpholine (20 mL). About 8 mL of dialyzed solution of protein **2** was collected and concentrated to 2 mL final volume using a centrifugal filter.

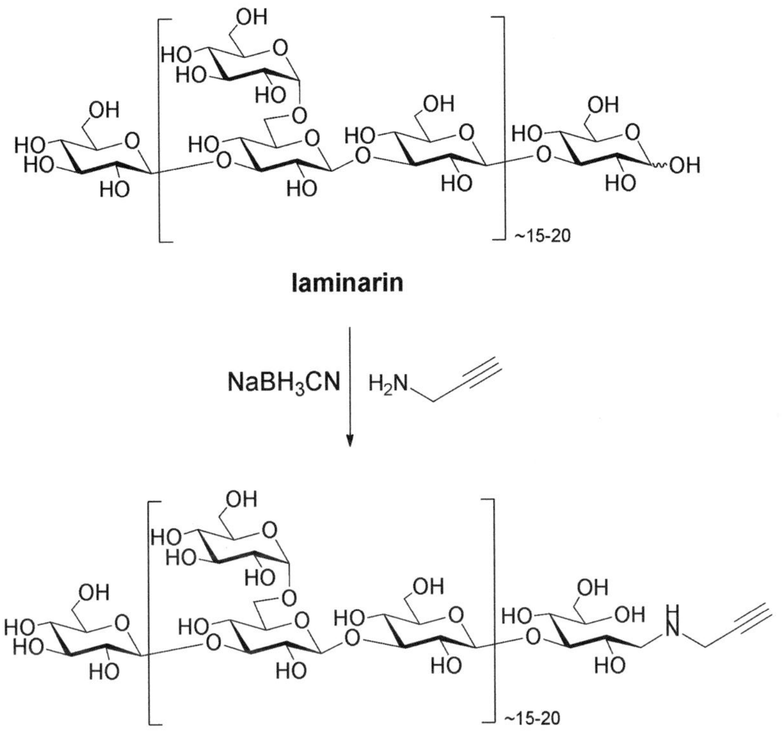

Scheme 1 Reductive amination of laminarin molecules possessing uncapped reducing glucose residues by propargylamine

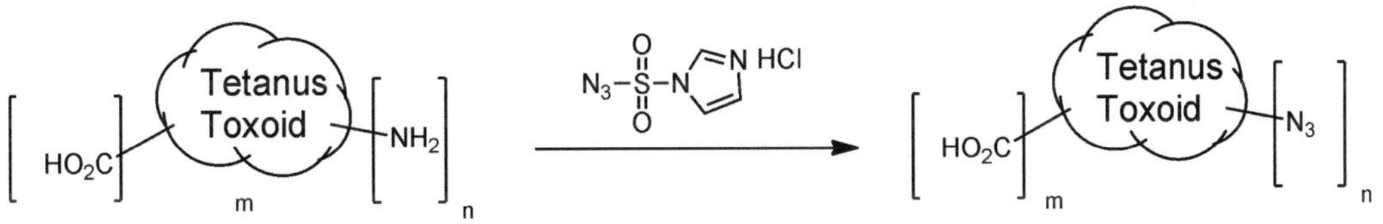

Scheme 2 Reaction of tetanus toxoid **1** with azo transfer reagent, imidazole-1-sulfonyl azide hydrochloride converts accessible terminal amino acid and lysine residues to the corresponding azidinated toxoid **2** with terminal azido amino acid and azido-lysines

3.3 Conjugation of the Amino Tether with Protein Carboxylic Acid Residues (See Scheme 3)

The solution of azidinated tetanus toxoid **2** (2 mL) is transfered to a 4 mL Kimball vial, and 5.9 mg of trisaccharide **3** (26 M equivalents) is added followed by DMTMM (16 mg). The consumption of trisaccharide and the progress of conjugation may be followed by TLC using the solvent mixture dichloromethane containing 5 % methanol. After ~5 h another portion of DMTMM (8 mg) is added and the pH adjusted to >8.0 by addition of 4-methylmorpholine. The reaction is assumed to reach completion when only trace amounts of unconjugated trisaccharide are detectable by TLC. The reaction mixture is then dialyzed against binding buffer.

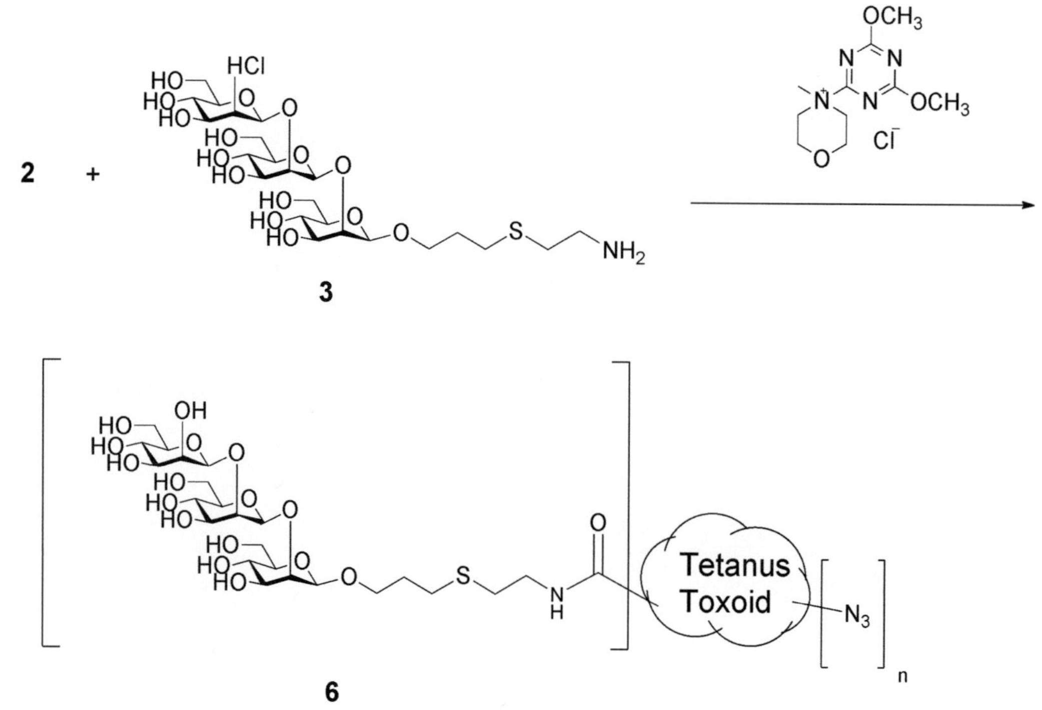

Scheme 3 Amide bond formation between the trimannoside ligand **3** equipped with a amino terminated tether and the Asp and Glu residues of azidinated tetanus toxoid **2**

To remove Tween the conjugate protein is loaded on a DEAE column equilibrated with the binding buffer. The conjugate is eluted with elution buffer, concentrated and then fractionated on Superdex S-200 in PBS. Material corresponding to a monomeric tetanus toxoid fraction is collected (~MW 170 kDaltons). Combined fractions are dialyzed against dialysis buffer and concentrated on a centrifugal filter unit to 4.4 mL of conjugate **6** (*see* **Note 6**).

3.4 Conjugation of Propargylated Laminarin with Mannotriose–Azidinated Tetanus Toxoid (See Scheme 4)

2.1 mL of a trisaccharide–tetanus toxoid conjugate **6** (10 mg) in cycloaddition buffer is transferred to a 4 mL Kimball vial containing 12.5 mg of propargylated laminarin **4**. Upon dissolution of laminarin, copper powder (~20 mg) and isobutanol (50 μL) are added and the vial is sealed with a septum, degassed under reduced pressure and purged with argon. The click reaction is initiated by addition of bathophenanthroline/Cu^{+1} catalyst (*see* **Note 2**) (25 μL per 1 mL of reaction mixture). After 12 h incubation the reaction mixture is filtered, dialyzed against PBS and purified on Superdex S-200 to give the azido derivative of the tricomponent conjugate vaccine **7** (*see* **Note 7**).

3.5 Reduction of Protein Azide Groups to Amine (See Scheme 5)

The azide groups of **7** are reduced to amines by reaction with trimethylphosphine. Conjugate **7** (10 mg/mL) is first subjected to buffer exchange from PBS to 0.5 M sodium carbonate via dialysis in Thermo Scientific Slide-A-Lyzer Dialysis Cassette (10K MWCO)

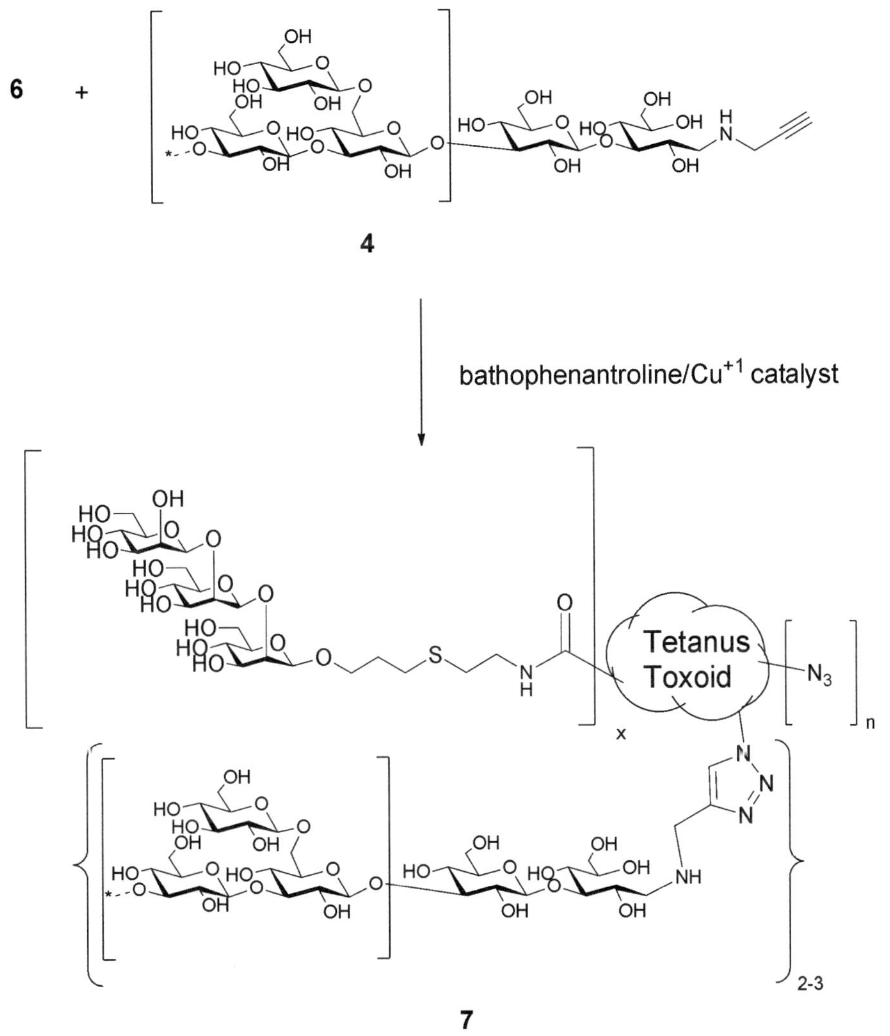

6 +

4

bathophenantroline/Cu^{+1} catalyst

Tetanus Toxoid

7

Scheme 4 Conjugation of propargylated laminarin **4** with tetanus toxoid trimannoside conjugate **6** by Huisgen cycloaddition conjugation, "click chemistry"

concentrated to ~1.5 mL and reacted with trimethylphosphine (50 μL of a 1 M solution in THF) at room temperature for 18 h in a 4 mL Kimball glass vial closed with a septa. Conjugate is then dialyzed against PBS, concentrated and sterile filtered through 0.22 μm filters to yield a solution of the tricomponent vaccine **8** (2.7 mg/mL).

3.6 Labeling Conjugates with Alexa Fluor 546 (See Scheme 6)

To a solution of the tricomponent tetanus toxoid conjugate vaccine **8** in 150 μL of PBS containing 400 μg of protein, 15 μL of 1 M sodium bicarbonate solution is added followed by Alexa Fluor 546 NHS ester (Molecular Probes, Life Technologies, Grand Island, NY) (20 μg) dissolved in DMSO (~30 μL). The tubes are wrapped in aluminum foil and left on an inverting mixer for 18 h.

7

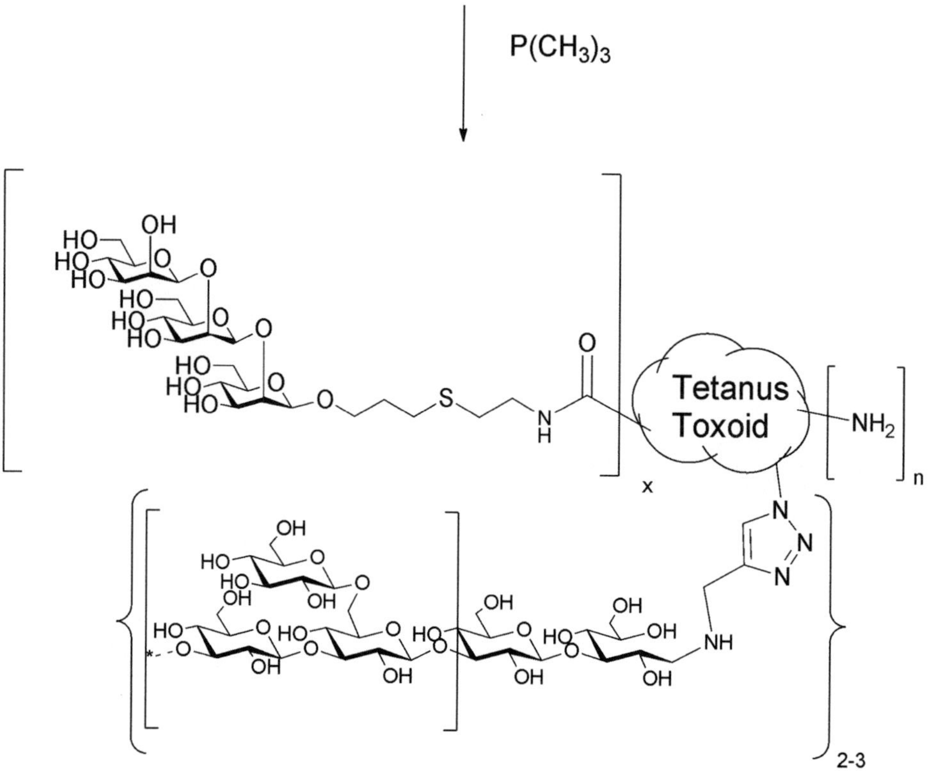

$$P(CH_3)_3$$

8

Scheme 5 Reduction of any unconjugated azide groups of **7** groups to amines by trimethylphosphine

Purification is performed on a PD-10 desalting column (GE Healthcare) equilibrated with PBS. Fractions containing labelled conjugates are collected and concentrated on an Amicon Ultra-4 Centrifugal Filter Unit (10 kDa) to give the fluorescently label vaccine **9** at a final concentration of 2.5 mg/mL.

4   Notes

1. Imidazole-1-sulfonyl azide hydrochloride is prepared by chemical synthesis according to a published method [14] and according to internet search is now commercially available from several sources including Sigma Aldrich and numerous Chinese companies. However, this reagent must be treated with extreme care. There has been one report of a serious explosion during its preparation and the authors of the original synthesis have discussed safe handling of the reagent and its salts [14, 15].

8

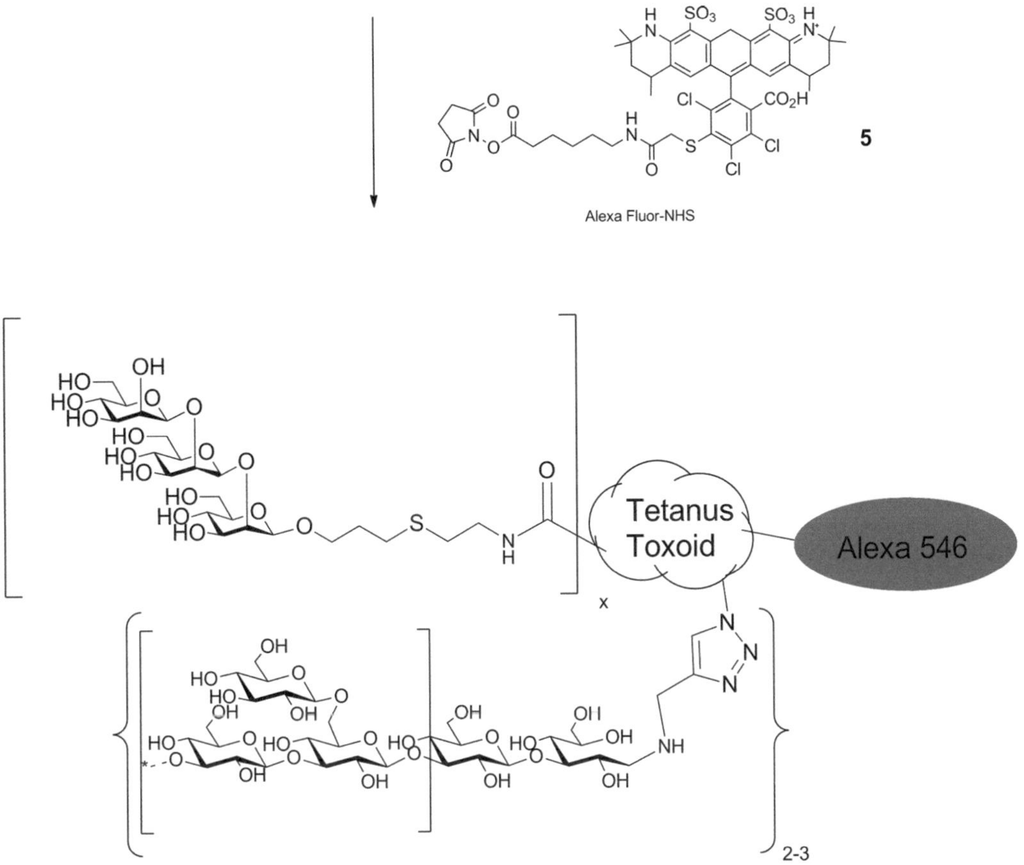

Alexa Fluor-NHS

5

Tetanus Toxoid

Alexa 546

9

Scheme 6 The tricomponent vaccine **8** is labelled by reaction with Alex 546 dye succinimide ester to give **9**

The reagent is crystalline and only the first crop of crystals should be collected. Further processing of mother liquors *should not be attempted*. The reagent is stored anhydrous in 100 mg aliquots in Kimble glass vials with a polypropylene cap at –20 °C. The authors of the original synthesis of this reagent have published a discussion of the dangers and recommended handling of this reagent and preparation of less hazardous salts [14, 15].

2. Bathophenanthroline/Cu^{+1} catalyst [2, 20]
   The catalyst is prepared by a slight modification [2] of a published procedure [20] as follows: $CuSO_4 \cdot 5H_2O$ (10 mg) and bathophenanthroline sulfonate (64.4 mg) (GFS Chemicals Inc.) are dissolved in 0.2 M Tris–HCl pH 8.0 buffer (1 mL) in a 4 mL Kimball glass vial. Copper powder (~50 mg) is added; the vial is closed with an open top screw cap with rubber septa

and purged with argon. The vial is then rotated for 2 h; the reduction of copper II to copper I by metallic copper is indicated by the appearance of a dark green color.

3. Detection of carbohydrate containing fractions from a G-25 column. This may be achieved by monitoring the eluent with a refractive index monitor or by simply spotting each fraction on a silica gel TLC plate and without development simply char by spraying with 5 % sulfuric acid in ethanol followed by heating to char any carbohydrate (which appears as a brown spot).

4. Approximately 15 % of laminarin molecules are substituted with a propargyl group, since only about 20 % of laminarin molecules in commercial preparations are available for derivatization via the reducing end since the majority of chains are capped by mannitol. The remainder of the laminarin molecules are unable to undergo reductive amination and therefore lack a propargyl group. Consequently these molecules do not undergo subsequent click reaction and are removed when the protein conjugate is purified (*see* Subheading 3.4).

5. Tetanus toxoid is partially N-methylated during inactivation with formaldehyde and different batches of toxoid may have varying numbers of available lysine residues with ξ side-chain amino groups. These may be determined by a colorimetric assay [21]. The concentration of tetanus toxoid and its conjugates can be estimated by UV absorbance at 280 nm. A solution of this protein at 1 mg/mL has an optical density of 1.24.

6. Approximate protein content measured by absorption at 280 nm should be 4.6 mg/mL

7. Mannitol capped laminarin molecules which are unreactive in the cycloaddition reaction are removed by the S-200 column (since laminarin elutes much later than the conjugate).

Acknowledgement

The research was made possible by grants awarded to D. R. B.; a Discovery grant from the Natural Science and Engineering Research Council of Canada and support from the Alberta Innovates Centers Program.

References

1. Lipinski T, Wu X, Sadowska J et al (2012) A trisaccharide conjugate vaccine induces high titer β-mannan specific antibodies that aid clearance of *Candida albicans* in immunocompromised rabbits. Vaccine 30:6263–6269

2. Lipinski T, Kitov P, Szpacenko A et al (2011) Synthesis and immunogenicity of a glycopolymer conjugate. Bioconjug Chem 22:274–281

3. Jiang J, Swiggard WJ, Heufler C et al (1995) The receptor DEC-205 expressed by dendritic cells and thymic epithelial cells is involved in antigen processing. Nature 375: 151–155

4. Banchereau J, Steinman RM (1998) Dendritic cells and the control of immunity. Nature 392:245–252

5. Steinman RM, Banchereau J (2007) Taking dendritic cells into medicine. Nature 449:419–426

6. van Kooyk Y, Rabinovich GA (2008) Protein-glycan interactions in the control of innate and adaptive immune responses. Nat Immunol 9: 593–601

7. Geijtenbeek TB, Gringhuis SI (2009) Signalling through C-type lectin receptors: shaping immune responses. Nat Rev Immunol 9: 465–479

8. Unger WW, van Kooyk Y (2011) "Dressed for Success" C-type lectin receptors for delivery of glyco-vaccines to dendritic cells. Curr Opin Immunol 23:131–137

9. van Kooyk Y, Unger WW, Fehres CM et al (2013) Glycan-based DC-SIGN targeting vaccines to enhance antigen cross-presentation. Mol Immunol 55:143–145

10. Fehres CM, Unger WW, Garcia-Vallejo JJ et al (2014) Understanding the biology of antigen cross-presentation for the design of vaccines against cancer. Front Immunol 5:149

11. Lipinski T, Fitieh A, St. Pierre J et al (2013) Enhanced immunogenicity of a tricomponent mannan tetanus toxoid conjugate vaccine targeted to DCs via Dectin-1 by incorporating β-glucan. J Immunol 190:4116–4128

12. Lönngren J, Goldstein IJ, Niederhuber JE (1976) Aldonate coupling, a simple procedure for the preparation of carbohydrate-protein conjugates for studies of carbohydrate-binding proteins. Arch Biochem Biophys 175:661–669

13. Svenson SB, Lindberg AA (1979) Coupling of acid labile *Salmonella* specific oligosaccharides to macromolecular carriers. J Immunol Methods 25:323–335

14. Goddard-Borger ED, Stick RV (2007) An efficient, inexpensive, and shelf-stable diazotransfer reagent: imidazole-1-sulfonyl azide hydrochloride. Org Lett 9:3797–3800, see also Additions and Corrections to the above paper (2011) *Org Lett* **13**, 2514–2514

15. Fischer N, Goddard-Borger ED, Greiner R et al (2012) Sensitivities of some imidazole-1-sulfonyl azide salts. J Org Chem 77: 1760–1764

16. Tornoe CW, Christensen C, Meldal M (2002) Peptidotriazoles on solid phase: [1,2,3]-triazoles by regiospecific copper(I)-catalyzed 1,3-dipolar cycloadditions of terminal alkynes to azides. J Org Chem 67:3057–3064

17. Wang Q, Chan TR, Hilgraf R et al (2003) Bioconjugation by copper (I)-catalyzed azide alkyne [3+2] cycloaddition. J Am Chem Soc 125:3192–3193

18. Wu X, Bundle DR (2005) Synthesis of glyco-conjugate vaccines for *Candida albicans* using novel linker methodology. J Org Chem 70: 7381–7388

19. Kunishima M, Kawachi C, Iwasaki F et al (1999) Synthesis and characterization of 4-(4,6-dimethoxy-1,3,5-triazin-2-yl)-4-methylmorpholinium chloride. Tetrahedron Lett 40:5327–5330

20. Sen Gupta S, Kuzelka J, Singh P et al (2005) Accelerated bioorthogonal conjugation: a practical method for the Ligation of diverse functional molecules to a polyvalent virus scaffold. Bioconjug Chem 16:1572–1579

21. Habeeb AFS (1966) Determination of free amino groups in proteins by trinitrobenzene-sulfonic acid. Anal Biochem 14:328–336

Chapter 10

Gold Nanoparticles as Carriers for Synthetic Glycoconjugate Vaccines

Fabrizio Chiodo and Marco Marradi

Abstract

Recent advances in the preparation and characterization of metal core-based nanoparticles have opened the way to their exploration as carriers for carbohydrate-based vaccines. Here we describe the protocols for the preparation and characterization of water dispersible gold nanoparticles (1–3 nm gold diameter) as carriers for carbohydrate antigens. We mainly refer to the protocols we used for the preparation of gold glyconanoparticles as carrier for an *S. pneumoniae* carbohydrate-antigen. The high number of ligands at the gold nanoparticles surface and the easiness of their one-pot preparation make these biocompatible nanomaterials an attractive tool for glyco-scientists.

Key words Gold nanoparticles, Glyconanoparticles, Carbohydrate-based vaccines, Nanomedicine

1 Introduction

Gold nanoparticles offer the possibility to combine in a controlled way different bioactive molecules required for the preparation of synthetic glycoconjugate vaccines [1, 2]. A high-carbohydrate loading and to overcome the risk of a carrier-induced epitopic suppression are the main characteristics of modern synthetic glycoconjugate vaccines-carrier that can be achieved by a full molecular-level control [3]. In this scenario, since 2005, different works have been published describing the attempt to exploit gold nanoparticles as carrier for synthetic glycoconjugate vaccines. The main idea was to functionalize the gold surface with antigenic carbohydrates and additional components, such as immunogenic peptides and adjuvants, which can aid at eliciting a significant immune response to the carbohydrates which are mostly T-independent antigens [3, 4].

The seminal works in this direction were reported by the groups of Barchi and Penadés: The group of Barchi reported the preparation of gold nanoparticles functionalised with a model glycopeptide mimicking the tumor-associated Thomsen–Friedenreich

(TF) antigen β-D-Gal$p$-(1→3)-α-D-Gal$p$NAc(1→)-OSer/Thr [5]. The group of Penadés described the preparation of different gold glyconanoparticles (GNPs) of 2 nm incorporating simultaneously the disaccharide antigen sialyl-Tn (α-Neu$p$5Ac-(2→6)-α-D-Gal$p$NAc(1→)), the tetrasaccharide Lewis Y antigen (Le^Y, α-L-Fucp-(1→2)-β-D-Gal$p$-(1→4)-[α-L-Fuc$p$-(1→3)]-β-D-Glc$p$NAc(1→)) and a T cell helper peptide [6]. In 2012 the same two groups showed the effectiveness of their synthetic approaches describing in vivo results after mice immunization with the GNPs. Gold nanoparticles bearing different ratios of synthetic carbohydrate-antigens and T-helper epitopes (fundamental to trigger an adaptive anti-carbohydrate immune response) were prepared as promising model for synthetic glycoconjugate vaccines. GNPs carrying the tetrasaccharide β-D-Gal$p$-(1→4)-β-D-Glc$p$-(1→6)-[β-D-Galp-(1→4)]-β-D-Glc$p$NAc(1→) (TetraPn14, repeating unit of S. pneumoniae capsular polysaccharide type 14) and OVA$_{323-339}$ peptide showed the ability to trigger in mice the production of specific and active IgGs against S. pneumoniae capsular polysaccharide [7]. GNPs carrying simultaneously tumor-associated glycopeptides and immunological adjuvants were effective in triggering specific IgGs in mice against tumor-associated carbohydrate-antigens [8]. In addition, the group of Scrimin reported the preparation of gold nanoparticles as carrier for synthetic analogues of the capsular polysaccharide repeating unit of serogroup A N. meningitidis [9]. Recently, Davis and Cameron proposed the synthesis of "multicopy-multivalent" gold nanoparticles covered with homopolymers bearing tumor-associated Tn antigen (α-D-Gal$p$NAc) [10].

Here we describe the protocols reported by the group of Penadés proposed to prepare water dispersible GNPs (1–3 nm gold diameter) as carriers for carbohydrate antigens. In designing the nanoparticles and reading the protocols, keep in mind that the easiness of the one-pot reaction to obtain gold nanoparticles allows also the simultaneous incorporation of different type of biomolecules and/or molecular imaging probes on the GNPs. Viral entry inhibitors [11], immune stimulating glycans [12] and MRI probes [13, 14] can be co-included on the gold nanoparticles surface in addition to carbohydrate entities.

2 Materials

Use analytical grade reagents and solvents. Prepare the reagent and GNP solutions using ultrapure water (electrical resistance of 18 MΩ cm at 25 °C).

2.1 GNP Preparation and Purification

1. Hydrogen tetrachloroaurate (III) hydrate (HAuCl$_4$) (Sigma-Aldrich or Strem Chemicals, Inc).

2. Sodium borohydride ($NaBH_4$).

3. Prepare the desired thiol-ending neoglycoconjugates via the conjugation of a thiol ending linker as aglycon to the antigenic carbohydrate under study.

4. The T cell epitopes can be synthesized by automatic peptide synthesis or ordered to suppliers.

5. SnakeSkin pleated dialysis tubing with a Molecular weight cut-off (MWCO) of 3500 Da (Thermo scientific pleated dialysis tubing, 22 mm × 35 ft dry diameter) or dialysis cassettes with a MWCO of 3500 Da.

6. Centrifugal filtering can also be used for small amount of GNPs.

7. Sephadex LH-20 (GE Healthcare) for gel-filtration chromatography on the unreacted ligands after nanoparticles formation.

2.2 GNPs Characterization

1. ^1H NMR spectra were recorded at 500 MHz at 25 °C with a Bruker AVANCE (500 MHz) spectrometer. Record the NMR spectra of the GNPs by setting the delay time (d1) at 15 s.

2. Deuterium oxide (99.9 atom % D), containing 0.05 wt% 3(trimethylsilyl)propionic-2,2,3,3-d4 acid (TSP), sodium salt.

3. For TEM characterization PELCO® ultrathin carbon film supported by a lacey carbon film on a 400 mesh copper grid (Product no. 01824, TED PELLA, INC.) were used. The micrographs were obtained using a JEOL JEM-2100F microscope working at 200 kV.

4. UV/Vis spectra were measured with a spectrophotometer.

5. Infrared spectra (IR) were recorded from 4000 to 400 cm^{-1} with a FT-IR spectrometer.

2.3 GNPs-ELISA

1. Coating buffer: 50 mM Na_2CO_3, pH = 9.7. Dissolve 2.65 g of Na_2CO_3 in 490 mL of H_2O. Adjust the pH to 9.7 and fill up to 500 mL. Store the buffer at 4 °C.

2. NUNC Maxisorp plate (Nunc MaxiSorp® flat-bottom 96-well plate).

3. BSA (lyophilized powder, ≥98 % measured by agarose gel electrophoresis) or other ELISA blocking agents can be used.

4. Substrate buffer: Acetate/Citrate buffer pH = 4. Dissolve 21.02 g of citric acid and 8.2 g of sodium acetate in 990 mL H_2O. Adjust the pH to 4.0 with acetic acid. Fill up to 1 L. Store at room temperature.

5. Substrate solution: Add 0.1 mL of 10 mg/mL of 3,3′,5,5′ Tetramethylbenzidine (TMB) in DMSO, and 2 μL of 30 % H_2O_2 in 10 mL of substrate buffer. Use this solution immediately.

6. Stop solution: 0.8 M H_2SO_4.

3 Methods

There are essentially two strategies for the preparation of multivalent gold glyconanoparticles: the direct gold salt reduction in the presence of a mixture of thiol-ending ligands in aqueous solution and the ligands-phase exchange on preformed gold nanoclusters. Both strategies are very well studied and applied [15].

For the preparation of the gold nanoparticles as carriers for synthetic glycoconjugate vaccines, we describe the direct strategy following a slightly modified Brust–Schiffrin method [16]. This procedure will give water-soluble GNPs from 1 to 3 nm in diameter carrying 70–150 molecules on their surface [2, 17]. In addition, a recent approach to bio-characterize the GNPs and to detect anti-carbohydrate antibodies will be described [18]. The simultaneous presence of different components on the gold nanoparticles is fundamental to trigger a specific anti-carbohydrate immune response. All these components need to have a thiol-ending linker for their inclusion on the gold surface (*see* Fig. 1). Mainly three components are fundamental for the correct design of gold nanoparticles as carrier for carbohydrate-antigens:

- a selected carbohydrate epitope against which the antibodies will be trigged.

- a T-cell epitope (murine or human) to activate the T cells [19].

- an inner component to modulate the ligand density on the gold nanoparticles [20].

Recently, Davis and Cameron reported a new and interesting approach based on the direct functionalization of gold

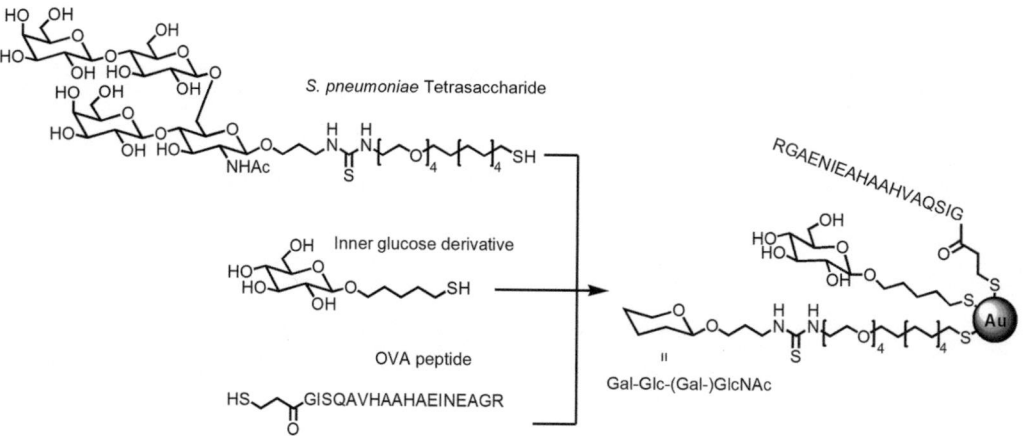

Fig. 1 General scheme for a three-component GNP carrying a synthetic antigen of *S. pneumoniae* capsular polysaccharide, an inner glucose derivative and the T-cell epitope OVA peptide [7]

nanoparticles with protected polymeric thioled-molecules [10]. Homopolymers bearing the Tn antigen and dithioester end groups were synthesized by the reversible addition-fragmentation chain transfer polymerization and then used directly as ligands for the gold nanoparticles. This new peptide-free platform showed a significant immune response in mice.

3.1 Gold Nanoparticles Preparation

Except for the approach presented by Davis and Cameron (*see* ref. 10), the preparation of GNPs as carrier for carbohydrate-antigens starts with the solubilization of the desired thiol-ending ligands in presence $HAuCl_4$ followed by a strong reduction in presence of $NaBH_4$.

1. Prepare a soluble mixture of the thiol-ending molecules that will be multimerized on the gold surface with the desired molar ratio (*see* Table 1 for a concrete example). MeOH, water or a mixture of the two is good solvents-system for the direct GNPs preparation. pH can also be changed to increase ligands solubility especially when peptides are included in the mixture. In the example showed in Table 1 a glucose derivative is used as inner component and we refer to the work described in ref. 7.

2. Perform a ^{1}H-NMR spectrum of the ligands mixture to confirm the desired ligand-ratio (*see* Subheading 3.2.1).

3. Adjust the solution with the thiol-ending molecules to 0.012 M with MeOH and/or H_2O and transfer it to an eppendorf or pear-shaped flask (*see* **Notes 1** and **2**).

4. Calculate the reagent-equivalents for the nanoparticles preparation as follow:

   – 1 eq. of $HAuCl_4$ at 0.025 M concentration in water.

   – 5 eq. of thiol-ending molecules 0.012 M in methanol/water (*see* **Note 3**).

   – 21 eq. of freshly prepared $NaBH_4$ 1 M in water (*see* **Note 4**).

   See Table 2 for a concrete example.

5. Add the $HAuCl_4$ solution to the thiol-ending molecules mixture.

Table 1
Calculation with the molar ratio of the soluble mixture of thiol-ending molecules that will be multimerized on the gold surface

Thiol-ending molecules	mg	MW	μmol	Eq.	Final %
S. pneumoniae Tetrasaccharide	1.89	1186.38	1.593	9	45
OVA$_{323-339}$ (T-cell epitope)	0.34	1919.08	0.177	1	5
Glucose derivative	0.5	282.35	1.77	10	50

Table 2
Calculation with the reagents needed for the in situ reduction of HAuCl$_4$ in presence of NaBH$_4$ for the preparation of GNPs

Reagents	µmol	Eq.	Conc. (M)	Volume (µL)
Thiol-ending molecules[a]	3.54	5	0.012	295
HAuCl$_4$	0.708	1	0.025	28.32
NaBH$_4$	14.87	21	1	14.87

[a]In this case we refer to a specific example reported in ref. 7

6. Mix slowly the reagents. White flocculate may appear.

7. Add NaBH$_4$ solution under shaking (1000 mot/min) in 3–4 portions. A dark/brown suspension will appear.

8. Shake vigorously the mixture for 2 h at room temperature (*see* **Note 5**).

9. After 2 h, leave the GNPs reaction rest for few minutes to have a precipitate on the bottom of the eppendorf. You can accelerate the GNPs precipitation by centrifugation (5′, 9600×g) and/or adding EtOH (*see* **Note 6**).

10. Wash the dark solid 4–5 times with MeOH to remove the reagent excess.

11. Collect the supernatants for ^{1}H NMR analysis and to recover them for another reaction (*see* **Note 7**).

12. Dissolve the nanoparticles (dark/brown solid) in the minimum volume of ultrapure water.

13. Load the GNPs solution into ~5–10 cm segment of SnakeSkin pleated dialysis tubing.

14. Change 7–8 times the dialysis-water during 72 h. For small amount of GNP-purification, use centrifugal filtering.

15. If some precipitate appears during dialysis filter them over cotton.

16. Freeze-dry the nanoparticles and store them (*see* **Note 8**).

17. To manage the GNPs and to perform cellular and/or in vivo experiments *see* **Notes 9** and **10**.

3.2 Gold Nanoparticles Characterization

A combination/correlation between different characterization techniques is needed to have a good, reliable and reproducible GNPs characterization. NMR is the main technique used, mainly combined with TEM. HPLC can also be used for the ligand ratio before and after the GNPs preparation. UV and IR can also be performed to have a better chemical characterization.

3.2.1 NMR

1. Record a ¹H NMR spectrum of the thiol-ending ligands mixture before the nanoparticles preparation (Fig. 2).

2. Record a ¹H NMR spectrum of the crude thiol-ending ligands mixture after the nanoparticles preparation. A desalting column (Sephadex LH-20) to remove the NaBH₄ excess is then also suggested before running another ¹H NMR. The ratio of the ligands after and before the nanoparticles preparation should be maintained.

3. Record a ¹H NMR spectrum of the purified nanoparticles: 2–3 mg/mL D₂O solution is a good concentration to perform this kind of analysis. Large number of scans is suggested (*see* Fig. 3 and **Note 11**).

4. Correlate the GNPs diameter (*see* Subheading 3.2.2) with the ligands ratio from ¹H NMR spectra pre/post GNPs preparation. From the literature [21] a correlation between the GNPs diameter, gold atoms and the number of ligands for each GNP can be found. Knowing the ligands ratio on the gold surface from the NMR spectra recorded before and after GNPs preparation, an

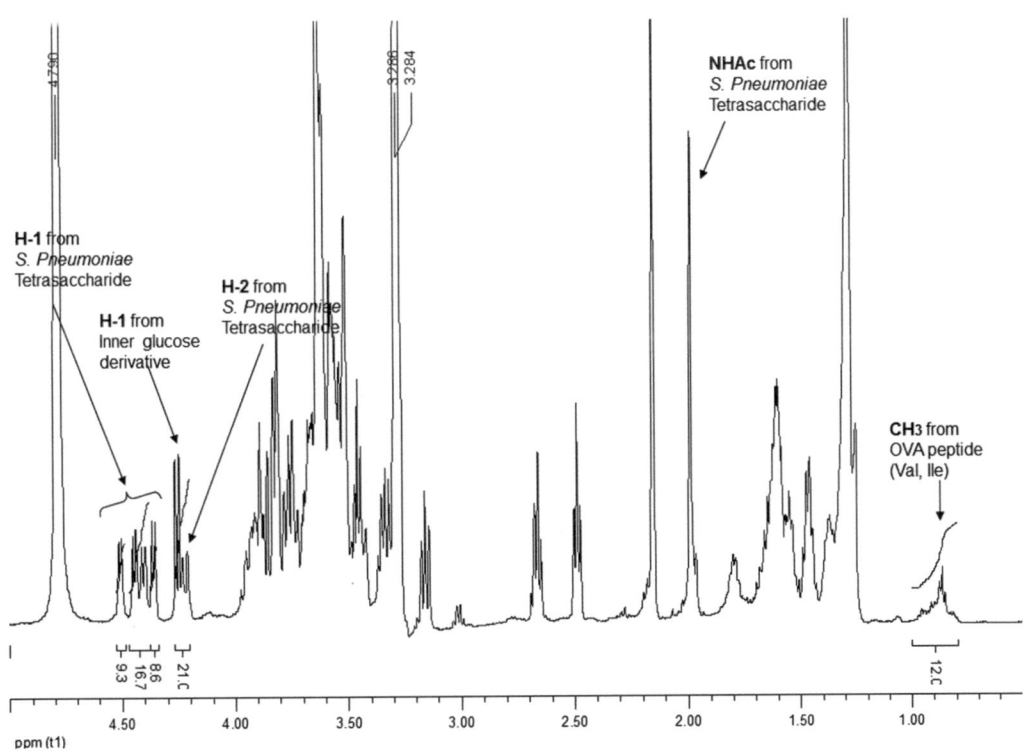

Fig. 2 ¹H-NMR spectrum of the thiol-ending ligands before the GNP preparation. The *arrows* indicate the different proton-signals from the three components of the GNP

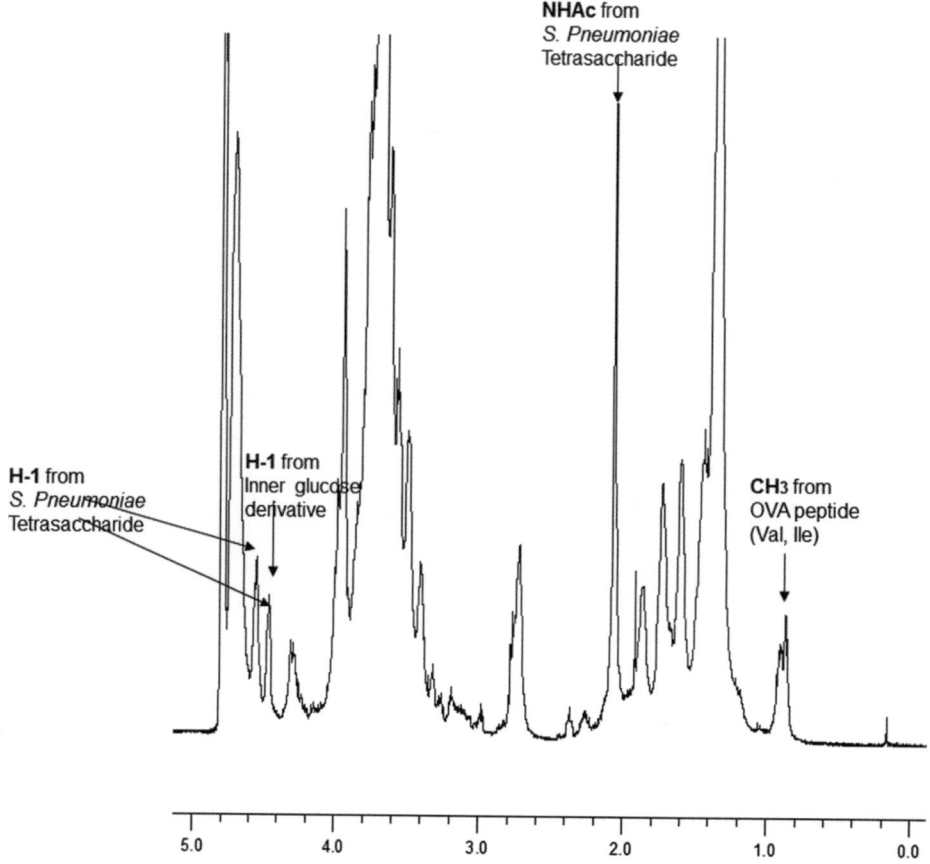

Fig. 3 ¹H-NMR spectrum of a three components GNP. As depicted in Fig. 2, the *arrows* indicate the different proton-signals from the different GNP-components

average molecular weight of the nanoparticles can be calculated and used for the ligands moles per mg of GNP [22]. In case the amount of nanoparticles in hand is sufficiently high and not limited by the cost/quantity of the starting neoglycoconjugates/peptides, an elemental analysis can be performed to obtain the percentage of the elements and adjust the average molecular formula of the GNPs.

5. Record a ¹H NMR dissolving the GNPs with D_2O containing 0.05 % (w/w) of TSP to perform a quantitative NMR (qNMR) on the intact nanoparticles with a similar approach described in ref. 9.

6. Integrate the signals of the molecules attached to the nanoparticle and compare them with internal standard.

7. Remove the TSP by dialysis after the qNMR quantification.

3.2.2   TEM

1. Put a single drop (5 μL) of the aqueous GNPs solution (ca. 0.1 mg/mL in Milli-Q water) onto a TEM grid.

2. Leave the grid dry in air for several hours at room temperature.

3. Analyze the grid with TEM microscope.

4. Evaluate the particle size distribution of the gold nanoparticles from several TEM micrographs by means of an image analyzer.

3.2.3   UV and IR

UV/vis on the intact GNP is not a good ligands characterization technique when you have not a UV/vis chromophore on the gold surface. However, the GNP dispersion color gives an indication of the gold core size. GNPs below 2 nm (gold diameter) afford usually a brownish solution; bigger nanoparticles afford reddish/purple colors. Turning to violet usually is an indication of nanoparticles aggregation. The UV spectra give a good indication of the GNPs dimensions: small GNPs, with gold core diameter below 2 nm, usually do not show the Plasmon absorption at 520 nm at 0.1 mg/mL in water. Typically, the absorbance spectra of gold nanoparticles show a maximum around 520 nm with peak shift towards higher wavelengths as the gold size increases. IR on KBr pellet is a qualitative technique that may help to confirm the presence of organic molecules especially when small amounts of ligands are attached on the gold surface. A GNPs spatula tip is enough to run IR spectra.

3.3   GNP-ELISA for Abs Detection and Bio-characterization

In order to detect the anti-carbohydrates antibodies trigged by the GNPs during the immunization studies, we have developed a sensitive ELISA-based assay exploiting the GNPs as coating for the ELISA plates [18]. In addition, the GNPs-functionalized surface, offers the opportunity to bio-characterize the GNPs with monoclonal antibodies and/or lectins (*see* Fig. 4 for a general scheme).

The GNPs described in ref. 7 have been used to coat ELISA plates and detect the IgG trigged in mice. GNPs bearing 10 % of a thiol-ending conjugate of dimannoside Man(α1-2) Man(α1→) (DiMan) or tetramannoside Man(α1-2) Man(α1-2) Man(α1-3) Man(α1→) (TetraMan) [22] were explored to detect the monoclonal antibody 2G12 and to evaluate the bioactivity of the glycans on the GNPs by a lectin recognition. GNPs fully covered by a glucose thiol-ending conjugate (Glc-GNPs) were used as negative control.

1. Coat the ELISA NUNC Maxisorp plate with 50 μL of GNP suspension (20 μg/mL in the coating buffer from the stock GNPs-solution). Coat the wells in triplicate for 2–3 h room temperature or overnight at 4 °C.

2. Discard the wells and wash them with PBS (2×200 μL) (*see* **Note 12**).

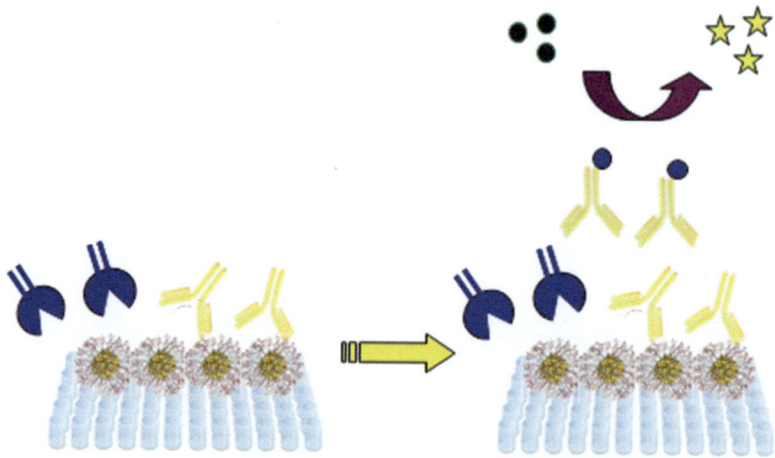

Fig. 4 Schematic picture to visualize the ELISA-GNP approach. ELISA plates are coated with the GNP and incubated with sera, monoclonal antibody or lectin (*left*). After washes, a secondary antibody coupled to horse-radish peroxidase (HRP) is then added followed by the addition of the chromogenic substrate TMB (*right*)

3. Block the wells with 200 μL of PBS with 1 % BSA. Leave the plate at room temperature for 30 min.

4. Discard the wells and do not wash them.

5. Add 100 μL of the solution containing the antibody or serum you need to test (*see* **Note 13**). For unknown samples, try different dilutions. Dilute them in PBS containing 0.5 % BSA.

6. Shake slowly the plate for 1 h at room temperature.

7. Discard the wells and wash them with PBS (3 × 200 μL).

8. Add 100 μL of the secondary antibody solution in PBS containing 0.5 % BSA. Use anti mouse or anti human secondary antibody coupled to horseradish peroxidase (HRP) at 1 μg/mL. Shake slowly the plate for 30 min at room temperature.

9. Discard the wells and wash them with PBS (3 × 200 μL).

10. Add 100 μL of substrate solution (*see* **Note 14**) and wait few minutes until a weak/strong blue color appears.

11. Stop the reaction with 50 μL of the stop solution (*see* **Note 15**).

12. Read the plate at 450 nm in ELISA reader.

Figure 5 shows two different applications of the GNP-ELISA assay: Fig. 5, left, shows the ability of GNPs to "capture" specific IgG from sera of immunized mice. These mice where immunized with TetraPnOv-GNP (the three components GNP described in Subheading 3.1) that was also use to coat the ELISA plate. Fig. 5, right, shows the biocharacterization of DiMan and TetraMan-GNPs (carrying α, 1-2 oligomannosides). ELISA plates were coated with these GNPs and then the binding to DC-SIGN was determined by ELISA in a calcium/magnesium containing buffer (TSM).

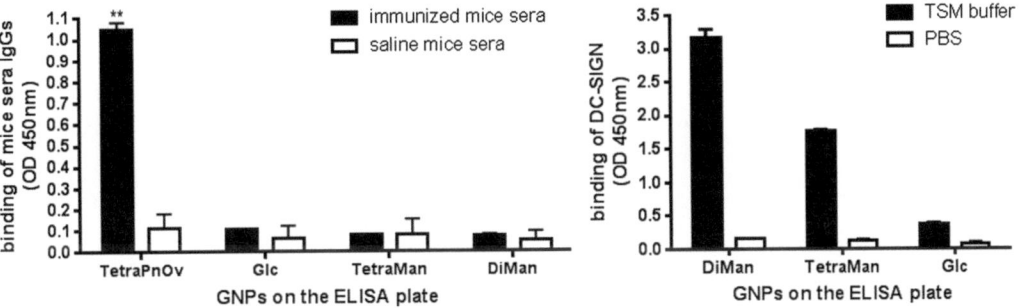

Fig. 5 *Left*: Detection of specific IgG by coating the ELISA plates with different GNPs. TetraPnOv shows strong binding to immunized mice serum. Glc-, TetraMan-, and DiMan-GNPs were not recognized by the sera's IgG. Sera of mice immunized with saline were used as negative control. Differences between sera from immunized mice and control samples are significant, two *asterisks* ($p < 0.01$). *Right*: The binding of lectin DC-SIGN to GNPs was determined using GNPs-ELISA in PBS and in calcium/magnesium containing buffer (TSM). DiMan and Te-GNPs show strong binding to DC-SIGN while Glc-GNP shows a weaker binding. Error bars indicate standard deviations

4   Notes

1. For high amount of GNPs (more than 2–4 mg) use pear-shaped flasks pretreated with aqua regia. For small amount of ligands (below 10 µmol of thiol-ending ligands), plastic 2.5 mL eppendorfs are good containers for the GNPs preparation.

2. The preparation of the GNPs in multiple eppendorfs (with max 1.5 mL solution) increases the yield of the GNPs.

3. For the preparation of GNPs carrying no peptides, 3 eq. of thiol-ending molecules respect to $HAuCl_4$ are suggested. When peptides should be included on the gold surface 5 eq. of the organic ligands are suggested.

4. $NaBH_4$ solution needs to be prepared freshly just before its addition in the reaction mixture.

5. Due to the fact that the reductive reaction is exothermic and H_2 will be produced from water reduction, it is suitable to perforate the eppendorfs on the top.

6. Sometimes the GNPs are soluble in the reaction mixture. After 2 h EtOH can be used to precipitate the GNPs followed by centrifugation (5 min, $9600 \times g$). Also in this case the supernatants need to be collected and the GNPs washed several times as explained before (*see* Subheading 3.1, **step 10**).

7. The un-reacted ligands can be reused for others GNPs preparations after purification. Size-exclusion column chromatography can be performed on Sephadex LH-20 in $MeOH/H_2O = 9/1$ to de-salt the organic mixture after GNPs preparation.

8. We find a better stability when the GNPs are stored as dry powder under Argon after lyophilization. For long time storage, keep GNPs at 4 °C. In general, depending on the components stability, GNPs are stable at room temperature for several weeks even in water dispersion.

9. Do not use gloves managing the dry GNPs in order to avoid electrostatic induction, weigh them with a Teflon spatula.

10. Make the GNP stock-solutions at 1–3 mg/mL in H_2O. For the biological experiments or other assays, dilute from the stock in the proper solvent/buffer/cellular medium.

11. NMR parameters like delay (d1) and number of scans (NS) can be adjusted to have a proper ratio of the ligands in the spectrum.

12. During the ELISA-GNPs washings, do not use surfactant-containing buffers.

13. Different antibody/serum dilution can be used in relation with the strength of the carbohydrate-interactions of interest. For pure monoclonal antibodies we suggest to use from 1 to 20 nM antibody solution. For mice sera we suggest to dilute it 1:10,000. For human serum we suggest to dilute it 1:500. For lectins we suggest to dilute them at 1–5 μg/mL. Dilute the samples in PBS with 0.5 % BSA.

14. The ELISA substrate solution needs to be prepared just before its use.

15. Add the ELISA stop solution simultaneously to the wells when a weak-mild blue color appears.

Acknowledgement

We thank Soledad Penadés for her scientific support throughout the last years. MM acknowledges COST Action CM1102 and the Department of Education, Universities and Research of the Basque Government (Project PI2012–46).

References

1. You CC, Chompoosor A, Rotello VM (2007) The biomacromolecule-nanoparticle interface. Nano Today 2:34–43

2. Marradi M, Chiodo F, García I et al (2013) Glyconanoparticles as multifunctional and multimodal carbohydrate systems. Chem Soc Rev 42:4728–4745

3. Berti F, Adamo R (2013) Recent mechanistic insights on glycoconjugate vaccines and future perspectives. ACS Chem Biol 8:1653–1663

4. Astronomo RD, Burton DR (2010) Carbohydrate vaccines: developing sweet solutions to sticky situations? Nat Rev Drug Discov 9: 308–324

5. Svarovskya SA, Szekely Z, Barchi JJ Jr (2005) Synthesis of gold nanoparticles bearing the Thomsen–Friedenreich disaccharide: a new multivalent presentation of an important tumor antigen. Tetrahedron Asymmetry 16: 587–598

6. Ojeda R, de Paz JL, Barrientos AG et al (2007) Preparation of multifunctional glyconanoparticles as a platform for potential carbohydrate-based anticancer vaccines. Carbohydr Res 342: 448–459

7. Safari D, Marradi M, Chiodo F et al (2012) Gold nanoparticles as carriers for a synthetic *Streptococcus pneumoniae* type 14 conjugate vaccine. Nanomedicine 7:651–662

8. Brinãs RP, Sundgren A, Sahoo P et al (2012) Design and synthesis of multifunctional gold nanoparticles bearing tumor-associated glycopeptide antigens as potential cancer vaccines. Bioconjug Chem 23:1513–1523

9. Manea F, Bindoli C, Fallarini S et al (2008) Multivalent, saccharide-functionalized gold nanoparticles as fully synthetic analogs of Type A *Neisseria meningitidis* antigens. Adv Mater 20:4348–4352

10. Parry AL, Clemson NA, Ellis J et al (2013) 'Multicopy multivalent' glycopolymer-stabilized gold nanoparticles as potential synthetic cancer vaccines. J Am Chem Soc 135: 9362–9365

11. Di Gianvincenzo P, Marradi M, Martinez-Avila OM et al (2010) Gold nanoparticles capped with sulfate-ended ligands as anti-HIV agents. Bioorg Med Chem Lett 20:2718–2721

12. Chiodo F, Marradi M, Park J et al (2014) Galactofuranose-coated gold nanoparticles elicit a pro-inflammatory response in human monocyte-derived dendritic cells and are recognized by DC-SIGN. ACS Chem Biol 9: 383–389

13. Irure A, Marradi M, Arnáiz B et al (2013) Sugar/gadolinium-loaded gold nanoparticles for labelling and imaging cells by magnetic resonance imaging. Biomater Sci 1:658–668

14. Frigell J, García I, Gómez-Vallejo V et al (2014) 68Ga-labeled gold glyconanoparticles for exploring blood-brain barrier permeability: preparation, biodistribution studies, and improved brain uptake via neuropeptide conjugation. J Am Chem Soc 136:449–457

15. Marradi M, Martín-Lomas M, Penadés S (2010) Glyconanoparticles polyvalent tools to study carbohydrate-based interactions. Adv Carbohydr Chem Biochem 64:211–290

16. Brust M, Walker M, Bethell D et al (1994) Synthesis of thiol derivatised gold nanoparticles in a two-phase liquid/liquid system. J Chem Soc Chem Commun 7:801–802

17. de la Fuente JM, Barrientos AG, Rojas TC et al (2001) Gold glyconanoparticles as water-soluble polyvalent models to study carbohydrate interactions. Angew Chem Int Ed 40:2257–2261

18. Chiodo F, Marradi M, Tefsen B et al (2013) High sensitive detection of carbohydrate binding proteins in an ELISA-solid phase assay based on multivalent glyconanoparticles. PLoS One 8:e73027

19. Avci FY, Li X, Tsuji M et al (2011) A mechanism for glycoconjugate vaccine activation of the adaptive immune system and its implications for vaccine design. Nat Med 17:1602–1609

20. Barrientos AG, de la Fuente JM, Rojas TC et al (2003) Gold glyconanoparticles: synthetic polyvalent ligands mimicking glycocalyx-like surfaces as tools for glycobiological studies. Chem Eur J 9:1909–1921

21. Hostetler MJ, Wingate JE, Zhong C-J et al (1998) Alkanethiolate gold cluster molecules with core diameters from 1.5 to 5.2 nm: core and monolayer properties as a function of core size. Langmuir 14:17–30

22. Martínez-Avila O, Hijazi K, Marradi M et al (2009) Gold manno-glyconanoparticles: multivalent systems to block HIV-1 gp120 binding to the lectin DC-SIGN. Chem Eur J 15: 9874–9888

Chapter 11

Identification and Characterization of Carbohydrate-Based Adjuvants

Timo Johannssen and Bernd Lepenies

Abstract

Modern vaccines such as recombinant proteins or nucleic acids are usually of pure origin, enhancing their tolerability and overall safety. However, this purity often renders them less immunogenic, creating the need for potent adjuvants. Carbohydrates are promising candidates to fulfill this role as they enable direct targeting of dendritic cells and modulation of adaptive immunity. C-type lectin receptors (CLRs) comprise a major group of carbohydrate binding receptors. As they are predominantly expressed by cells of innate immunity, CLR targeting can enhance or dampen early stages of cytokine secretion and antigen presentation, thus modulate the activation and differentiation of T cells. Here, we provide a protocol for the identification of novel CLR ligands by glycan array using recombinant CLR-Fc chimeras followed by the covalent conjugation of carbohydrate CLR ligands to the model antigen ovalbumin (OVA). The resulting glycoconjugates are subsequently used to evaluate T cell activation in vitro and immunomodulation in vivo.

Key words C-type lectin receptor, Fc fusion protein, Glycan array, Cell targeting, Carbohydrate adjuvants, Immunomodulation

1 Introduction

Vaccination constitutes a major means of prophylactic medication against infectious diseases. Older generations of vaccines contain comparably impure attenuated or killed pathogens. While these agents are capable of inducing strong and long lasting immune responses, they also bear an increased risk of local inflammation and systemic reactions. In contrast, modern vaccines are usually of defined and pure origin due to chemical synthesis (carbohydrate or DNA vaccines) or biotechnological production (recombinant proteins or subunits), leading to better overall tolerability. Ironically, the defined and pure state of those molecules may also favor reduced immunogenicity, rendering them less potent than their predecessors. To ensure vaccine-induced protection combining both tolerability and immunogenicity, adjuvants are needed to efficiently boost immune responses towards administered antigens [1].

Bernd Lepenies (ed.), *Carbohydrate-Based Vaccines: Methods and Protocols*, Methods in Molecular Biology, vol. 1331,
DOI 10.1007/978-1-4939-2874-3_11, © Springer Science+Business Media New York 2015

Carbohydrate adjuvants are a promising new class of agents which may be used for the directed delivery of vaccines to antigen presenting cells as well as the modulation of T cell responses. Carbohydrates can significantly impact immune responses and usually possess a low risk for toxicity. Compared to alum, carbohydrate adjuvants are easily excreted, hindering the formation of tissue deposits [2]. Further, next to the induction of humoral responses, glycan-based adjuvants are capable of enhancing cell-mediated immunity, which is crucial for efficient clearance of intracellular pathogens [3].

Many effects of carbohydrates on our immune system are mediated by C-type lectin receptors (CLRs). Engagement of a myeloid CLR may lead to diverse immune functions such as cell adhesion, phagocytosis, secretion of cytokines, and antigen presentation. Most CLRs are expressed by antigen presenting cells, most notably macrophages and dendritic cells [4]. Depending on the receptor addressed, the immunological outcome of CLR targeting can involve either cellular activation for host defense, or immunoregulation to ensure tolerance towards autoantigens [5].

The common feature of C-type lectins is the presence of a so-called C-type lectin-like domain (CTLD) within the carbohydrate recognition domain (CRD), which is responsible for ligand recognition. Although not all C-type lectin-like domain containing proteins bind carbohydrates, most CLRs exhibit binding of glycans in a Ca^{2+}-dependent manner. Ligand specificity is determined by amino acid motifs present in the CRD of the receptor. Two major motifs include the EPN motif, displaying affinity for mannose-containing glycans, and the QPD motif, which exhibits specificity for galactose-type carbohydrates [6]. Based on the exposed glycan profile, CLRs are capable of sensing an array of distinct pathogen-associated molecular patterns (PAMPs). Until now, several species of viruses, fungi, bacteria, and parasites have been described to be recognized by CLRs. For example, pathogens bound by the CLR Dendritic cell-specific intercellular adhesion molecule-3-grabbing non-integrin (DC-SIGN) include HIV via gp120 [7], *Candida albicans* via mannan [8, 9], and *Mycobacterium tuberculosis* via lipoarabinomannan [10]. Mycobacterial species are further detected by the Macrophage-inducible C-type lectin (Mincle) via trehalose-6,6′-dimycolate (TDM) [11, 12]. On the other hand, CLRs were also shown to modulate immune function in response to danger-associated molecular patterns (DAMPs) released by necrotic cells. Reported self-antigens include SAP130 detected by Mincle [13], F-actin bound by Clec9a [14], and uric acid crystals recognized by Clec12a [15].

Depending on the intracellular motif, engagement of a CLR may initiate multifaceted biological responses. Two major signaling motifs present in the intracellular domain of CLRs are immunoreceptor tyrosine-based activation motifs (ITAMs) and inhibitory motifs (ITIMs), which mediate cellular activation by recruitment of Syk family kinases and participate in dampening of cellular responses by recruitment of tyrosine phosphatases, respectively [16]. Cellular functions induced or altered upon CLR ligation include phagocytosis, antigen presentation, production of reactive oxygen species and nitric oxide, and secretion of proinflammatory or regulatory cytokines, as well as chemokines [17].

Since several CLRs display endocytic capacity and are involved in the initiation and modulation of immune responses, they are attractive targets for both cell targeting and antigen delivery [18, 19]. However, since protein/carbohydrate interactions are usually of low affinity, multivalent presentation is employed to enhance binding avidity [20]. Glycan-based targeting of CLRs has been achieved using various carrier systems such as nanoparticles, dendrimers, and liposomes [21]. Apart from multivalent presentation, detailed knowledge about the targeted receptor's ligand recognition profile is needed to enable specific delivery to host cells. This bears two major challenges: First, although various receptors have been identified so far, there is limited knowledge about respective glycan ligands. Second, known ligands may potentially bind to several CLRs associated with different signaling pathways and subsequently divergent cellular responses. In this context, glycan arrays have proven a beneficial technique since a large amount of different carbohydrates can be screened for recognition by multiple receptors in parallel.

Here, we present a protocol for the identification of novel CLR ligands and their subsequent evaluation regarding immunomodulatory capabilities. Initially, CLRs of interest are expressed as soluble fusion proteins consisting of the extracellular domain (ECD) and the Fc portion of human IgG1, enabling bivalent ECD presentation and facilitated detection using secondary anti-Fc antibodies. The resulting fusion proteins are used to screen potential carbohydrate ligands using the glycan array technology. Following ligand screening, selected hits are conjugated to the model antigen ovalbumin (OVA) and used for cell targeting. To assess the capability of T cell activation in vitro, CD11c+ dendritic cells are pulsed with the glycoconjugates and cocultured with T cells. The glycoconjugates are further subjected to in vivo immunization studies followed by the evaluation of cytokine levels and antibody titers [22] (Fig. 1).

2 Materials

All buffers need to be prepared in ultrapure water (18 MΩ cm, 25 °C), passed through a 0.2 μm filter and stored at RT. Protein containing solutions are freshly prepared.

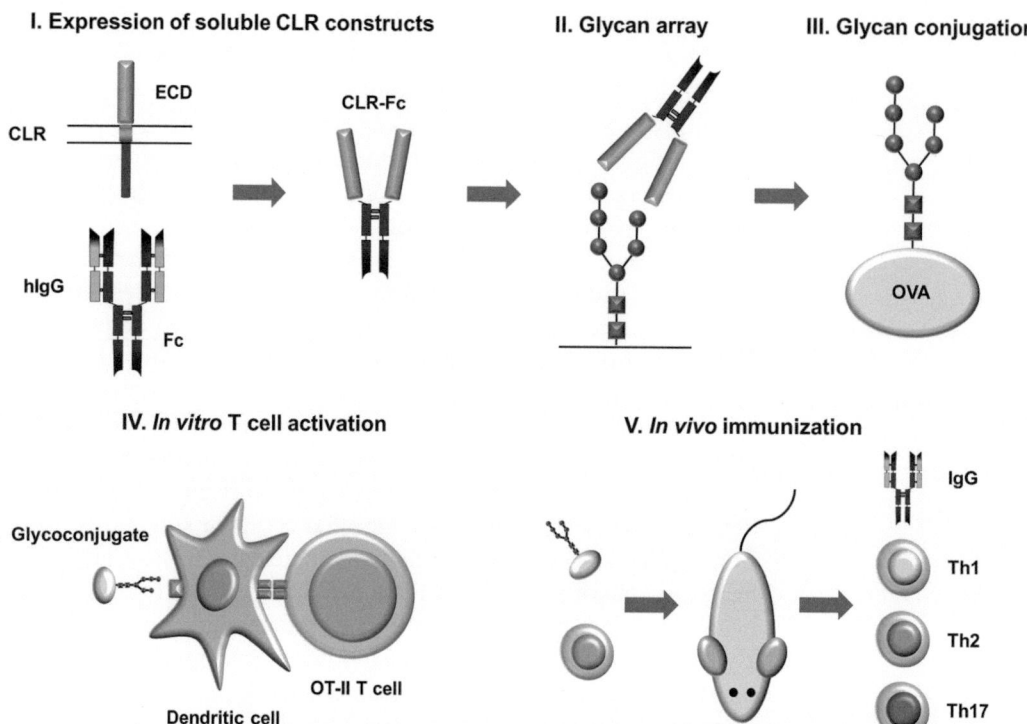

Fig. 1 Major steps for screening and characterization of CLR targeting immunomodulatory carbohydrates. CLRs are expressed as soluble Fc fusion proteins and used to identify novel carbohydrate ligands by glycan array. Resulting ligands are conjugated to the model antigen OVA and tested for immunomodulatory properties using T cell activation assays and immunization studies

2.1 Generation of Recombinant Fc Fusion Proteins

1. Fc expression vector: pFUSE-hIgG1-Fc2 (InvivoGen, San Diego, CA, USA).

2. FreeStyle™ CHO-S cells (Life Technologies, Carlsbad, CA, USA).

3. Culture medium: Gibco® FreeStyle™ CHO expression medium (Life Technologies) supplemented with 8 mM L-glutamine.

4. Transfection medium: OptiPRO™ SFM (Life Technologies).

5. Transfection reagent: FreeStyle™ MAX reagent (Life Technologies).

6. PE-conjugated anti-human IgG Fc.

7. HRP-conjugated anti-human IgG Fc.

8. HiTrap Protein G HP, 1 ml (GE Healthcare, Freiburg, Germany).

9. Binding buffer: 20 mM sodium phosphate, pH 7.0.

10. Elution buffer: 0.1 M glycine-HCl, pH 2.7.

11. Neutralization buffer: 1 M Tris–HCl, pH 9.0.

12. Ultrafiltration concentrator, MWCO 10 kDa.

2.2 Screening of Amino-Functionalized CLR Ligands Using Glycan Array

1. Immobilization buffer: 50 mM NaH_2PO_4, pH 8.5.

2. Phosphate buffered saline (PBS): 137 mM NaCl, 2.7 mM KCl, 10 mM NaH_2PO_4, 1.8 mM KH_2PO_4, pH 7.4.

3. Thio linker immobilization buffer: 1 mM tris(2-carboxyethyl) phosphine hydrochloride in PBS, pH 7.4.

4. Microarray printing device: sciFlexarrayer (Scienion, Berlin, Germany).

5. Epoxy-functionalized microarray slides.

6. Quenching buffer: 50 mM NaH_2PO_4, 100 mM ethanolamine, pH 9.0.

7. Lectin buffer: 10 mM HEPES, 1 mM $CaCl_2$, 1 mM $MgCl_2$, pH 7.4.

8. Blocking buffer: 2 % bovine serum albumin (BSA) in lectin buffer.

9. Secondary antibody: fluorophore-conjugated anti-human IgG1.

10. Antibody dilution buffer: 0.5 % BSA in lectin buffer.

11. Microarray scanning device: Genepix scanner 7 (Molecular Devices, Sunnyvale, CA, USA).

2.3 Conjugation of CLR Ligands to OVA

1. *N*-succinimidyl adipate (DSAP) linker. DSAP can be synthesized from NHS (*N*-hydroxysuccinimide), adipoyl dichloride, and triethylamine or alternatively purchased from commercial sources (e.g., Synchem UG & Co. KG, Felsberg/Altenburg, Germany).

2. DMSO.

3. Triethylamine.

4. Phosphate buffer: 100 mM NaH_2PO_4 pH 7.4.

5. Chloroform.

6. Ovalbumin.

7. Ultrafiltration concentrator, MWCO 10 kDa.

2.4 In Vitro T cell Activation Assay

1. C57BL/6 mice.

2. OT-II mice.

3. Complete RPMI: RPMI 1640, 10 % FCS, 2 mM L-glutamine, 100 U/ml penicillin, 100 µg/ml streptomycin.

4. Cell strainer, 40 µm.

5. RBC lysis buffer: 144 mM NH_4Cl, 10 mM Tris, pH 7.5.

6. MACS buffer: PBS, 0.5 % BSA, 2 mM EDTA.

7. Anti-mouse CD16/32 antibody.

8. Complete IMDM: IMDM, 10 % FCS, 2 mM L-glutamine, 100 U/ml penicillin, 100 µg/ml streptomycin.

9. CD11c microbeads (Miltenyi Biotech, Bergisch Gladbach, Germany).

10. Pan T cell isolation kit II (Miltenyi Biotech).

11. LS/LD columns (Miltenyi Biotech).

12. MidiMACS or QuadroMACS cell separator (Miltenyi Biotech).

2.5 Immunization Studies

1. C57BL/6 mice.

2. OT-II mice.

3. Complete RPMI: RPMI 1640, 10 % FCS, 2 mM L-glutamine, 100 U/ml penicillin, 100 µg/ml streptomycin.

4. Cell strainer, 40 µm.

5. RBC lysis buffer: 144 mM NH_4Cl, 10 mM Tris, pH 7.5.

6. *Optional*: cell proliferation dye eFluor® 670 (eBioscience, San Diego, CA, USA).

7. Phosphate buffered saline (PBS).

8. Aluminum hydroxide: Alhydrogel (Brenntag, Mühlheim an der Ruhr, Germany).

9. Ovalbumin.

3 Methods

3.1 Generation of Recombinant Fc Fusion Proteins

To screen for novel CLR ligands by glycan array, corresponding receptors are expressed as soluble Fc fusion proteins. The extracellular domain is cloned into an expression vector encoding human Fc, thereby fusing the receptor's carbohydrate recognition domain to the CH2, CH3, and hinge region of IgG1. Resulting constructs are transfected into CHO-S cells and purified from culture supernatants using affinity chromatography.

Template cDNA is generated from a cell subset expressing the CLR of interest (*see* **Note 1**). First, RNA is isolated followed by digestion of genomic DNA and reverse transcription. The resulting cDNA is used to amplify the extracellular domain by PCR, ligated in-frame into the pFUSE expression vector and amplified in *E. coli*. Isolated plasmid DNA should be passed through a 0.22 µm filter before transfection of CHO-S cells. CHO-S cells have been adapted to suspension growth, enabling facilitated scale-up of the production volume. Further, these cells grow under serum-free conditions, avoiding the need for IgG depletion of FCS when producing Fc fusion proteins.

1. Culture CHO-S cells in expression medium at 37 °C and 8 % CO_2 on an orbital shaker (125 rpm) according to the manufacturer's instructions. Passage every 2–3 days and do not let

the culture grow above 1.5×10^6 cells/ml, as this may decrease transfection efficiency. Seed the cells in 30 ml at 5×10^5 cells/ml 24 h before transfection.

2. On the day of transfection, count cells and adjust to 1.0×10^6 cells/ml.

3. Dilute 40 µg of the expression vector and 37.5 µl of transfection reagent to 600 µl with transfection medium each.

4. Mix both dilutions and incubate at RT for 10 min.

5. Add transfection mix to the cells while swirling.

6. Culture transfected cells for at least 3 days. When expressing a fusion protein for the first time, daily supernatant samples should be collected and analyzed for target protein expression by Western blot (using HRP-conjugated anti-Fc antibody) to determine the optimal time of harvest (*see* **Note 2**).

7. To verify target protein expression, perform intracellular staining 24 h post transfection using an antibody directed against the Fc tag and the extracellular domain, if available (Fig. 2, *see* **Note 3**).

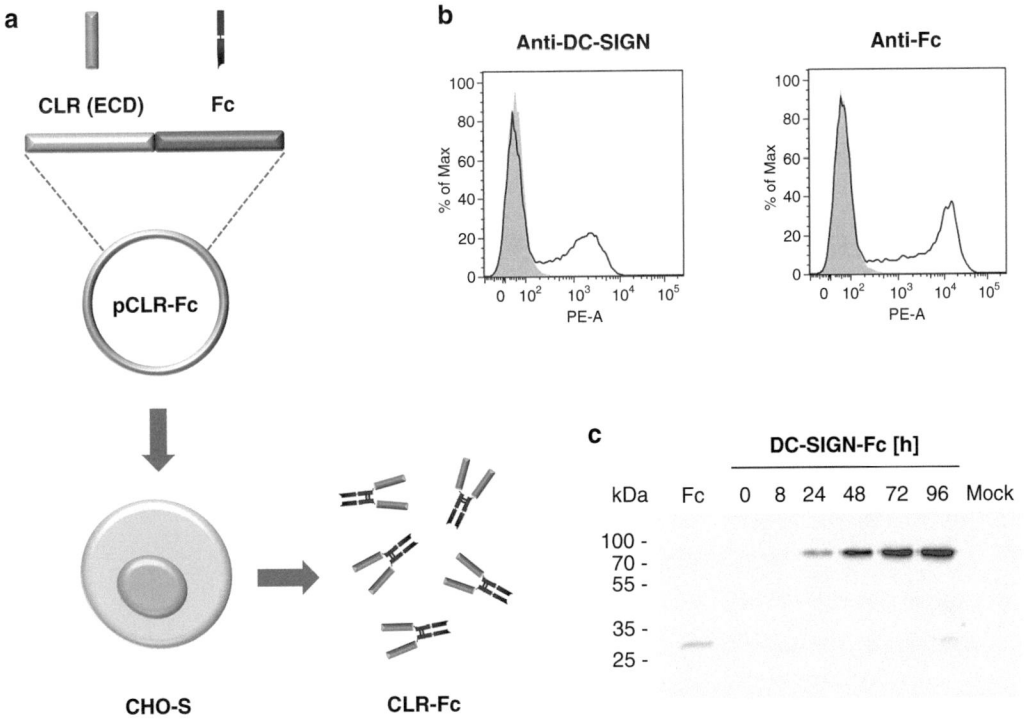

Fig. 2 Transient expression of DC-SIGN as Fc chimera in CHO-S cells. (**a**) The extracellular domain of DC-SIGN was fused to the Fc portion of human IgG1 and transfected into CHO-S cells. (**b**) 24 h post transfection, flow cytometric analysis revealed the presence of the DC-SIGN domain as well as the Fc part (*black line*: DC-SIGN-Fc transfected vs. filled: mock transfected control). (**c**) Supernatants were collected at indicated time points and target protein expression was analyzed by Western blot (anti-Fc-HRP, mock control corresponds to 96 h post transfection)

8. Centrifuge culture at $300 \times g$ for 5 min and pass supernatant through a 0.22 μm filter to remove debris. Perform all following steps at 4 °C or on ice.

9. Purify the Fc-tagged protein by affinity chromatography on a protein G column using a preparative protein chromatography system of choice. Equilibrate the column with binding buffer and apply the culture supernatant. After washing, the protein is eluted in elution buffer followed by immediate neutralization by adding 10 % (v/v) neutralization buffer.

10. Exchange the elution buffer with PBS using an ultrafiltration concentrator. We recommend using a molecular weight cutoff (MWCO) of 10 kDa. Pre-rinse the membrane with sterile PBS and concentrate the protein by centrifugation at $3200 \times g$ and 4 °C. Refill to initial volume with PBS and repeat twice.

11. Repeat the centrifugation step and adjust the final volume of the conjugate with sterile PBS. Determine protein concentration, aliquot and store at −80 °C until further use. Every batch should be tested regarding size, purity, and functionality (*see* **Note 4**).

3.2 Screening of Amino-Functionalized CLR Ligands Using Glycan Array

Generated fusion proteins are used for identification of novel C-type lectin ligands using the glycan array platform. This widely used technique is based on immobilization of functionalized glycan candidates on microarray slides. The array is probed with Fc fusion proteins which are in turn detected using a fluorophore-conjugated secondary antibody (Fig. 3).

1. For glycan array printing, amine-functionalized carbohydrates are dissolved in immobilization buffer. Dissolved carbohydrates (25 μl each) are transferred into a 384-well plate which serves as printing reservoir. Briefly centrifuge the plate at

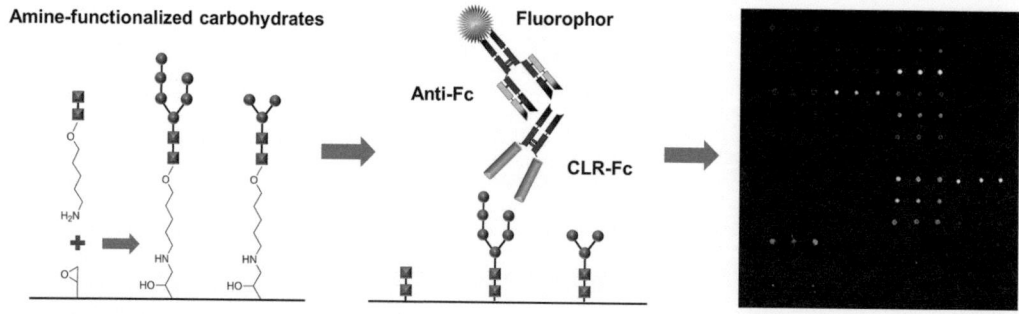

Fig. 3 Screening of novel CLR carbohydrate ligands using the glycan array platform. Functionalized carbohydrates are immobilized on epoxy slides. Following quenching and blocking, arrays are probed with CLR-Fc fusion proteins. Bound proteins are subsequently detected using a fluorophore-conjugated secondary antibody directed against the Fc tag and visualized using a microarray scanning device

$800 \times g$ to remove bubbles. Print glycans at 10, 1, and 0.1 mM in triplicates. Plates can be sealed and stored at –20 °C for later use.

2. Place epoxy functionalized microarray slide into the printing device. Adjust relative humidity to 60 % and print the glycans at RT following the desired pattern (*see* **Note 5**).

3. Incubate the printed slide in a humidified chamber overnight, allowing covalent glycan binding.

4. Gently rinse the slide three times with ultrapure water to remove unbound material.

5. Remove remaining liquid by centrifugation at $800 \times g$, 5 min in a 50 ml conical tube. If multiple slides are printed in advance, slides can be stored at RT under anhydrous conditions.

6. Incubate the slide in quenching buffer to block non-occupied epoxy groups at 50 °C for 1 h.

7. Gently rinse slide three times with water. Centrifuge slide at $800 \times g$, for 5 min.

8. Incubate slide in blocking buffer at RT for 1 h to prevent non-specific adsorption of proteins.

9. Gently wash slide three times in lectin buffer for 5 min each.

10. Attach a multiwell grid of desired size to the slide. We commonly use 16- or 64-well grids, depending on the amount of proteins screened. Take care not to touch the printed surface.

11. Thaw CLR-Fc fusion proteins on ice and dilute them in lectin buffer. Samples should include multiple concentrations typically ranging from 10 to 50 ng/μl in at least duplicate values. Use 100 μl per well for 16-well grids. Add Tween 20 at a final concentration of 0.01 % to minimize nonspecific interactions.

12. Incubate the slide at RT for 1 h with gentle agitation.

13. Carefully wash slides with 100 μl lectin buffer per well to remove unbound protein, three times 5 min each. From this point, make sure to pipet in the same corner of the well and avoid shear stress.

14. Dilute secondary antibody directed against the human Fc tag 1:200 in antibody dilution buffer.

15. Add 100 μl antibody per well for 16-well grids. Incubate at RT for 1 h with gentle agitation.

16. Wash slide to remove unbound antibody. Wash the slide twice with lectin buffer for 5 min each followed by one washing step with water for a couple of seconds.

17. Analyze binding of fusion proteins by evaluating the mean fluorescent intensity (MFI) of each printed spot using a microarray scanning device.

3.3 Conjugation of CLR Ligands to OVA

To covalently conjugate amine-functionalized glycans to the model antigen OVA, a disuccinimido adipate (DSAP) linker is used. DSAP reacts with primary amines via an amine-reactive *N*-hydroxysuccinimide (NHS) ester to form stable amide bonds. The functionalized glycan is subsequently linked to lysine residues within OVA (Fig. 4).

1. Dissolve DSAP in DMSO and activate by addition of 10 µl triethylamine.

2. Dissolve carbohydrates in DMSO and add dropwise to a ten-fold molar excess of DSAP.

3. Incubate at RT for 1.5 h, stirring.

4. Add 250–500 µl phosphate buffer.

5. Extract excess DSAP: Add 10 ml chloroform and centrifuge at $3000 \times g$, for 5 min. Recover carbohydrate-linker conjugates from the aqueous phase and repeat twice.

6. Add OVA to the glycan-linker conjugate in a total volume of 250–500 µl phosphate buffer and incubate overnight.

7. Separate conjugated OVA and free glycans using an ultrafiltration concentrator as described in Subheading 3.1, **step 10**.

8. Adjust the final volume of the conjugate using sterile PBS, determine protein concentration and store at ≤ -20 °C until further use. We recommend analyzing the glycan conjugate for successful conjugation by SDS-PAGE compared to unfunctionalized OVA. Mass spectrometric analysis should be employed to determine the carbohydrate/protein ratio.

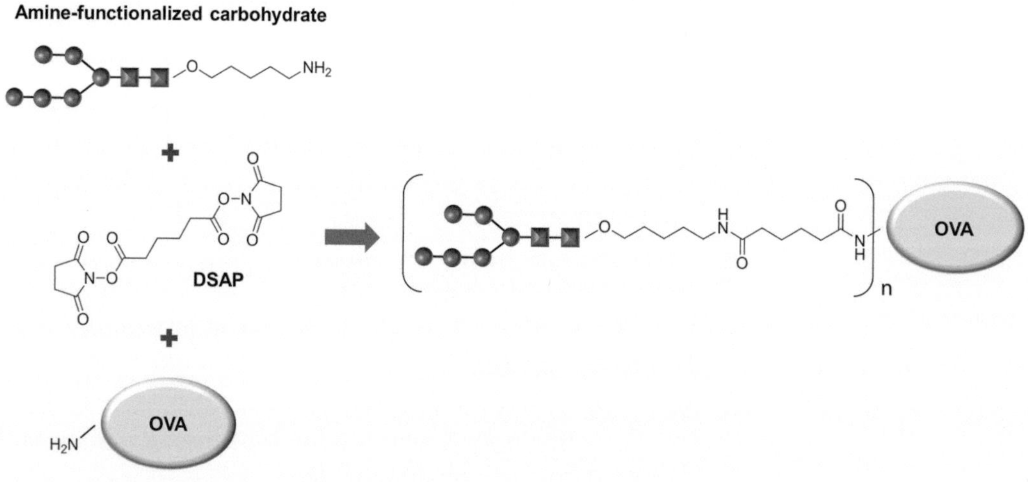

Fig. 4 Conjugation of carbohydrates to the model antigen OVA. Amine functionalized carbohydrates identified in previous screenings are conjugated to *N*-succinimidyl adipate (DSAP). After extraction of unbound linker, OVA is added to the reaction resulting in covalent glycan/protein conjugation

3.4 In Vitro T cell Activation Assay

Since many CLRs are endocytic receptors mediating uptake of antigens and subsequently presentation to T cells, generated neo-glycoconjugates are tested in a cell-based assay for activation of T cells. Dendritic cells are first isolated and pulsed with neoglycoconjugates, leading to uptake and presentation of OVA peptides on MHC-II molecules. In a second step, T cells from OT-II transgenic mice, displaying a T cell receptor specific for the MHC-II-presented ovalbumin 323–339 peptide, are added to the culture followed by analysis of activation markers and cytokine production (Fig. 5). Alternatively, T cells from OT-I mice can be used that recognize the ovalbumin 257-264 peptide presented on MHC-I molecules, thus allow to evaluate cross-presentation.

1. Isolate murine splenocytes. Sacrifice C57BL/6 and OT-II mice. Mount mice on styrofoam and sterilize the abdomen with 70 % ethanol. Open the abdominal cavity and carefully remove the spleen using scissors and forceps. Flush the spleen with 5 ml of complete RPMI using a syringe and needle. Collect cells and filter through a 40 μm cell strainer.

2. Centrifuge cell suspension at $300 \times g$ for 5 min. Aspirate the supernatant and resuspend the cell pellet in 5 ml RBC lysis buffer. Incubate at RT for 5 min and wash once with complete RPMI. Determine total cell number using a hemocytometer.

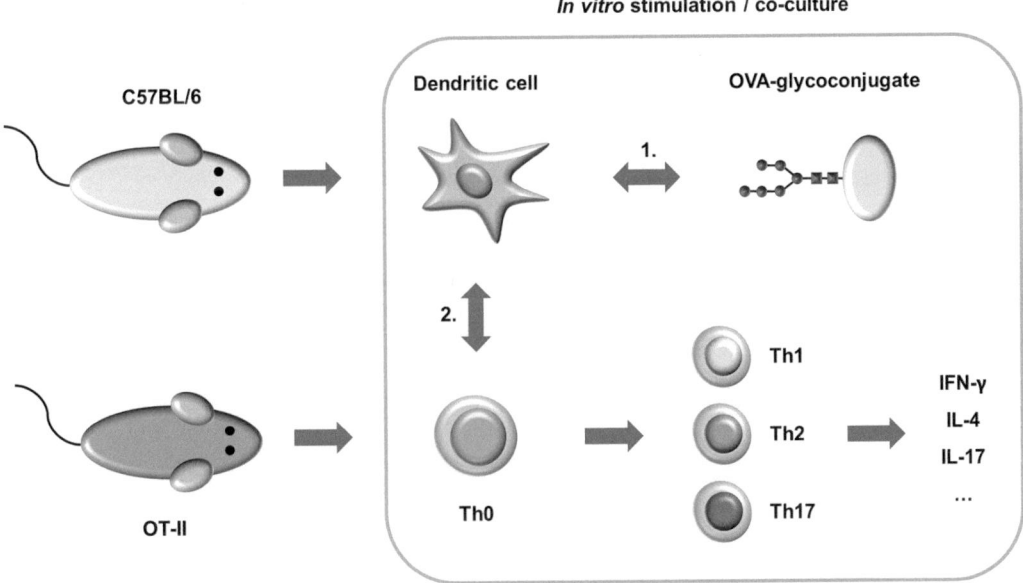

Fig. 5 Workflow for analysis of subsequent T cell activation in vitro. Dendritic cells are isolated from C57BL/6 mice by magnetic sorting of CD11c+ cells and pulsed with the neoglycoconjugates. Following uptake, transgenic OT-II T cells are added to the culture and subsequently analyzed for secretion of cytokines

3. Resuspend splenocytes in MACS buffer. Block Fc receptors by adding anti-CD16/32 (1:100) and incubate at 4 °C for 15 min.

4. Isolate CD11c+ dendritic cells using CD11c microbeads according to the manufacturer's instructions. In brief, resuspend cells in 400 µl MACS buffer per 10^8 cells, add 100 µl microbeads and incubate at 4 °C for 15 min. Add 10 ml MACS buffer, centrifuge at $300 \times g$ for 5 min and resuspend the pellet in 500 µl MACS buffer. Place the LS column in a cell separator and equilibrate with 3 ml MACS buffer. Apply cells, wash column three times with 3 ml MACS buffer each and elute in 5 ml MACS buffer.

5. Determine cell number, centrifuge, and resuspend for a suspension of 2×10^5 cells/ml in complete IMDM. Add 100 µl per well to a 96-well round bottom plate. Let the cells settle at 37 °C and 5 % CO_2 for 30 min.

6. Add neoglycoconjugates to the dendritic cell culture. We use a final concentration of 30 µg/ml next to unfunctionalized OVA as control. Incubate at 37 °C and 5 % CO_2 for 1 h.

7. Purify OVA-specific T cells from OT-II transgenic mice. Spleen cells are obtained by flushing and purified by magnetic-activated cell sorting as described above. Since this is a negative selection, the flow-through from the column is used.

8. Determine cell number, centrifuge, and resuspend cells in complete IMDM for a cell count of 1×10^6 cells/ml. Add 100 µl per well, corresponding to a DC–T cell ratio of 1:5.

9. Incubate at 37 °C and 5 % CO_2 for 48–72 h. Glycan-dependent activation of T cells can be evaluated by flow cytometry, ELISpot, and ELISA. We commonly determine concentrations of corresponding T cell cytokines such as IFN-γ and IL-2 in the culture supernatant.

3.5 Immunization Studies

To analyze the impact of CLR targeting in vivo, splenocytes are isolated from OT-II mice and adoptively transferred into C57BL/6 mice, enabling an OVA-specific response at low antigen doses. Mice are subsequently immunized with neoglycoconjugates and analyzed for T cell activation and anti-OVA antibody titers (Fig. 6).

1. Isolate splenocytes from OT-II mice as described above: Remove spleen, flush with complete RPMI, filter through a 40 µm cell strainer, lyse erythrocytes and count cells.

2. Inject 1.5×10^7 splenocytes in 100 µl PBS intravenously into C57BL/6 mice. Labeling of splenocytes may be performed to verify successful transfer (*see* **Note 6**). Let the mice sit overnight.

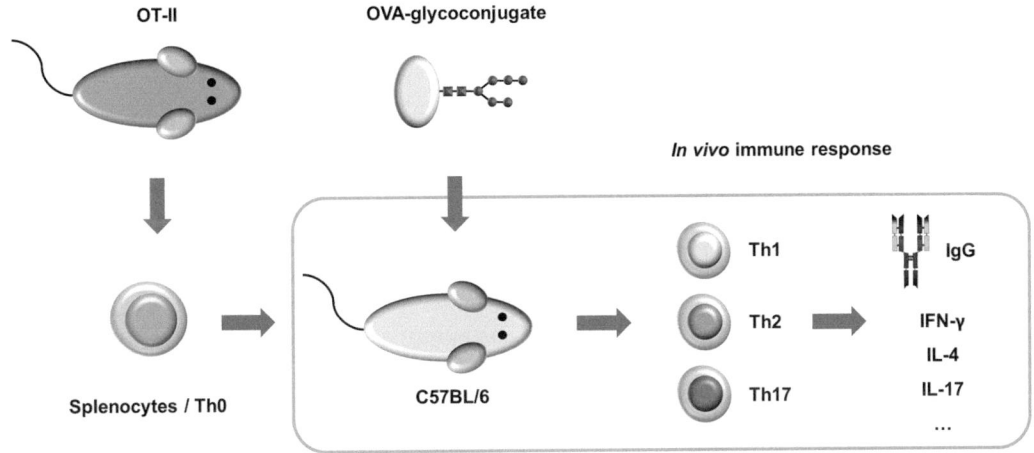

Fig. 6 Workflow for immunization of mice with neoglycoconjugates. Splenic T cells derived from OT-II mice are adoptively transferred into C57BL/6 mice followed by immunization with neoglycoconjugates. Serum samples are analyzed for cytokines and anti-OVA antibodies using ELISA. Mice are sacrificed after 4 weeks and splenic T cells are analyzed for differentiation status

3. The next day, immunize mice with at least 20 μg of neoglyco-conjugate per animal using intraperitoneal injection. Controls should include the use of unfunctionalized OVA alone and OVA supplemented with adjuvants such as alum. Perform a boost immunization after 2–3 weeks. Isolate serum samples weekly, centrifuge, and store at –80 °C until analysis of anti-OVA antibodies by ELISA.

4. Sacrifice mice 4 weeks after initial immunization. Isolate serum and splenocytes as described above and analyze by flow cytometry and ELISpot for cytokine expression (e.g., IFN-γ, IL-2, IL-4, IL-10, IL-17).

4   Notes

1. Splenocyte-derived cDNA is usually sufficient for most murine CLRs expressed by dendritic cells. For cloning of human CLRs, cDNA can be generated from CD14 monocytes after in vitro differentiation into dendritic cells or macrophages.

2. We found that peak protein levels occur after 3–5 days for most CLR-Fc fusion constructs. Extended expression times rather favor protein degradation and formation of cleaved fragments.

3. Since untransfected cells maintain their proliferation rate, analysis at later time points results in a lower percentage of CLR-Fc positive cells. However, this does not affect the absolute number of CLR-Fc expressing cells.

4. We characterize generated CLR-Fc fusion proteins by BCA assay, SDS-PAGE, Western blot, mass spectrometry, and binding assays using known CLR ligands.

5. It is recommended to use at least 16 identical fields to allow for enough proteins to be screened in parallel.

6. We usually use mice at 8–12 weeks of age and confirm success of the transfer by analysis of blood samples 1 h after injection using flow cytometry. To enable verification of a successful adoptive transfer, isolated splenocytes are labeled with eFluor® 670 according to the manufacturer's instructions. In brief, wash cells twice with PBS to remove serum and resuspend at 1×10^7 cells/ml in PBS. Add the same volume of 10 µM dye in PBS while vortexing for a final staining of 5×10^6 cells/ml in 5 µM dye. Incubate for 10 min at 37 °C. Stop the reaction by adding 5 volumes of RPMI and incubation on ice for 5 min. Wash cells three times with PBS and adjust to 1.5×10^8 cells/ml.

Acknowledgements

This work was supported by the Max Planck Society. We further acknowledge funding by the German Federal Ministry of Education and Research (Fkz. 0315446 to B. L.) and the SFB 765 (to B.L.).

References

1. Reed SG, Orr MT, Fox CB (2013) Key roles of adjuvants in modern vaccines. Nat Med 19:1597–1608

2. Petrovsky N, Cooper PD (2011) Carbohydrate-based immune adjuvants. Expert Rev Vaccines 10:523–537

3. Petrovsky N, Aguilar JC (2004) Vaccine adjuvants: current state and future trends. Immunol Cell Biol 82:488–496

4. Robinson MJ, Sancho D, Slack EC et al (2006) Myeloid C-type lectins in innate immunity. Nat Immunol 7:1258–1265

5. Geijtenbeek TB, Gringhuis SI (2009) Signalling through C-type lectin receptors: shaping immune responses. Nat Rev Immunol 9:465–479

6. Zelensky AN, Gready JE (2005) The C-type lectin-like domain superfamily. FEBS J 272:6179–6217

7. Geijtenbeek TB, Kwon DS, Torensma R et al (2000) DC-SIGN, a dendritic cell–specific HIV-1-binding protein that enhances *trans*-infection of T cells. Cell 100:587–597

8. Cambi A, Gijzen K, de Vries IJM et al (2003) The C-type lectin DC-SIGN (CD209) is an antigen-uptake receptor for Candida albicans on dendritic cells. Eur J Immunol 33:532–538

9. Cambi A, Netea MG, Mora-Montes HM et al (2008) Dendritic cell interaction with Candida albicans critically depends on N-linked mannan. J Biol Chem 283:20590–20599

10. Tailleux L, Schwartz O, Herrmann JL et al (2003) DC-SIGN is the major Mycobacterium tuberculosis receptor on human dendritic cells. J Exp Med 197:121–127

11. Ishikawa E, Ishikawa T, Morita YS et al (2009) Direct recognition of the mycobacterial glycolipid, trehalose dimycolate, by C-type lectin Mincle. J Exp Med 206:2879–2888

12. Schoenen H, Bodendorfer B, Hitchens K et al (2010) Mincle is essential for recognition and adjuvanticity of the mycobacterial cord factor and its synthetic analog trehalose-dibehenate. J Immunol 184:2756–2760

13. Yamasaki S, Ishikawa E, Sakuma M et al (2008) Mincle is an ITAM-coupled activating receptor that senses damaged cells. Nat Immunol 9:1179–1188

14. Zhang JG, Czabotar PE, Policheni AN et al (2012) The dendritic cell receptor Clec9A

binds damaged cells via exposed actin fila-ments. Immunity 36:646–657

15. Neumann K, Castiñeiras-Vilariño M, Höckendorf U et al (2014) Clec12a is an inhibitory receptor for uric acid crystals that regulates inflammation in response to cell death. Immunity 40:389–399

16. Figdor CG, van Kooyk Y, Adema GJ (2002) C-type lectin receptors on dendritic cells and Langerhans cells. Nat Rev Immunol 2:77–84

17. Sancho D, Reis e Sousa C (2012) Signaling by myeloid C-type lectin receptors in immunity and homeostasis. Annu Rev Immunol 30:491–529

18. Tacken PJ, de Vries IJM, Torensma R et al (2007) Dendritic-cell immunotherapy: from ex vivo loading to in vivo targeting. Nat Rev Immunol 7:790–802

19. Lepenies B, Yin J, Seeberger PH (2010) Applications of synthetic carbohydrates to chemical biology. Curr Opin Chem Biol 14:404–411

20. Eriksson M, Serna S, Maglinao M et al (2014) Biological evaluation of multivalent lewis X-MGL-1 interactions. Chembiochem 15:844–851

21. Lepenies B, Lee J, Sonkaria S (2013) Targeting C-type lectin receptors with multivalent carbo-hydrate ligands. Adv Drug Deliv Rev 65:1271–1281

22. Maglinao M, Eriksson M, Schlegel MK et al (2014) A platform to screen for C-type lectin receptor-binding carbohydrates and their potential for cell-specific targeting and immune modulation. J Control Release 175:36–42

Chapter 12

Characterization of Carbohydrate Vaccines by NMR Spectroscopy

Francesco Berti and Neil Ravenscroft

Abstract

Physicochemical techniques are a powerful tool for the structural characterization of carbohydrate-based vaccines. High-field Nuclear Magnetic Resonance (NMR) spectroscopy has been established as an extremely useful and robust method for tracking the industrial manufacturing process of these vaccines from polysaccharide bulk antigen through to the final formulation. Here, we describe the use of proton NMR for structural identity and conformity testing of carbohydrate-based vaccines.

Key words Nuclear magnetic resonance spectroscopy, Capsular polysaccharide, Carbohydrates, Antigens, Vaccines

1 Introduction

In the past 60 years, NMR spectroscopy has undergone a revolution, and today it is undoubtedly a technique of the utmost importance in studies of the structure, dynamics, and function of many molecules, including those related to carbohydrate chemistry and biochemistry. Physicochemical techniques are a powerful tool for the structural characterization of vaccine antigens at the level of both the bulk and final formulation, and NMR spectroscopy is now crucial for vaccine characterization and quality control [1, 2]. Qualitative and quantitative NMR methods have been proposed and developed for a number of applications related to the characterization of polysaccharide and glycoconjugate vaccines; these include determination of the identity of polysaccharide antigens and their combination vaccines [3–7], quantification of labile groups which might be important for immunogenicity (e.g., O-acetyl content) [3, 5–7], identification of end groups as markers of depolymerization of the carbohydrate chains [8–12], polysaccharide identification and monitoring of the conjugation process to assess the production process consistency [12, 13], determination

Bernd Lepenies (ed.), *Carbohydrate-Based Vaccines: Methods and Protocols*, Methods in Molecular Biology, vol. 1331,
DOI 10.1007/978-1-4939-2874-3_12, © Springer Science+Business Media New York 2015

of polysaccharide-protein ratio [1, 2], and quantification of NMR-sensitive residual process contaminants [6, 14].

In accordance with regulatory requirements specified by World Health Organization (WHO) Recommendations and Pharmacopeia, NMR spectroscopy provides an appropriate option as a routine release test. This test is performed together with other methodologies applied during the discovery and development phase of the product and follows principles established for drug analysis [15]. Structural characterization by NMR is also important for product comparability studies and predicting and evaluating the quality of the vaccines in the absence of reliable animal models for potency testing [16].

Here, we describe an NMR method for polysaccharide identity and conformity testing for the bulk monovalent polysaccharide, blended polysaccharide bulk, activated polysaccharide intermediate, bulk monovalent conjugate, blended conjugate bulks, and final fills.

2   Materials

Prepare all the analytical samples using analytical grade solvents and reagents. Store all the reagents and analytical samples at the recommended and appropriate temperatures respectively.

2.1 Solvents and Reagents

1. Deuterated solvents and reagents (i.e., deuterium oxide, D_2O; deuterated dimethyl sulfoxide, DMSO-d_6; 40 % sodium deuteroxide (NaOD) in deuterium oxide) with a high proportion of deuterium (e.g., >99.9 atom % D).

2. Chemical shift reference compound (i.e., sodium 2,2-dimethyl-2-silapentane-5-sulfonate, DSS; sodium trimethylsilylpropionate, TSP; or a deuterated analogue, TSP-d_4 to set the reference to zero ppm for the methyl signals).

3. High quality and qualified NMR reference standards to monitor the instrumental performance of NMR spectrometer (e.g., 0.1 % ethylbenzene in chloroform-d to perform the proton sensitivity test, a 1 % solution of chloroform in acetone-d_6 to evaluate the resolution, spectral line shape test).

2.2 Polysaccharide Samples

1. Solid or liquid state 0.5–20 mg aliquot of polysaccharide sample, as saccharide content, for the bulk monovalent polysaccharide, blended polysaccharide bulk, activated polysaccharide (if isolated), bulk monovalent conjugate, blended conjugate bulks, and final fills.

2.3 Devices and Instrumentation

1. Freeze-drier (lyophilizer) or other solvent evaporator.

2. NMR glass tubes (i.e., 5 mm diameter) of a quality suitable for use in high field spectrometers with caps (i.e., rubber caps).

3. NMR spectrometer with a minimal nominal field strength corresponding to a proton resonance frequency of 400 MHz, equipped with a high precision temperature controller (i.e., ±1 K) and a probe for proton detection (e.g., 5 mm).

4. Suitable host computer and software for instrument control, data collection, and processing.

5. Plastic or ceramic spinner for NMR tube (i.e., 5 mm diameter). Ceramic spinners are used for temperatures >323 K.

3 Methods

Carry out all procedures at room temperature unless otherwise specified.

3.1 Preparation of Analytical NMR Samples

1. Dry under vacuum (using a freeze-drier, lyophilizer, or other solvent evaporator) the relevant amount (0.5–20 mg) of polysaccharide material in liquid solution to obtain a solid aliquot. The procedure is not required if the material is already in solid state (*see* **Note 1**).

2. Add a low amount (0.5 % has been found to be appropriate) of chemical shift reference compound (i.e., DSS, TSP, TSP-d4) to the deuterated solvent used to dissolve the polysaccharide sample. DMSO (0.01 %) may be added as an internal intensity standard.

3. Dissolve the solid aliquot, contained in appropriate vial (i.e., 1–15 mL vial/tube) in ca. 0.7 mL of deuterated solvent with chemical shift reference compound, and mix (e.g., by vortex agitator) the solution to obtain a uniform concentration. Any particulate matter held in suspension will severely compromise field homogeneity and thus line shape. Low speed centrifugation (e.g., $7200 \times g$ for 10 min) can be used to pellet out any undissolved material [6].

4. Transfer the solution in NMR tube (i.e., 5 mm diameter) using a pipette (i.e., Pasteur pipette) and fix the cap.

5. For *O*-acetylated samples, record a second spectrum after performing de-*O*-acetylation in the NMR tube. Add 40 % sodium deuteroxide (NaOD) in deuterated water (15 µL into 0.7 mL of sample) which corresponds to a final concentration of approximately 200 mM NaOD in the NMR tube. The rate of de-*O*-acetylation varies with polysaccharide structure; for example heating at 310 K for 1 h is recommended for complete de-*O*-acetylation of meningococcal group C polysaccharide whereas the other polysaccharides are more readily de-O-acetylated [7].

3.2 NMR Data Collection and Processing

1. Login as a workstation user and run the software for NMR spectrometer control, data collection, and processing (e.g., TopSpin™ Bruker; VNMR™ Agilent Varian; Delta™ Jeol).

2. Create a new dataset or open a dataset listed in the browser to be saved with a different name.

3. Set the desired sample temperature (e.g., 298 ± 1 K).

4. Inserting sample in the NMR magnet (Fig. 1):

   (a) Hold the sample tube by the top (**step 1**), place it in the plastic or ceramic spinner (**step 2**), the spinner in the sample depth gauge and keeping locked the spinner push the sample tube to touch the bottom (**step 3**).

   (b) Remove the black cap from the top of the magnet bore. Press the LIFT button and wait for the airflow (hissing sound that can be heard) and insert the sample tube with the spinner into the magnet bore (**step 4**). Press the LIFT button again to toggle the airflow off: the sample tube with spinner gently drop to the magnet bore where it positioned at the top of the probe. In recent version of NMR instrumentation equipped with an autosampler (i.e., installed on the top of the magnet and controlled by the software), the sample tube and spinner are positioned into the autosampler holder (i.e., 20, 96 positions) and directly inserted in the magnet.

5. Locking:

   (a) Open the lock display window.

   (b) Lock the signal by selecting the appropriate solvent in the table window (i.e., D_2O) and wait for the message which confirms the end of process.

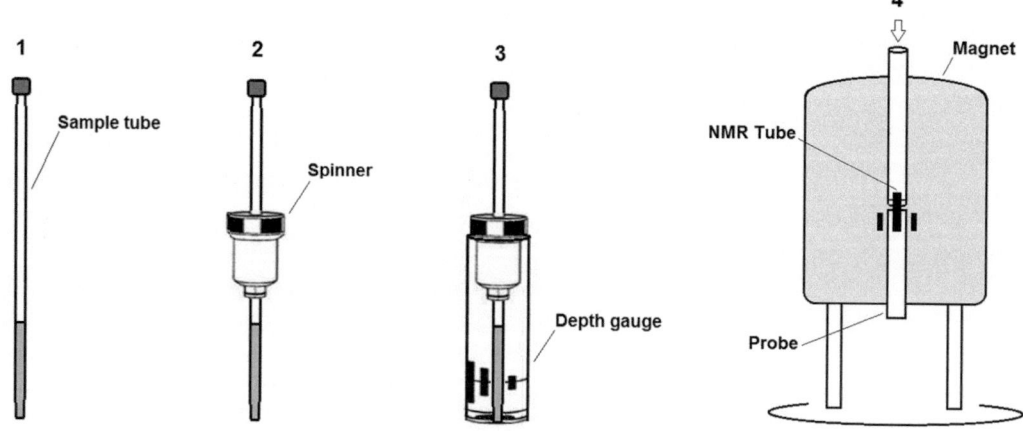

Fig. 1 Inserting sample in the NMR magnet

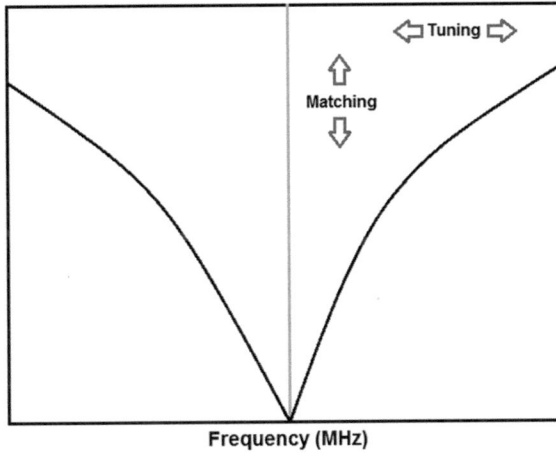

Fig. 2 Example of wobble curve with good matching and tuning

6. Tuning and matching the probe (*see* **Note 2**):

    (a) For probes not equipped with automatic tuning and matching, open the control window and adjust the proper buttons to adjust the reflected radiofrequency power at the minimum (Fig. 2).

    (b) For probes equipped with automatic tuning and matching, type the proper command and wait for the end of process.

7. Shimming the magnetic field (*see* **Note 3**):

    (a) For NMR spectrometers not equipped with automated procedures, adjust the proper buttons to shim in all the directions (i.e., Z1, Z2, Z3, X, Y, etc.) to obtain the highest value of lock signal.

    (b) For modern NMR spectrometers, the shimming can be performed by automated procedures: type the proper commands and wait for the end of process. Typically the procedure is complete within 1–2 min, although it can be shorter or longer to get convergence depending on the initial homogeneity.

    (c) If the temperature is changed, then re-tune and re-shim once the target temperature is reached.

8. Setting up the experiment:

    (a) Acquisition:

    • Create a new dataset by defining the nucleus (i.e., proton), the sample temperature (e.g., 298 ± 0.1 K or 343 ± 0.1 K, *see* **Note 4**), the pulse-program (i.e., standard mono-dimensional spectrum), the data points (i.e., 16–32 k), the spectral width (i.e., 10–16 ppm), the transmitter frequency (e.g., water signal at 4.79 and 4.48 ppm for 298 and 343 K respectively), the total recycle time (i.e., fivefold Longitudinal Relaxation Time T_1 to ensure a full recovery of each signal and obtain spectrum in quantitative manner). For instance, the T_1 values detected for methyl groups of Glc*p*NAc and Neu*p*NAc of GBS polysaccharides ranged from ~0.7 to ~1.9 s [17]. Typical total recycle time is approximately 10 s.

- Spinning the sample can improve spectral resolution by canceling out field inhomogeneities, but may lead to the presence of spinning sidebands. Typically samples are recorded without spinning.

- Determine the 90° proton pulse at high power (i.e., pulse calibration procedure) and set the pulse length and power and the appropriate receiver gain for the experiment.

- Collect data until the signal-to-noise (S/N) ratio in the anomeric-proton region of the spectrum is approximately 5 or better; this will depend on the amount of saccharide available. Typically 64 or 128 scans are recorded for a 1 mg polysaccharide sample.

(b) Processing (Fig. 3):

- Apply a weighting function (e.g., 0.2–0.3 line broadening function) to the Free Induction Decay (FID) and Fourier transform.

- After transforming the FID, adjust the phase to pure adsorption phase.

- If required for the experiment (e.g., quantitation), apply a baseline correction.

9. Monitor the instrumental performance of the NMR spectrometer: high quality and qualified NMR reference standards are required. A sealed NMR tube containing 0.1 % ethylbenzene in chloroform-d is used to perform ^1H sensitivity tests for the instrument/probe combination, whilst a 1 % solution of chloroform in acetone-d6 is used to evaluate spectral line shape (resolution), and compared to manufacturer's specifications.

10. Record and process the NMR experiments of analytical samples by performing the procedures already described. If needed, apply additional processing steps such as peaks integration for method quantification. In general "identity" requires that the peaks in the spectra of the test and reference sample, acquired using the same procedure and operating conditions, should correspond in position, intensity, and multiplicity.

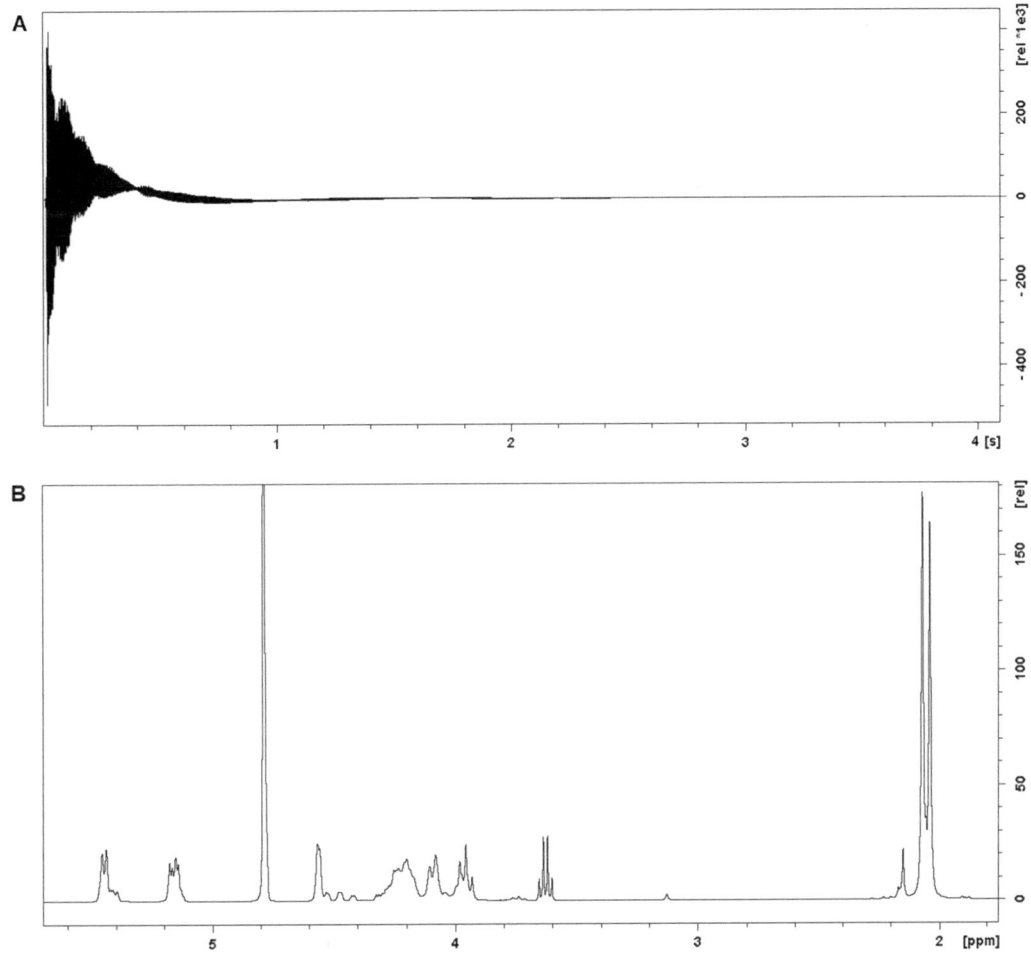

Fig. 3 Example of (**a**) FID (time domain signal) and (**b**) proton NMR spectrum

3.3 Spectral Interpretation

The NMR techniques described in this chapter have been used to facilitate the development, licensure, and quality control of carbohydrate-based vaccines against *Haemophilus influenzae* type b, *Salmonella enterica* serotype Typhi, and multiple strains of *Neisseria meningitidis* and *Streptococcus pneumoniae*. The methodologies established are being applied to vaccines being developed against other encapsulated pathogens including *Streptococcus pneumoniae*, *Staphylococcus aureus*, and enteric bacteria as well as fungi such as *Candida albicans* and *Cryptococcus neoformans*. Some examples of the application of NMR spectroscopy to the analysis and control of carbohydrate-based vaccines are provided in this section.

3.3.1 Haemophilus influenzae Type b (Hib) Vaccines

The structure of the Hib polysaccharide repeating unit is → 3)-β-D-Ribf-(1 → 1)-D-Ribitol-5-(P→. The proton NMR spectrum (Fig. 4a) constitutes a "fingerprint" of the Hib polysaccharide and can be used to confirm the identity and purity of different batches.

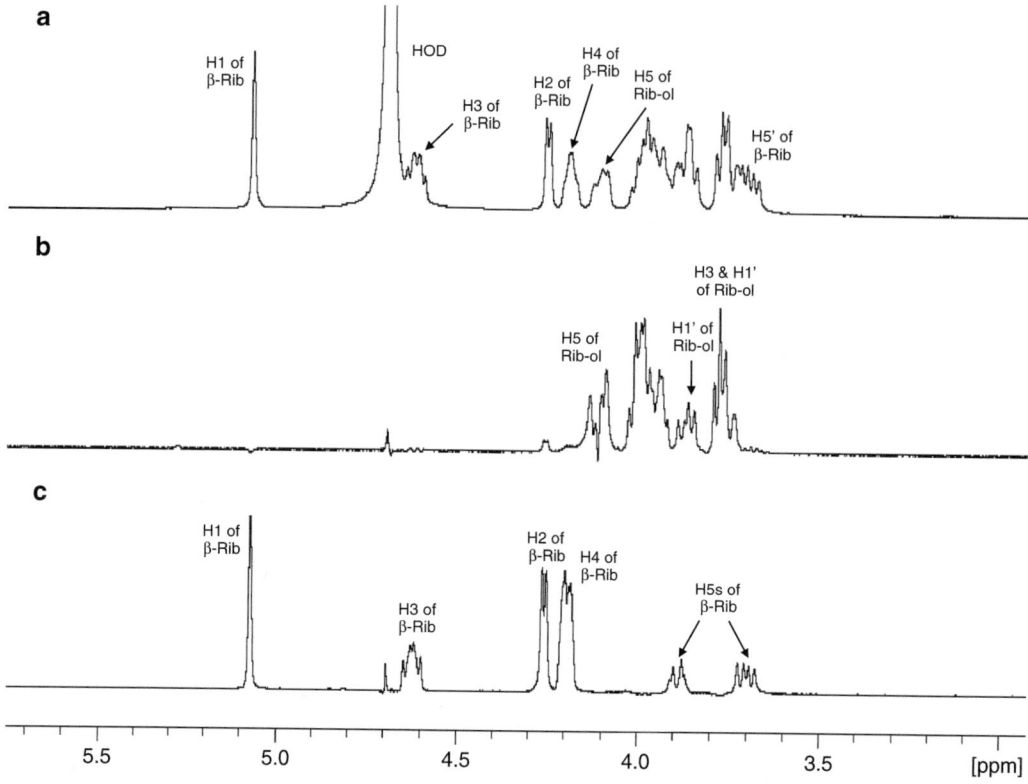

Fig. 4 ^{1}H-NMR spectra of Hib polysaccharide recorded at 400 MHz (303 K). (**a**) ^{1}H NMR spectrum, (**b**) 1D TOCSY (250 ms) correlations from H5 of ribitol (4.10 ppm), (**c**) 1D TOCSY (250 ms) correlations from H2 of Rib$f$ (4.25 ppm). Key proton assignments are labeled

Establishment of a "fingerprint" NMR spectrum follows from the full NMR characterization of the polysaccharide antigen by use of 1D and 2D ^{1}H, ^{13}C and ^{31}P experiments (*see* **Note 5**). Validation of the NMR identity test for the Hib polysaccharide has been published [4]. The ^{1}H NMR spectrum of Hib polysaccharide and those of the ribitol and ribose spin systems revealed by 1D TOCSY experiments are shown in Fig. 4.

Degradation of the Hib polysaccharide can be detected by the presence of new signals due to end groups formed. At neutral pH cleavage occurs through formation of 2,3-cyclophosphate ribose whereas the Rib$f$-(1 → 1)-D-Ribitol bond is hydrolyzed under acidic conditions to generate ribose at the reducing end [9, 10, 18, 19]. Several Hib glycoconjugate vaccines are licensed and they consist of either a long-chain polysaccharide or an oligosaccharide which is activated and attached to the protein carrier directly or via a spacer [20, 21]. NMR analysis can be used to track Hib antigen integrity from the starting polysaccharide, through activated intermediates to the final conjugate although the applicability depends on the type of vaccine and the coupling chemistry [1, 2, 22]. Detailed NMR characterization of oligosaccharide-based vaccines prepared by acid hydrolysis or periodate oxidation of the polysaccharide has been described [2, 13, 23]. For polysaccharide-based vaccines, the degree of activation/derivatization of a Hib polysaccharide intermediate generated by hydroxyl activation, reaction with excess butanediamine followed by acetylation with bromoacetyl chloride, has been quantified by NMR spectroscopy [24].

3.3.2 Salmonella enterica serotype Typhi (Vi) Vaccines

Vi polysaccharide vaccines are available and several conjugate vaccines are being developed for which draft WHO guidelines have been prepared [25]. The structure of the Vi repeating unit is →)-α-D-GalpNAcA(3OAc)-(1→ and *O*-acetylation is considered to be important for immunogenicity. The use and validation of an NMR test for the identity and *O*-acetyl content for the Vi polysaccharide has been described [5]. The method involves recording the NMR spectrum of Vi before and after de-*O*-acetylation. The de-O-acetylation is performed in the NMR tube by the addition of NaOD to a final concentration of 200 mM (Fig. 5). The final spectrum

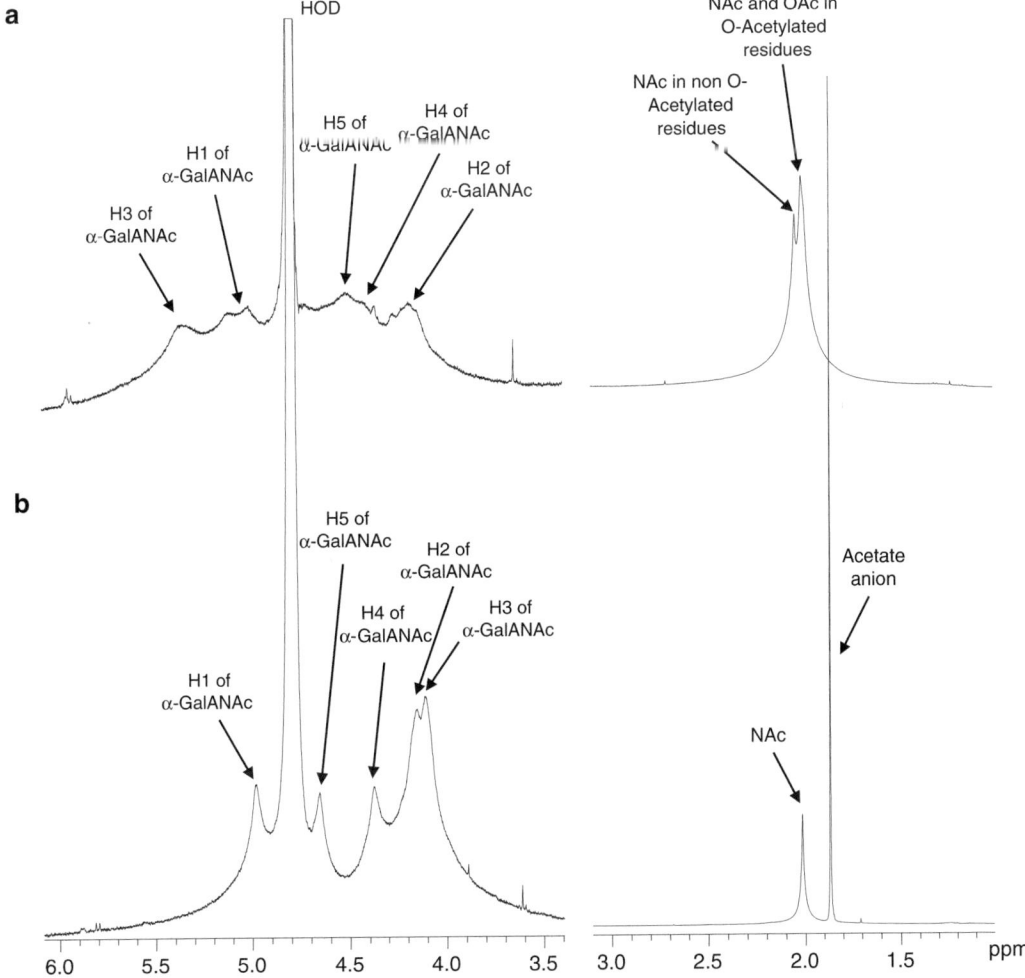

Fig. 5 ¹H-NMR profiles of the Vi polysaccharide (**a**) before and (**b**) after de-*O*-acetylation; some assignments are labeled

yields a simpler spectrum with sharper lines, and the degree of
O-acetylation may be determined by comparison of the integrals of
well-resolved N-acetyl and acetate resonances.

*3.3.3 Neisseria
meningitidis (Mn) Vaccines*

The structures of the most important meningococcal polysaccharide repeating units are shown in Table 1.

Polysaccharide vaccines against Mn groups A, C, Y, and W are
available; however, they are largely becoming replaced by the more
immunogenic glycoconjugate vaccines [26]. Several Mn X conjugate vaccines are in development [27]. Full NMR assignments for
the meningococcal polysaccharides have been published [3, 28,
29]. The proton NMR spectra are shown in Fig. 6; the spectral
complexity due to the presence of O-acetylation for groups A, C,
Y, and W complicates the NMR fingerprinting approach.
Furthermore the NMR spectrum can also change over time due to
O-acetyl migration [3]. As for the Vi polysaccharide, this problem
has been solved by recording a second spectrum after performing
de-O-acetylation in the NMR tube (shown for Mn A in Fig. 7).
This method has been validated and can be used to confirm the
identity of the polysaccharide backbone and the degree of
O-acetylation [7].

Degradation may be detected by the presence of new signals
due to end groups formed, e.g., β-NeuNAc reducing ends groups
for Mn C, Y, and W [8–12]. In addition, NMR analysis can determine the position and degree of O-acetylation and thus indicate if
there is loss or migration of O-acetyl groups [3]. Mn A polysaccharide was shown to be more labile than the structurally similar
Mn X polysaccharide due to the axial orientation of the man-
nosamine N-acetyl group that assists in the cleavage of the α phosphate group [30]. Monovalent (Mn A and Mn C) and tetravalent
(Mn A, C, Y, and W) glycoconjugate vaccines have been licensed.
The structure of each intermediate and the conjugate of
oligosaccharide-based vaccines can be characterized by NMR

Table 1
Repeating unit structures of some meningococcal polysaccharides

Polysaccharide	Repeat unit structure
Group A	→6)-α-D-Man*p*NAc(3/4OAc)-(1→*P*→
Group B	→8)-α-D-Neu*p*NAc-(2→
Group C	→9)-α-D-Neu*p*NAc(7/8OAc)-(2→
Group W	→6)-α-D-Gal*p*-(1→4)-α-D-Neu*p*NAc(7/9OAc)-(2→
Group Y	→6)-α-D-Glc*p*-(1→4)-α-D-Neu*p*NAc(7/9OAc)-(2→
Group X	→4)-α-D-Glc*p*NAc-(1→*P*→

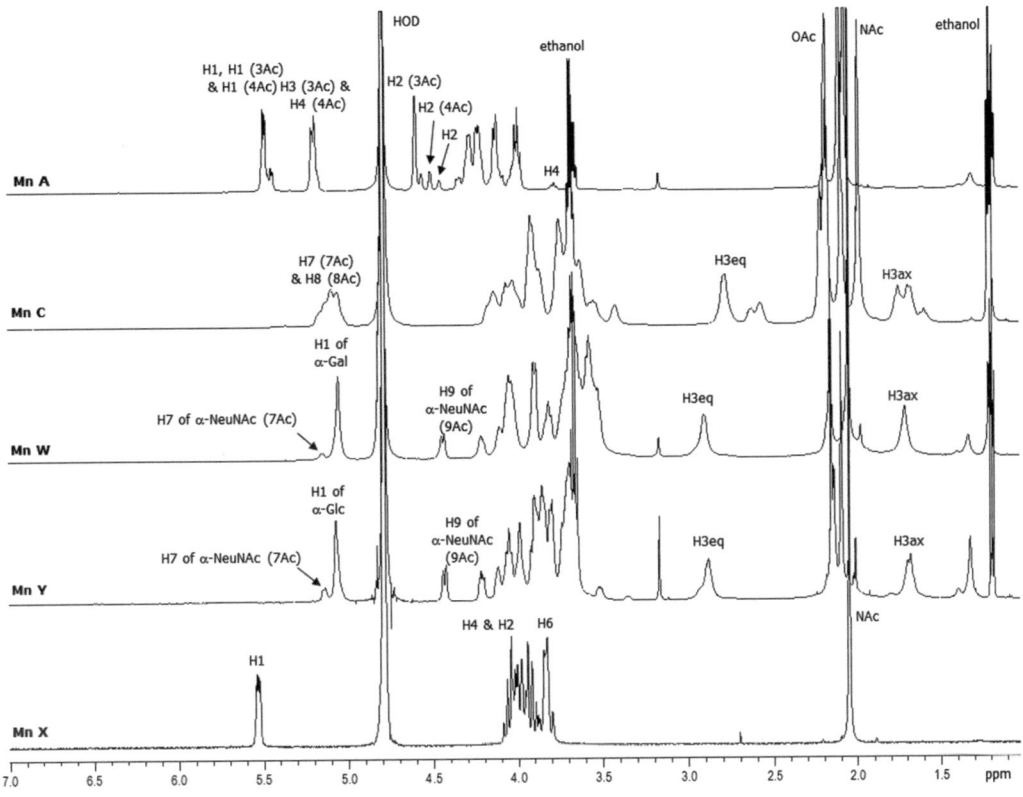

Fig. 6 ¹H-NMR profiles of the meningococcal polysaccharides groups A, C, W, Y, and X recorded at 400 MHz (298 K); some assignments are labeled

spectroscopy (*see* Fig. 8), and thereby confirming that the structural integrity of the carbohydrate antigen (including *O*-acetylation) is maintained throughout the manufacturing process [12].

3.3.4 Streptococcus pneumoniae (Pn) Vaccines

Over 90 serotypes have been recognized; polysaccharide vaccines against 23 serotypes (Table 2) are available, and 7, 10, and 13 valent glycoconjugate vaccines have been licensed with higher valency conjugates in development [31].

An overlay of the anomeric region of the NMR spectra for polysaccharides of 23 Pn serotypes recorded at 600 MHz (323 K) is shown in Fig. 9 [6]. The line width depends on the structure of the repeating unit and molecular weight (*see* **Note 1**). Relatively sharp lines are obtained for polysaccharides containing flexible linkages such as phosphodiesters and alditols whereas highly charged polysaccharides and those with many large substituents give poorer spectra [32].

This spectral region was used for polysaccharide identity through the calculation of correlation coefficients between test and reference spectra [6]. The test was validated by calculating a matrix of the correlation coefficients from all possible combinations of the

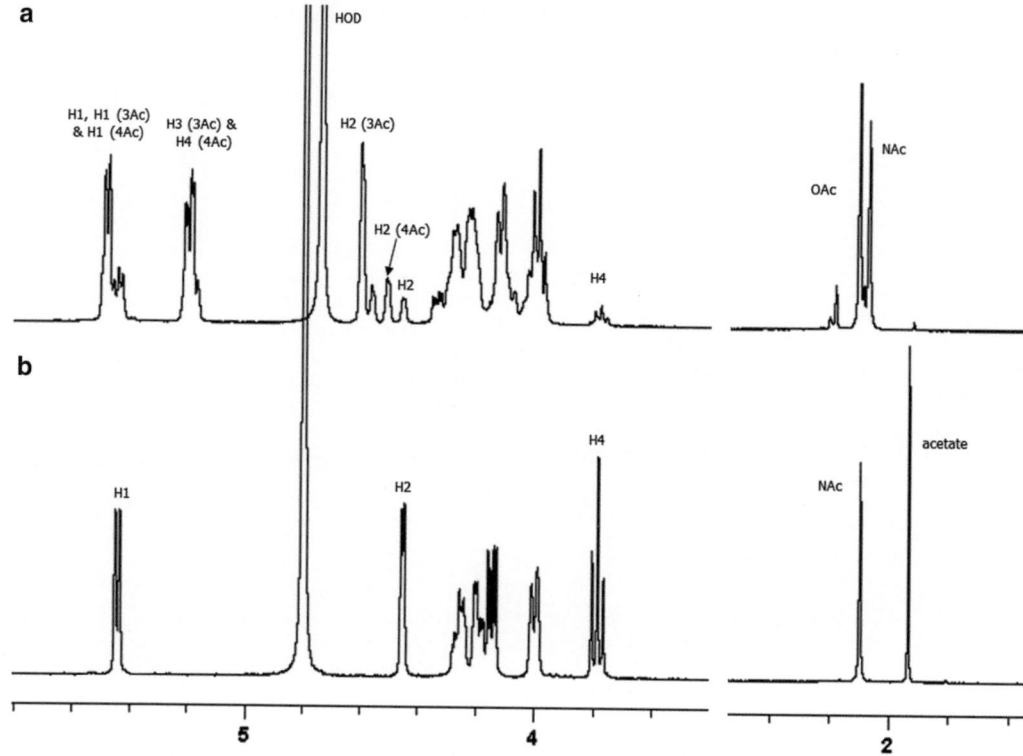

Fig. 7 ¹H-NMR profiles of the Mn A polysaccharide recorded at 500 MHz (303 K); (**a**) before and (**b**) after de-O-acetylation; some assignments are labeled

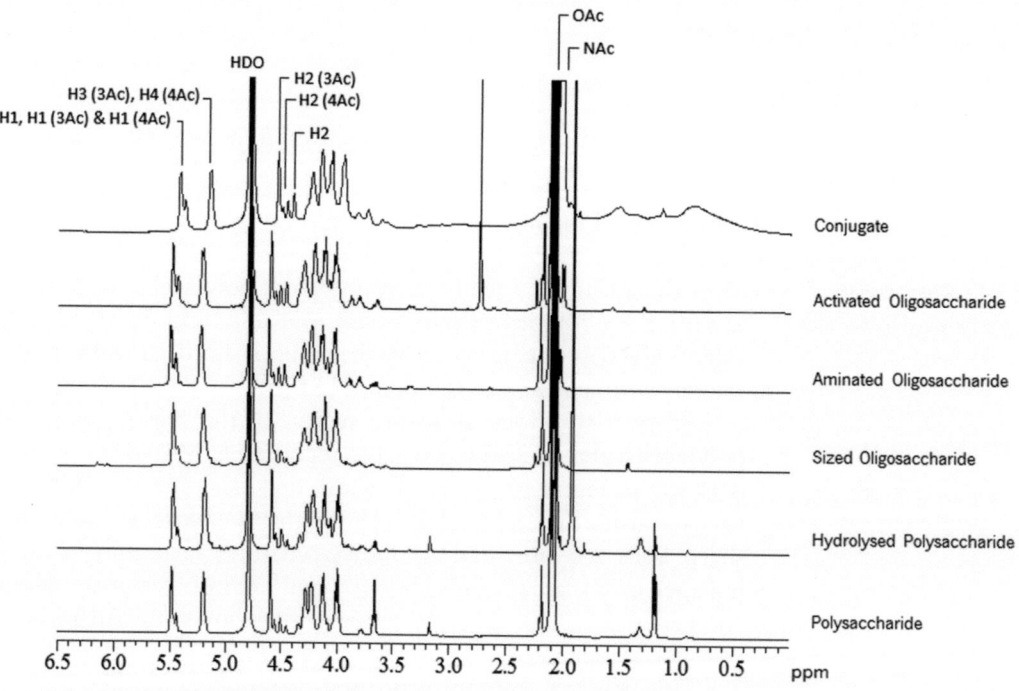

Fig. 8 An example of ¹H-NMR tracking of the conjugation process from polysaccharide to oligosaccharide intermediates to bulk conjugate vaccine (Mn A)

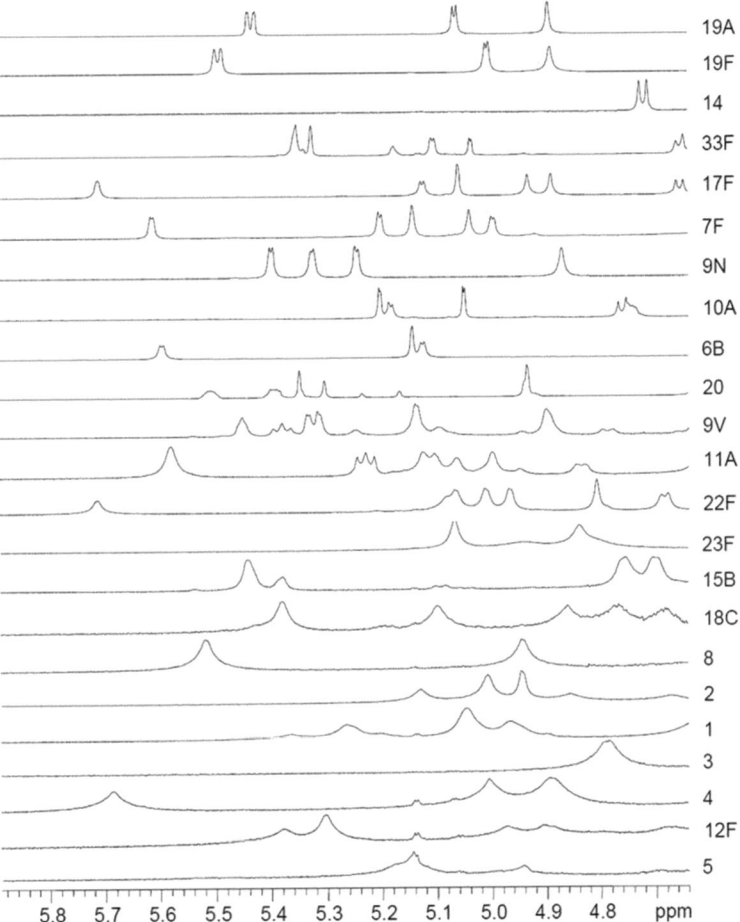

Fig. 9 ¹H-NMR overlay of the anomeric region of 23 pneumococcal polysaccharides recorded at 600 MHz (323 K). Spectra are displayed according to the average line width of their anomeric resonances. Reproduced from Ref. [6] with permission from Elsevier

23 serotypes. In addition to polysaccharide identity, the addition of DMSO (0.01 %) permitted the potential quantification of other organic materials present in the polysaccharide preparation such as contaminants and process residuals as well as product residuals such as cell wall polysaccharide (CWPS) [6, 14]. The NMR method used for quality control of the diverse structures of Pn polysaccharides is simple, specific, and reproducible compared to the elemental and colorimetric assays traditionally performed [6, 33]. Although no validation has been published, similar methodologies for determining the degree of O-acetylation as described for the Vi polysaccharides can be applied to O-acetylated serotypes (e.g., Pn 1, 7F, 9V, 11A, 15B, 17F, 18C, 20, 22F and 33F); an example is shown in Fig. 10.

Table 2
Repeating unit structures of some pneumococcal polysaccharides

Polysaccharide	Repeat unit structure
Type 1	→3)-D-AAT-α-Galp-(1→4)-α-D-GalpA(2/3OAc)-(1→3)-α-D-GalpA-(1→
Type 2	→4)-β-D-Glcp-(1→3)-[α-D-GlcpA-(1→6)-α-D-Glcp-(1→2)]-α-L-Rhap-(1→3)-α-L-Rhap-(1→3)-α-L-Rhap-(1→
Type 3	→3)-β-D-GlcpA-(1→4)-β-D-Glcp-(1→
Type 4	→3)-β-D-ManpNAc-(1→3)-α-L-FucpNAc-(1→3)-α-D-GalpNAc-(1→4)-α-D-Galp2,3(S)Py-(1→
Type 5	→4)-β-D-Glcp-(1→4)-[α-L-PnepNAc-(1→2)-β-D-GlcpA-(1→3)]-α-L-FucpNAc-(1→3)-β-D-Sugp-(1→
Type 6A	→2)-α-D-Galp-(1→3)-α-D-Glcp-(1→3)-α-L-Rhap-(1→3)-D-Rib-OL-(5→P→
Type 6B	→2)-α-D-Galp-(1→3)-α-D-Glcp-(1→3)-α-L-Rhap-(1→4)-D-Rib-OL-(5→P→
Type 7F	→6)-[β-D-Galp-(1→2)]-α-D-Galp-(1→3)-β-L-Rhap(2OAc)-(1→4)-β-D-Glcp-(1→3)-[α-D-GlcpNAc-(1→2)-α-L-Rhap-(1→4)]-β-D-GalpNAc-(1→
Type 8	→4)-β-D-GlcpA-(1→4)-β-D-Glcp-(1→4)-α-D-Glcp-(1→4)-α-D-Galp-(1→
Type 9N	→4)-α-D-GlcpA-(1→3)-α-D-Glcp-(1→3)-β-D-ManpNAc-(1→4)-β-D-Glcp-(1→4)-α-D-GlcpNAc-(1→
Type 9V	→4)-α-D-Glcp(2/3OAc)-(1→4)-α-D-GlcpA-(1→3)-α-D-Galp-(1→3)-β-D-ManpNAc(4/6OAc)-(1→4)-β-D-Glcp-(1→
Type 10A	→5)-β-D-Galf-(1→3)-β-D-Galp-(1→4)-[β-D-Galp-(1→4)]-[β-D-Galf-(1→)]-β-D-GalpNAc-(1→3)-α-D-Galp-(1→2)-D-Rib-OL-(5→P→
Type 11A	→3)-β-D-Galf-(1→4)-β-D-Glcp-(1→6)-[Gro-(1→P→4)]-α-D-Glcp(2/3OAc)-(1→4)-α-D-Galp-(1→
Type 12F	→4)-[α-D-Galp-(1→3)]-α-L-FucpNAc-(1→3)-β-D-GalpNAc-(1→4)-[α-D-Glcp-(1→2)-α-D-Glcp-(1→3)]-β-D-ManpNAcA-(1→
Type 14	→4)-β-D-Glcp-(1→6)-[β-D-Galp-(1→4)]-β-D-GlcpNAc-(1→3)-β-D-Galp-(1→
Type 15B	→6)-[α-D-Galp(2/3/4/6OAc)-(1→2)-[Gro-(2→P→3)]-β-D-Galp-(1→2)]-β-D-GlcpNAc-(1→3)-β-D-Galp-(1→4)-β-D-Glcp-(1→
Type 17F	→3)-β-L-Rhap-(1→4)-β-D-Glcp-(1→3)-α-D-Galp-(1→3)-β-L-Rhap(2OAc)-(1→4)-α-L-Rhap-(1→2)-D-Ara-OL-(1→P→
Type 18C	→4-)-β-D-Glcp-(1→4)-[α-D-Glcp(6OAc)-(1→2)][Gro-(1→P→3)]-β-D-Galp-(1→4)-α-D-Glcp-(1→3)-β-L-Rhap-(1→
Type 19A	→4)-β-D-ManpNAc-(1→4)-α-D-Glcp-(1→3)-α-L-Rhap-(1→P→
Type 19F	→4)-β-D-ManpNAc-(1→4)-α-D-Glcp-(1→2)-α-L-Rhap-(1→P→
Type 20	→6)-α-D-Glcp-(1→6)-β-D-Glcp-(1→3)-β-D-Galf-(1→3)-β-D-Glcp-(1→3)-[β-D-Galf-(1→4)]-α-D-GlcpNAc-(1→P→
Type 22F	→4)-β-D-GlcpA-(1→4)-[α-D-Glcp-(1→3)]-β-L-Rhap(2OAc)-(1→4)-α-D-Glcp-(1→3)-α-D-Galf-(1→2)-α-L-Rhap-(1→
Type 23F	→4)-β-D-Glcp-(1→4)-[α-L-Rhap-(1→2)]-[Gro-(2→P→3)]-β-D-Galp-(1→4)-β-L-Rhap-(1→
Type 33F	→3)-β-D-Galp-(1→3)-[α-D-Galp-(1→2)]-α-D-Galp-(1→3)-β-D-Galf-(1→3)-β-D-Glcp-(1→5)-β-D-Galf(2OAc)-(→

AAT is 2-acetamido-4-amino-2,4,6-trideoxygalactose, Gro is glycerol, Pne is 2-acetamido-2,6-dideoxytalose, Sug is 2-acetamido-2,6-deoxyhexose-4-ulose, and *P* is phosphate in a phosphodiester linkage

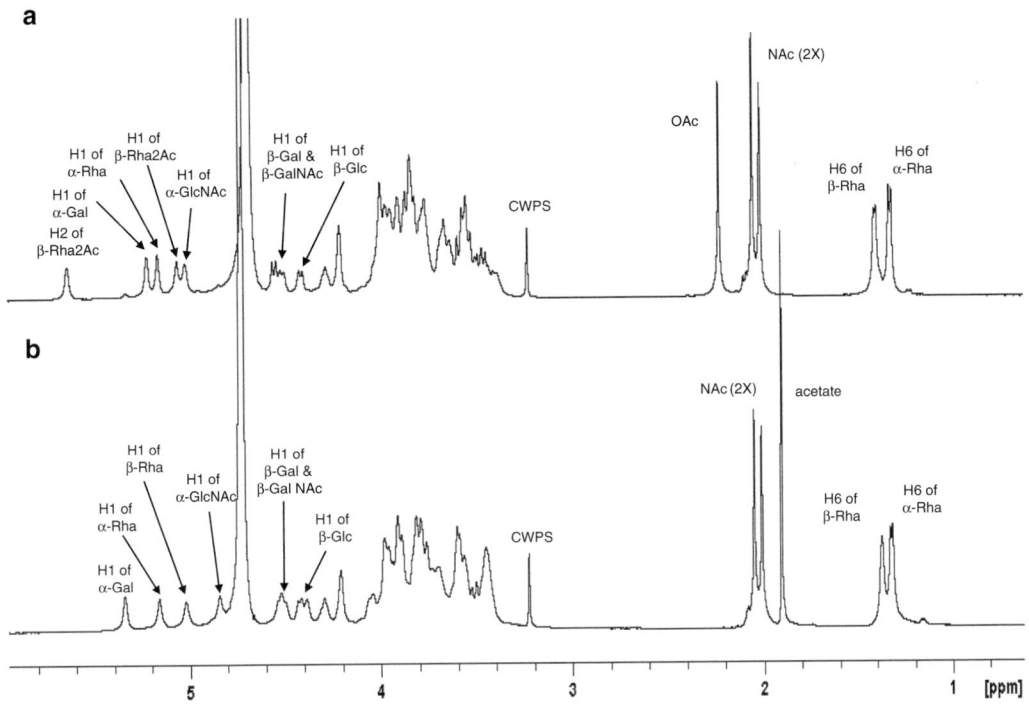

Fig. 10 ¹H-NMR profiles of the Pn 7 F polysaccharide recorded at 400 MHz (303 K); (**a**) before and (**b**) after de-O-acetylation; some assignments are labeled

3.3.5 Group B Streptococcus (GBS) Vaccines

Ten GBS serotypes have been recognized based on their expression of distinct capsular polysaccharides (Table 3) of which five are the current targets of glycoconjugate vaccines in development [34].

The structural diversity provides the basis for development of a proton NMR-based identity assay [17]; the proton NMR profiles are shown in Fig. 11.

4   Notes

1. Some polysaccharide samples are viscous and poorly soluble and may need to be vortexed and left overnight to achieve full dissolution. Alternatively the sample can be heated or subjected to sonication to improve solubility and spectral resolution. Dilute samples contain a large solvent (HOD) signal which appears in the anomeric region (~4.4–4.8 ppm) of the spectrum depending on temperature, pH, and concentration of the solution. This may overlap with peaks of interest and interfere with the analysis. The intensity of the HOD signal can be reduced by several cycles of deuterium exchange (i.e., dissolve the sample in the minimum amount of D₂O, freeze, and lyophilize) prior to analysis. This treatment may result in the loss of volatiles such as residual ethanol.

Table 3
Repeating unit structures of some GBS polysaccharides

Polysaccharide	Repeating unit structure
Type Ia	→4)-[α-D-NeupNAc-(2→3)-β-D-Galp-(1→4)-β-D-GlcpNAc-(1→3)]-β-D-Galp-(1→4)-β-D-Glcp (1→
Type Ib	→4)-[α-D-NeupNAc-(2→3)-β-D-Galp-(1→3)-β-D-GlcpNAc-(1→3)]-β-D-Galp-(1→4)-β-D-Glcp-(1→
Type II	→3)-β-D-Glcp-(1→2)-[α-D-NeupNAc-(2→3)]-β-D-Galp-(1→4)-β-D-GlcpNAc-(1→3)-[β-D-Galp-(1→6)]-β-D-Galp-(1→4)-β-D-Glcp-(1→
Type III	→6)-[α-D-NeupNAc-(2→3)-β-D-Galp-(1→4)]-β-D-GlcpNAc-(1→3)]-β-D-Galp-(1→4)-β-D-Glcp-(1→
Type IV	→4)-α-D-Glcp-(1→4)-[α-D-NeupNAc-(2→3)-β-D-Galp-(1→4)-β-D-GlcpNAc-(1→6)]-β-D-Galp-(1→4)-β-D-Glcp-(1→
Type V	→4)-[α-D-NeupNAc-(2→3)-β-D-Galp-(1→4)-β-D-GlcpNAc-(1→6)]-α-D-Glcp-(1→4)-[β-D-Glcp-(1→3)]-β-D-Galp-(1→4)-β-D-Glcp-(1→
Type VI	→6)-[α-D-NeupNAc-(2→3)-β-D-Galp-(1→3)]-β-D-Glcp-(1→3)-β-D-Galp-(1→4)-β-D-Glcp-(1→
Type VII	→4)-[α-D-NeupNAc-(2→3)-β-D-Galp-(1→4)-β-D-GlcpNAc-(1→6)]-α-D-Glcp-(1→4)-β-D-Galp-(1→4)-β-D-Glcp-(1→
Type VIII	→4)-[α-D-NeupNAc-(2→3)]-α-D-Galp-(1→4)-β-l-Rhap-(1→4)-β-D-Glcp-(1→
Type IX	→4)-[α-D-NeupNAc-(2→3)-β-D-Galp-(1→4)-β-GlcpNAc-(1→6)]-β-D-GlcpNAc-(1→4)-β-Galp-(1→4)-β-Glcp-(1→

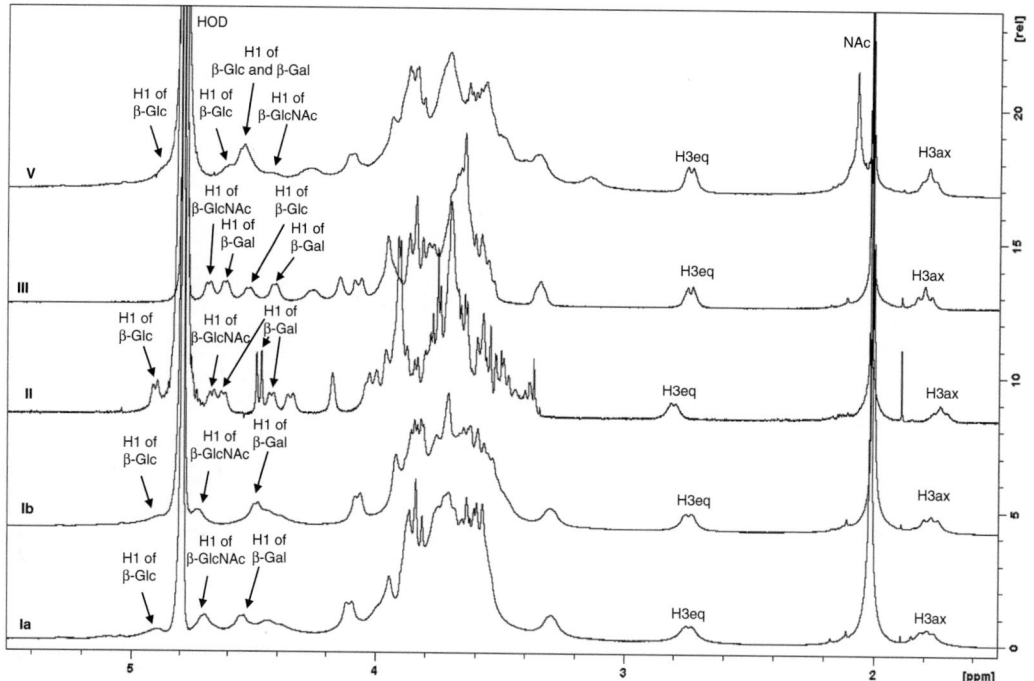

Fig. 11 ¹H-NMR profiles of the GBS polysaccharides recorded at 400 MHz (298 K); some diagnostic peaks labeled

2. The resonance frequency of the radiofrequency coil varies depending on the content of the individual sample tube (i.e., salt concentration, etc.). Consequently, the radiofrequency coil has to be tuned by adjusting two variable capacitors (called "tune" and "match") on the probe at the correct value (the reflected radiofrequency power is minimized at the proper frequency) to yield the correct resonance frequency for the magnet field strength.

3. Shimming is a process in which minor adjustments are made to the magnetic field until a uniform magnetic field is achieved around the sample.

4. Peaks of interest under the HOD signal can be revealed by recording the spectrum at higher temperatures that shift the HOD signal upfield, e.g., 323 or 343 K. Performing long experiments at high temperatures may result in spectral changes such as O-acetyl migration or formation of end groups due to hydrolysis of polysaccharides containing acid labile linkages. In addition or alternatively, the intensity of the HOD signal can be reduced by use of solvent suppression programs such as presaturation, WATERGATE, or 1D NOESY presaturation; care must be taken that the intensity of nearby signals of interest is not perturbed.

5. The ^1H, ^{13}C, and ^{31}P NMR spectra constitute fingerprints of the Hib antigen and can be assigned using standard 1D and 2D NMR experiments [35, 36]. Discernable signals in the ^1H NMR spectrum (Fig. 12a), typically anomeric (and deoxy for other antigens) signals, serve as the starting point for the ^1H-^1H 2D NMR scalar chemical shift correlation spectroscopy (COSY) and total correlation (TOCSY) experiments (Fig. 12b, c). Elucidation of the entire proton spin system for each residue may require the use of a range of TOCSY mixing times depending on the magnitude of the coupling constants involved, as well as the use of 1D variants that give better resolution and permit longer mixing times to be used (Fig. 4b, c). Alternatively NOESY experiments may provide correlations when the scalar couplings are "blocked" by small coupling constants. If sufficient material is available, ^{13}C NMR spectra can be recorded directly. Assignment of the groups present in the ^{13}C NMR spectrum follows from the chemical shift value and the DEPT experiment which identifies methyl, methylene, methine, and quaternary carbons (Fig. 12d, e). Full assignment of the ^{13}C NMR spectrum can be made from the assigned protons by use of ^1H-^{13}C 2D NMR scalar chemical shift correlation experiments such as HSQC (Fig. 12f). Complete assignment of the spin systems may require the use of hybrid experiments such as HSQC-TOCSY (a ^{13}C-dispersed TOCSY spectrum, Fig. 13a) or HSQC-NOESY, as well as

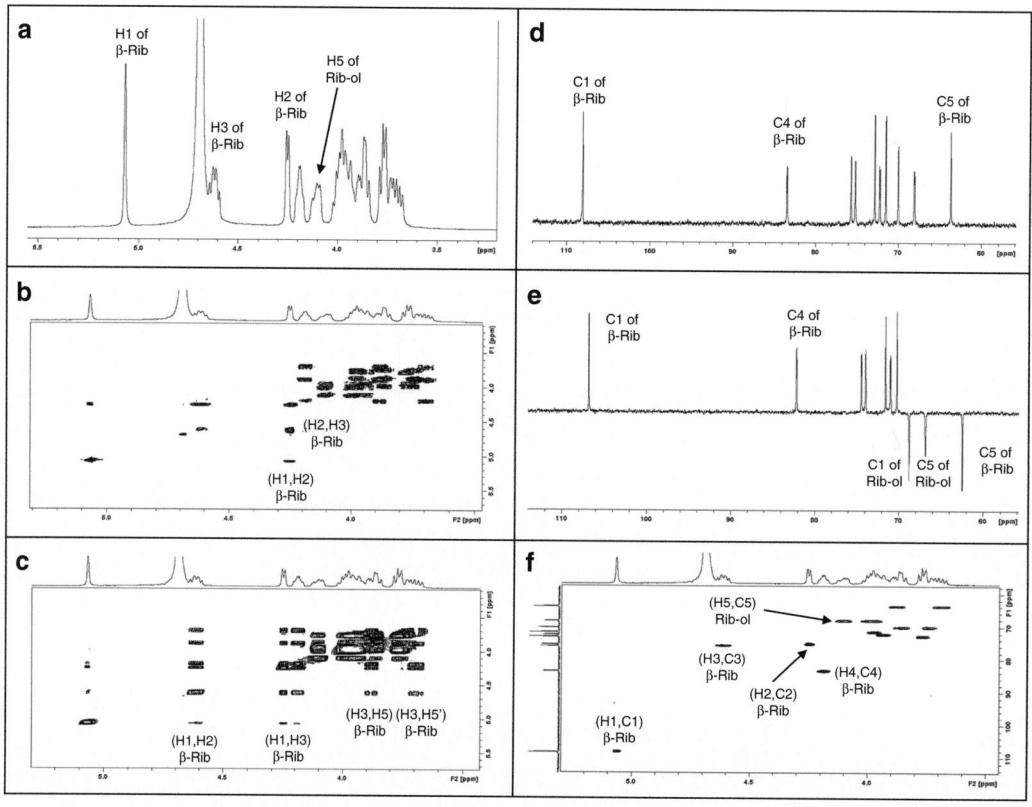

Fig. 12 1D and 2D NMR spectra of the Hib polysaccharide recorded at 400 MHz (303 K): (**a**) ¹H-NMR spectrum, (**b**) COSY, (**c**) TOCSY (120 ms), (**d**) ¹³C NMR spectrum, (**e**) ¹³C DEPT-135 NMR spectrum, (**f**) HSQC

long range ^{1}H-^{13}C experiments (HMBC and H2BC). Once the full set of ^{1}H and ^{13}C NMR data is obtained, the identity and linkages of the constituent sugars can be inferred from the magnitude of the glycosylation shifts facilitated by use of a carbohydrate chemical shift prediction program CASPER [37]. The sequence follows from correlations established between the sugar spin systems by use of long range ^{1}H-^{13}C experiments (HMBC, Fig. 13b) and ^{1}H-^{1}H dipolar experiments (NOESY). The ^{31}P NMR spectrum is less informative and for Hib it contains a single peak assigned to the phosphodiester group (Fig. 13c). The presence of the phosphodiester linkage between C3 of the ribose and C5 of the ribitol is indicated by the glycosylation shifts and characteristic splitting of the signals of the adjacent protons (H3 ribose and H5's of the ribitol) and the carbon signals of C2, C3, and C4 of the ribose and C4 and C5 of the ribitol (Fig. 12c). This linkage can be confirmed by performing a long range ^{1}H-^{31}P HMBC experiment (Fig. 13d). Thus the structure of the Hib polysaccharide is elucidated and full assignment of the ^{1}H, ^{13}C, and ^{31}P NMR fingerprint spectra achieved.

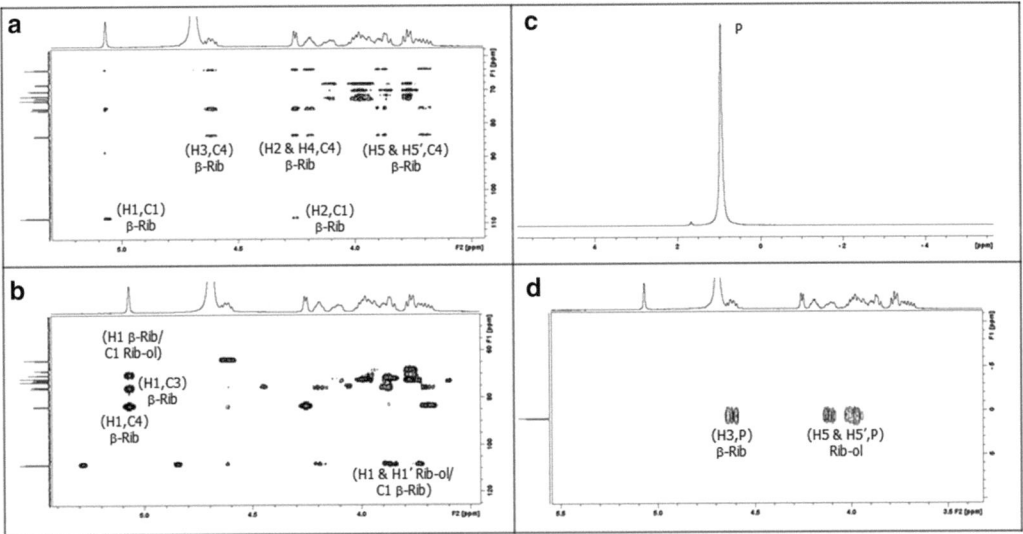

Fig. 13 1D and 2D NMR spectra of the Hib polysaccharide recorded at 400 MHz (303 K): (**a**) HSQC-TOCSY (120 ms), (**b**) ^{1}H-^{13}C HMBC (J=6 Hz), (**c**) ^{31}P NMR spectrum, (**d**) ^{1}H-^{31}P HMBC (J=10 Hz)

Acknowledgements

We thank Dr. Chris Jones (Laboratory for Molecular Structure, National Institute for Biological Standards and Control) for years of useful discussions on NMR and vaccines. One of us (N.R.) would like to thank PATH and all the vaccine manufacturers who have made available samples of polysaccharides and glycoconjugate vaccines.

References

1. Jones C (2005) NMR assays for carbohydrate-based vaccines. J Pharm Biomed Anal 38: 840–850
2. Jones C, Ravenscroft N (2008) NMR assays for carbohydrate-based vaccines. In: Holtzgrabe U, Wawer I, Diehl B (eds) NMR spectroscopy in pharmaceutical analysis. Elsevier, Oxford, UK, pp 341–368
3. Lemercinier X, Jones C (1996) Full ^{1}H NMR assignment and detailed O-acetylation patterns of capsular polysaccharides from *Neisseria meningitidis* used in vaccine production. Carbohydr Res 296:83–96
4. Lemercinier X, Jones C (2000) An NMR spectroscopic identity test for the control of the capsular polysaccharide from *Haemophilus influenzae* type b. Biologicals 28:175–183
5. Lemercinier X, Martinez-Cabrera I, Jones C (2000) Use and validation of an NMR test for

the identity and O-acetyl content of the *Salmonella* typhi Vi capsular polysaccharide vaccine. Biologicals 28:17–24
6. Abeygunawardana C, Williams TC, Sumner JS et al (2000) Development and validation of an NMR-based identity assay for bacterial polysaccharides. Anal Biochem 279:226–240
7. Jones C, Lemercinier X (2002) Use and validation of NMR assays for the identity and O-acetyl content of capsular polysaccharides from *Neisseria meningitidis* used in vaccine manufacture. J Pharm Biomed Anal 30: 1233–1247
8. Ravenscroft N, D'Ascenzi S, Proietti D et al (2000) Physicochemical characterisation of the oligosaccharide component of vaccines. Dev Biol (Basel) 103:35–47
9. Ravenscroft N, Averani G, Bartoloni A et al (1999) Size determination of bacterial capsular

oligosaccharides used to prepare conjugate vaccines. Vaccine 17:2802–2816

10. Jones C, Lemercinier X, Crane DT et al (2000) Spectroscopic studies of the structure and stability of glycoconjugate vaccines. Dev Biol (Basel) 103:121–136

11. Bardotti A, Averani G, Berti F et al (2005) Size determination of bacterial capsular oligosaccharides used to prepare conjugate vaccines against *Neisseria meningitidis* groups Y and W135. Vaccine 23:1887–1899

12. Bardotti A, Averani G, Berti F et al (2008) Physicochemical characterization of glycoconjugate vaccines for prevention of meningococcal diseases. Vaccine 26:2284–2296

13. Ravenscroft N (2000) The application of NMR spectroscopy to track the industrial preparation of polysaccharide and derived glycoconjugate vaccines. Pharmeuropa, Special Edition, pp 131–144

14. Xu Q, Abeygunawardana C, Ng AS et al (2005) Characterization and quantification of C-polysaccharide in *Streptococcus pneumoniae* capsular polysaccharide preparations. Anal Biochem 336:262–272

15. Holzgrabe U, Deubner R, Schollmayer C et al (2005) Quantitative NMR spectroscopy—applications in drug analysis. J Pharm Biomed Anal 38:806–812

16. Holliday MR, Jones C (1999) Meeting report: WHO-Co-sponsored informal Workshop on the use of physicochemical methods for the characterization of *Haemophilus influenzae* type b conjugate vaccines. Biologicals 27:51–53

17. Pinto V, Berti F (2014) Exploring the Group B *Streptococcus* capsular polysaccharides: the structural diversity provides the basis for development of NMR-based identity assays. J Pharm Biomed Anal 98:9–15

18. Egan W, Schneerson R, Werner KE et al (1982) Structural studies and chemistry of bacterial capsular polysaccharides. Investigations of phosphodiester-linked capsular polysaccharides isolated from *Haemophilus influenzae* types a, b, c, and f: NMR spectroscopic identification and chemical modification of endgroups and the nature of base-catalyzed hydrolytic depolymerization. J Am Chem Soc 104:2898–2910

19. Sturgess AW, Rush K, Charbonneau RJ et al (1999) *Haemophilus influenzae* type b conjugate vaccine stability: catalytic depolymerization of PRP in the presence of aluminium hydroxide. Vaccine 17:1169–1178

20. Lindberg AA (1999) Glycoprotein conjugate vaccines. Vaccine 17(Suppl 2):S28–S36

21. Verez-Bencomo V, Fernández-Santana V, Hardy E et al (2004) A synthetic conjugate polysaccharide vaccine against *Haemophilus influenzae* type b. Science 305:522–525

22. Aubin Y, Jones C, Freedberg DI (2010) Using NMR Spectroscopy to obtain the higher order structure of biopharmaceutical products. BioPharm Int (Suppl):28–38

23. D'Ambra AJ, Baugher JE, Concannon PE et al (1997) Direct and indirect methods for molar-mass analysis of fragments of the capsular polysaccharides of *Haemophilus influenzae* type b. Anal Biochem 250:228–236

24. Xu Q, Klees J, Teyral J et al (2005) Quantitative nuclear magnetic resonance analysis and characterization of the derivatized *Haemophilus influenzae* type b polysaccharide intermediate for PedvaxHIB. Anal Biochem 337:235–245

25. Anonymous (2013) Recommendations to assure the quality, safety and efficacy of typhoid conjugate vaccines. World Health Organization. http://www.who.int/biologicals/areas/vaccines/TYPHOID_BS2215_doc_v1.14_WEB_VERSION.pdf. Accessed 27 June 2014

26. Cohn AC, MacNeil JR, Clark TA et al (2013) Prevention and control of meningococcal disease: recommendations of the Advisory Committee on Immunization Practices (ACIP). MMWR Recomm Rep 62(RR-2):1–28

27. Micoli F, Romano MR, Tontini M et al (2013) Development of a glycoconjugate vaccine to prevent meningitis in Africa caused by meningococcal serogroup X. Proc Natl Acad Sci U S A 110:19077–19082

28. Xie O, Bolgiano B, Gao F et al (2012) Characterization of size, structure and purity of serogroup X *Neisseria meningitidis* polysaccharide, and development of an assay for quantification of human antibodies. Vaccine 30:5812–5823

29. Garrido R, Puyada A, Fernández A et al (2012) Quantitative proton nuclear magnetic resonance evaluation and total assignment of the capsular polysaccharide *Neisseria meningitides* serogroup X. J Pharm Biomed Anal 70:295–300

30. Berti F, Romano MR, Micoli F et al (2012) Relative stability of meningococcal serogroup A and X polysaccharides. Vaccine 30:6409–6415

31. Ginsburg AS, Alderson MR (2011) New conjugate vaccines for the prevention of pneumococcal disease in developing countries. Drugs Today 47:207–214

32. Jones C, Mulloy B (1993) The application of nuclear magnetic resonance to structural studies of polysaccharides. In: Jones C, Mulloy B, Thomas AH (eds) Spectroscopic methods and analyses. Humana, Totowa, NJ, pp 149–167

33. Anonymous (2005) Recommendations for the production and control of pneumococcal conjugate vaccines. WHO Tech Rep Series 927: 64–98

34. Madhi SA, Dangor Z, Heath PT et al (2013) Considerations for a phase-III trial to evaluate a group B *Streptococcus* polysaccharide-protein conjugate vaccine in pregnant women for the prevention of early-and late-onset invasive disease in young-infants. Vaccine 31:D52–D57

35. Bubb WA (2003) NMR spectroscopy in the study of carbohydrates: characterizing the structural complexity. Concept Magn Reson A 19:1–19

36. Duus JØ, Gotfredsen CH, Bock K (2000) Carbohydrate structural determination by NMR spectroscopy: modern methods and limitations. Chem Rev 100:4589–4614

37. Lundborg M, Widmalm G (2011) Structure analysis of glycans by NMR chemical shift prediction. Anal Chem 83:1514–1517

Chapter 13

Characterization of Capsular Polysaccharides and Their Glycoconjugates by Hydrodynamic Methods

Stephen E. Harding, Ali Saber Abdelhameed, Richard B. Gillis, Gordon A. Morris, and Gary G. Adams

Abstract

Hydrodynamic methods are relevant for the characterization of carbohydrates such as capsular bacterial polysaccharides or glycoconjugates in solution. This chapter focuses on the following hydrodynamic methods: sedimentation velocity analytical ultracentrifugation (SV AUC), dynamic light scattering (DLS), sedimentation equilibrium analytical ultracentrifugation (SE AUC), size exclusion chromatography coupled to multi-angle light scattering (SEC-MALS), and capillary viscometry—intrinsic viscosity measurement. The chapter highlights the general principle of these five methods, describes experimental details, and specifies advances in the last years.

Key words Hydrodynamic methods, Sedimentation velocity analytical ultracentrifugation, Dynamic light scattering, Sedimentation equilibrium analytical ultracentrifugation, Size exclusion chromatography, Multi-angle light scattering, Capillary viscometry, Intrinsic viscosity measurement

1 Introduction

This chapter will cover not just one method but a collection of five methods which fall under the umbrella of "hydrodynamics," a Greek derived word meaning "water movement":

- Sedimentation velocity analytical ultracentrifugation (SV AUC).
- Sedimentation equilibrium analytical ultracentrifugation (SE AUC).
- Dynamic Light Scattering (DLS).
- Size exclusion chromatography coupled to multi-angle light scattering (SEC-MALS).
- Capillary Viscometry—Intrinsic viscosity measurement.

Bernd Lepenies (ed.), *Carbohydrate-Based Vaccines: Methods and Protocols*, Methods in Molecular Biology, vol. 1331, DOI 10.1007/978-1-4939-2874-3_13, © Springer Science+Business Media New York 2015

Since in many applications a complex carbohydrate (including capsular bacterial polysaccharides or glycoconjugates of proteins with these polysaccharides) will be in solution, all these methods are relevant. Each method has its own particular strengths and limitations but used collectively they present a powerful library of methods.

1.1 Information Being Sought

The types of information that these methods can provide are:

1. Absolute molar mass (molecular weight) information (principally the weight average, but also the number and z-averages).

2. Heterogeneity information (in terms of molar mass or sedimentation coefficient distributions).

3. Interaction information (self-association, reversibility, complex interactions).

4. Conformation and conformational flexibility (in terms of power law parameters or persistence lengths).

For each method we will outline the principles avoiding almost all mathematical detail, give important experimental points, and indicate the limitations. Further information can be found from the key references indicated. Much of the methodology is similar to when the original Methods in Molecular Biology articles were written in 1994 [1–4] and (viscosity) in 1997 [5]. What is greatly different has been the advances in software for analysis rendering the application of hydrodynamic methodologies more powerful and user friendly. As we are essentially covering five techniques in this single chapter, we will cross-refer to those original articles where appropriate.

2 Sedimentation Velocity in the Analytical Ultracentrifuge (SV AUC)

2.1 Principle

The analytical ultracentrifuge is a high speed centrifuge (top speed 60,000 rpm) with an optical system to monitor the change in concentration distribution of a solution of macromolecules with time. Traditionally a sedimenting boundary would be detected and followed by the optical system from which a sedimentation coefficient (rate of movement per unit centrifugal field), s (units seconds s or Svedbergs S, where $1S = 10^{-13}$ s) is determined. Nowadays with the on-line detection system permitting the recording of the change in the whole concentration distribution with time, a distribution of sedimentation coefficients can be recorded.

2.2 Experimental Details

Approximately 400 μl solution is required at a concentration of at least 0.03 mg/ml. For unconjugated polysaccharides the refractometric (Rayleigh interference) optical system is selected. For glycoconjugates the UV/vis absorption optical system can also be selected if the macromolecule has sufficient chromophore content—in the case of a glycoprotein such as a tetanus toxoid conjugate, if there is sufficient tyrosine/tryptophan content, absorption at a wavelength $\lambda = 280$ nm can be used, with regular scans (~100 between the start and end of the experiment). Experiments are conducted at high vacuum and at precisely controlled temperatures (to ±0.1 °C). The Optima XL-I (Beckman-Coulter Instruments, Palo-Alto, CA, USA) allows a range from 4.0 to 40.0 °C.

Much of the experimental detail about cell filling and running the instrument for a sedimentation velocity experiment is similar to when the original Methods in Molecular Biology article was written [1], although the software is now vastly different. Gone are the days of simple boundary analysis. The computer program SEDFIT [6–8] can be used to analyze the change in the whole concentration profiles with time and yields a sedimentation coefficient distribution plot g(s) versus s. An example for a capsular polysaccharide is given in Fig. 1a [9]. The number and shape of the peaks can be

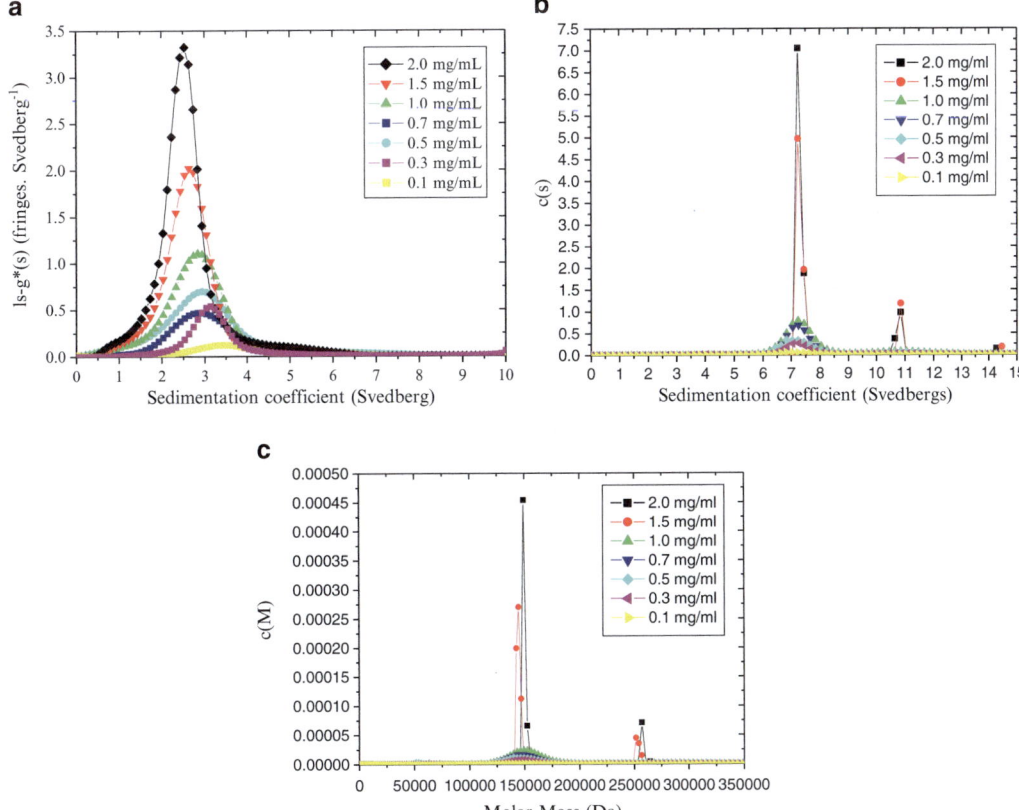

Fig. 1 Sedimentation coefficient distributions (**a**) $g^*(s)$ profile for capsular *Streptococcal* polysaccharides SP (5) at different concentrations. Reproduced from Ref. 9 with permission from Elsevier. (**b**) c(s) vs. s profile for tetanus toxoid showing monomer and ~14 % of dimer at higher s. (**c**) corresponding molar mass distribution plot of c(M) versus M. Reproduced from Ref. 10 with permission from Elsevier

used to provide/quantify the homogeneity or heterogeneity of a formulation. For polysaccharides diffusion is relatively slow but for proteins it is more significant, and for the latter a diffusion corrected sedimentation coefficient distribution plot of $c(s)$ versus s is more appropriate, and an example is given for a tetanus toxoid protein in Fig. 1b [10]. With the latter, clear resolution of a second component—"dimer" is seen. Furthermore the $c(s)$ plot can be converted into a molar mass distribution plot (Fig. 1c) provided all species have the same or very similar conformations [10]. SEDFIT will also provide the weight average sedimentation coefficients s of resolvable components and a good estimate of their relative proportions. If s values are to be interpreted further (e.g., for conformation analyses), then they need to be standardized to the case of the density and viscosity of water $s_{20,w}$. This is done using another useful algorithm SEDNTERP [11] where the user enters the buffer/solvent details and the normalization is done automatically. As polysaccharides and glycoconjugates are usually large and nonideal (arising from large co-exclusion volumes and, for polyelectrolytes, large repulsive interactions), it is necessary to correct for these effects as well. This can be done either working at very low concentrations (where nonideality effects are a minimum) or by measuring $s_{20,w}$ at a series of concentrations and extrapolating $s_{20,w}$ (or better $1/s_{20,w}$) to $c=0$, to yield $s^o_{20,w}$ (Fig. 2), and another useful parameter known as the Gràlen coefficient k_s (ml/g) from the slope.

Both $s^o_{20,w}$ and k_s can be related to conformation as we consider in Subheading 7 below (nb: if k_s is to be used then the concentrations c used in the evaluation have to be true sedimenting concentrations, i.e., corrected for radial dilution effects). For interactions,

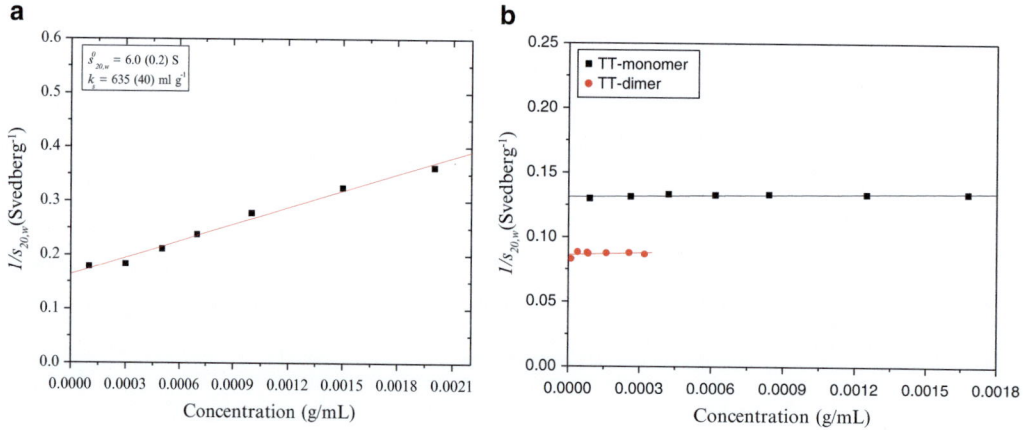

Fig. 2 Example of concentration dependence (reciprocal) of sedimentation coefficient plots (**a**) for capsular polysaccharide SP (1) (**b**) tetanus toxoid protein monomers (*upper*) and dimers (*lower*). Note the higher concentration dependence for the much more nonideal polysaccharide. Reproduced from Refs. 9 and 10 with permission from Elsevier

self-association manifests itself as providing the opposite contribution to nonideality in a sedimentation concentration dependence plot—i.e., an increase in s with c. Complex associations with other molecules can be observed from the appearance of higher s peaks and also co-sedimentation—particles moving with the same sedimenting coefficient (especially if the ligand has a characteristic chromophore that can be detected by the UV/visible absorption optical system) [1].

2.3 Information Provided	Heterogeneity (sedimentation coefficient distribution and, with additional information, molar mass distribution—*see* Subheading 7), conformation information, self-association, and complex formation with ligands.
2.4 Advantage	Highly resolving without the need for a separation matrix
2.5 Limitations	Expensive. Can take several hours, although samples can be run in multiples (up to seven samples simultaneously using a multi-hole rotor).

3 Sedimentation Equilibrium Analytical Ultracentrifugation (SE AUC)

3.1 Principle	At lower rotational speeds compared with sedimentation velocity, centrifugal forces and opposing diffusive forces become comparable and after a period of time (>24 h) come to equilibrium. The final steady state pattern is a function only of molar mass and related parameters, and not on shape effects as there are no friction forces.
3.2 Experimental Details	Approximately 100 μl solution is required at a concentration of at least 0.3 mg/ml, using the same optical system for sedimentation velocity. Prior to filling the ultracentrifuge cells solutions need to be dialyzed against the solvent particularly if the Rayleigh optical system is being used, and dialysate used in the reference channel. As with sedimentation velocity, much of the experimental detail about cell filling and running the instrument for a sedimentation equilibrium experiment is similar to when the original Methods in Molecular Biology article was written [2]. One huge difference is in the software used to analyze the data. The M* function of Creeth and Harding [12] is still central to the analysis but its implementation is now much more powerful and user friendly: the computer program SEDFIT-MSTAR [8] can be used to analyze the steady state profile produced of concentration $c(r)$ as a function of radial displacement (r) from the center of rotation. This profile is transformed into a plot of a function known as $M^*(r)$ versus r. The value of $M^*(r)$ at the base of the cell (the position in the cell furthest from the center of rotation) = $M_{w,app}$, the molar mass (or to be precise, the apparent weight average molar mass over all the macromolecular species in the solution). SEDFIT-MSTAR also

provides a plot of local or point average molar masses $M_{w,app}(r)$ as a function of local concentration $c(r)$ in the cell. At the "hinge point," i.e., the value $c(r)$ which corresponds to the initial loading concentration c^o, $M_{w,app}(r)$ also $= M_{w,app}$ the apparent weight average over the whole cell: this provides an internal check on the estimate from the $M^*(r)$ extrapolation. SEDFIT-MSTAR also provides an estimate for the molecular weight distribution, $c(M)$ versus M. By analogy with the sedimentation coefficient, to correct for nonideality either working at a concentration low enough so that $M_{w,app} \sim M_w$ or if not by measuring at several concentrations and extrapolating $(1/M_{w,app})$ back to $c = 0$ to obtain M_w is required. Unfortunately the minimum concentration required for a sedimentation equilibrium experiment is about 10× higher than that for sedimentation velocity, rendering a concentration extrapolation as more necessary. This is discussed in considerable detail in Schuck et al. [8].

3.3 Information Provided

An absolute (i.e., not requiring calibration standards) estimate for the weight (and z-) average molar mass. Molar mass distribution.

3.4 Advantage

Absolute molar masses without the need for reference standards or a separation matrix.

3.5 Limitations

Expensive. Can take 2–3 days, although samples can be run in multiples (up to seven at a time in a multi-hole rotor). Solutions have to be dialyzed against the solvent before the experiment. The user needs to supply an estimate of a parameter known as the partial specific volume, which can be either calculated from the composition or measured using a density meter [13]. Thermodynamic nonideality needs to be taken into consideration.

4 Dynamic Light Scattering (DLS)

4.1 Principle

If a beam of monochromatic and coherent laser light is incident on a solution of particles, the light will be scattered—the intensity of the light scattered at a given angle from the direction of the incident beam will rapidly fluctuate due to the Brownian motions of the particles [3]. The smaller and more compact particles are, generally the faster they move and the quicker the oscillations occur. A device known as an autocorrelator compares or correlates these intensity fluctuations, and associated computer algorithms evaluate the translational diffusion coefficient D. The diffusion coefficient is usually then converted to an "equivalent hydrodynamic radius" r_h, based on the Stokes relation for a sphere:

$$r_h = kT / \left(6\pi\eta_o \cdot D\right) \tag{1}$$

where η_o is the viscosity of the solvent, T is the absolute temperature, and k is the Boltzmann constant. If the solution is polydisperse, algorithms such as CONTIN [14]—now fully integrated into modern instrument software (such as Malvern Instrument's Zetasizer software; *see* Ref. 15)—will provide a size distribution f(r_h) versus r_h (or equivalently f(D) versus D) and also an estimate for the "Polydispersity factor" PF (normalized z-average variance of the distribution of diffusion coefficients [16]).

One important feature that is sometimes overlooked—but is nonetheless critical for work on polysaccharides and glycoconjugates—is that rotation diffusion effects (tumbling motions of the macromolecules) also contribute to the fluctuations in intensity. For spherical or near-significant particles these contributions will be insignificant irrespective of the scattering angle of the detector, but for nonspherical particles negligence of these effects can lead to error in estimates for D or r_h: since polysaccharides and large glycoconjugates are generally nonspherical, this is important. Burchard [17] described how these effects can be negated by making measurements at a series of angles and extrapolating back to zero angle where these effects vanish. When angular extrapolation is not possible, measurement at a single fixed low angle may suffice.

4.2 Experimental

Standard size cuvettes can be used or injection into an internal scattering cell depending on the design of the instrument. If a study requiring multiple angles is necessary, then cylindrical cuvettes can be used (square ones can still be used so long as the angles at the corners are avoided). Cuvettes have to be scrupulously clean and free of dust—injections of solutions have to be done via filters (conventionally less than 500 nm pore size)—and special cuvettes with no exposure to atmospheric dust can be employed (*see* ref. 4). Modern instrumental software is usually much better at computationally filtering out large particle contributions than at the time the original Methods in Molecular Biology article was written [4]. The diffusion coefficient, like the sedimentation coefficient, will be strongly affected by the temperature of the solution so accurate temperature control is essential (to at least ±0.1 °C). If D values are to be interpreted further (e.g., for conformation analyses), then they need to be standardized to the case of the viscosity of water to yield $D_{20,w}$ using standard formulae. This normalization is usually done automatically by the instrumental software. One difference with the sedimentation coefficient is that nonideality effects are much less significant for $D_{20,w}$ so extrapolations to $c = 0$ are normally not necessary. At higher concentrations, although nonideality is not such an issue, problems through multiple scattering events (particles scattering light already scattered by other particles) can become an issue. When nonideality is an issue, then it is possible to perform the extrapolation to $c = 0$ and, if measurement at a single low angle is not possible, also the angular extrapolation to zero scattering angle, on the same set of axes—called a dynamic Zimm plot [17].

4.3 Information Provided   Heterogeneity (diffusion coefficient or size distribution), conformation information (via the translation frictional ratio).

4.4 Advantage   No need for a separation matrix although it can be coupled to a SEC-MALS system. Relatively rapid measurement (~ few minutes) so can be used to follow changes.

4.5 Limitations   Not as resolving as AUC, although simpler to apply and less expensive. Nonsphericity of most polysaccharides and glycoconjugates means angular extrapolation or low angle measurement is essential. High dependence on sample clarification from supramolecular particles. Experiments on nonclarified material are not useful.

5   Size Exclusion Chromatography Coupled to Multi-angle Light Scattering (SEC-MALS)

5.1 Principle   The very first determination of the molar mass distribution $f(M)$ vs M of a polysaccharide by SEC-MALS (formerly "SEC-MALLS")—size exclusion chromatography coupled to laser light scattering—was reported by John Horton, Steve Harding, and John Mitchell [18] for a sodium alginate. The rapidity and convenience of the method was clearly illustrated and the effects of nonideality—and how to correct for them—were demonstrated. Encouragingly, results for the average weight average molar mass M_w were in agreement with results from the independent method of sedimentation equilibrium in the ultracentrifuge. Since that demonstration nearly a quarter of a century ago, it has become the method of choice for polysaccharide molar mass characterization. The columns not only provided separation of a polydisperse distribution of materials prior to analysis but also acted as online filters helping circumvent the age-old problem of light scattering on solutions of macromolecules, namely the crippling effects of the presence of trace amounts of dust and other supramolecular contaminants. It also replaced the hitherto laborious procedure of taking fractions from preparative columns, measuring their molar masses by either conventional light scattering or sedimentation equilibrium procedures to calibrate the columns before conversion to a distribution. The method could also be applied to glycoconjugates and the first demonstration of its application to mucin glycoproteins (which are over 80 % glycosylated) was by Jumel et al. [19].

5.2 Experimental   The method is an ingenious combination—pioneered by Philip Wyatt and colleagues in the 1980s at Wyatt Technology (Santa Barbara, CA, USA)—of the separating power of SEC coupled online via a flow cell to the absolute molar mass ability of MALS [20]. The MALS works by registering the time averaged intensity of laser light scattered by a solution of molecules as a function of angle: as the solution moves through the flow cell, the angular scattered intensity envelope is instantaneously recorded by a series of photodetectors. The presence online also of a concentration detector (refractive index and/or ultra-violet detector based) means that the apparent molar mass M_{app} can be measured as a function of elution volume V_e. Since the concentration is known also as a function of V_e a molar mass distribution f(M) vs. M can be specified as well as (principally) the (apparent) weight average $M_{w,app}$, together with the number $M_{n,app}$ and z-averages $M_{z,app}$ for the distribution. Because of the sensitivity of the detectors and the low concentrations after dilution from the columns, nonideality can usually be assumed to be negligible and $M_w \sim M_{w,app}$; $M_n \sim M_{n,app}$ and $M_z \sim M_{z,app}$. If the macromolecules have a molar mass generally >150,000 (i.e., if the particles are essentially "Rayleigh Gans Debye scatterers") it is possible from the change in intensity with scattering angle to estimate a conformation parameter known as the radius of gyration, R_g. Below a molar mass of ~150,000 g/mol generally the variation of intensity with angle is too small for accurate measurement.

Although the software is greatly improved, much of the practical requirements in terms of calibration and clarity is the same as in the 1994 Methods in Molecular Biology article [3]:

1. Choice of appropriate combination columns with separation range and including a guard column.

2. Selection of appropriate inert tubing minimizing the dead volumes between detectors.

3. Checking all optical components are clean and free of supramolecular contamination.

4. The requirement of accurate calibration of the system using a small Rayleigh scatterer (e.g., highly purified toluene).

5. Requirement of an accurate value of the refractive index increment dn/dc, and if necessary (higher concentration work) a value for the second thermodynamic virial coefficient B.

5.3 Information Provided

Heterogeneity (molar mass distribution), average molar masses, conformation information (via the radius of gyration).

5.4 Advantage

Gives directly a molar mass distribution without calibration standards or assumptions on conformation. Can be coupled to an online viscometer.

5.5 Limitations

For large molar mass polysaccharides such as xanthan (M_w ~3–4×10^6 g/mol), the columns give poor separation, and for others such as chitosans noninertness or interactions with the columns give anomalous results. The separation problem can be circumvented by the use of field-flow fractionation systems although problems of noninertness through anomalous interactions with the membranes can also lead to erroneous results.

6   Capillary Viscometry: Intrinsic Viscosity Measurement

6.1   Principle

Although a lot simpler—and a loss less expensive—than the other methods considered above, the information content from its careful implementation is still very large. There are three main types of capillary viscometer—the simple Ostwald or U-tube viscometer, involving comparison of times of flow under gravity through a capillary for solutions at various concentrations and solvent. Specially constructed extended Ostwald viscometers—with extra capillary length—give better resolution and allow materials at lower concentrations to be measured (as low as 0.1 mg/ml, with 1–2 ml of sample required). As described in detail in 1997 in the Progress in Biophysics and Molecular Biology article [5], the ratio of flow times (together with a straightforward correction involving the ratio of densities) for solution to solvent gives the relative viscosity η_r and if the concentration c (g/ml) is known, the reduced viscosity $\eta_{red} = (\eta_r - 1)/c$ can be defined. Extrapolation of η_{red} to $c = 0$ gives a parameter—the intrinsic viscosity $[\eta]$—which is an intrinsic function of the size, hydration, and shape characteristics of the molecule. The density correction is not necessary if concentrations at ~0.1 mg/ml or less (e.g., for polysaccharides) can be employed. The 1997 *Progress* article compares the various forms of the extrapolation, and a how a combination of two or three can prove useful.

A newer development—not covered in the 1997 article—is the Rolling Ball viscometer: a steel ball rolls along a capillary at a rate depending on the viscosity of the solution. Flow times can be measured as before for solution and solvent (with density correction as appropriate), defining η_r. The tilt angle of the viscometer can be varied—this means the shear rate can also be varied facilitating the (1) checking for non-Newtonian behavior and (2) extrapolation to zero shear rate if such behavior is present.

Differential pressure viscometry (DPV) has also developed considerably since 1997. Instead of comparing flow times of solution and solvent flowing through a capillary sequentially, differential pressure viscometers use a pressure transducer to measure the pressure difference between solute flowing through a capillary and pure solvent moving through a capillary. This pressure drop can be related via Poiseuille's equation to the relative viscosity and hence the reduced viscosity η_{red} if the concentration is known. The unit can be connected in between the MALS and concentration detector in a SEC-MALS setup enabling the reduced viscosity η_{red} to be measured as a function of elution volume V_e. Because of the high sensitivity of the transducer system, very low concentrations can be measured and hence the approximation $\eta_{red}(V_e) \sim [\eta](V_e)$ can be made. If there is more than one discrete component present $[\eta]$ can then be measured for each component—this gives the methodology a huge advantage compared with conventional capillary or rolling ball viscometers, and is very useful for investigating the presence and amount of unbound protein or polysaccharide in a glycoconjugate: intrinsic viscosities for globular proteins are generally in the range 3–5 ml/g, whereas polysaccharides are usually >20 ml/g.

6.2 Information Provided

Conformation—for proteins capillary viscometry can provide information about the axial dimensions. For polysaccharides and intact glycoconjugates, it can provide information about particle flexibility, especially if combined with molar mass and sedimentation data. For nonglobular particles it is also dependent on molar mass and can be used as an assay for stability.

6.3 Advantage

The intrinsic viscosity is a sensitive function of conformation. Coupling to SEC allows the conformations of mixed systems to be assessed.

6.4 Limitations

Capillaries have to be scrupulously clean. Besides nonideality, non-Newtonian behavior can cause problems, as well as molecular overlap at higher concentrations, so all measurements need to be in the dilute regime.

7 Conformation Assessment

1. *Wales van-Holde ratio.* The ratio of the Grálen concentration dependence of the sedimentation coefficient k_s to the intrinsic viscosity $[\eta]$ (*see* ref. 21) is perhaps the simplest guidance of a polysaccharide/glycoconjugate's conformational flexibility. The limits appear to be ~1.6 for a compact sphere or a nondraining random coil, and ~0.1 for a stiff rod [22, 23]. In the evaluation of k_s (plot of $1/s_{20,w}$ *vs* c), the concentrations c need to be corrected for radial dilution effects in the centrifuge cell. The aqueous solvent needs to be of sufficient ionic strength to suppress polyelectrolyte effects. Examples for a variety of *Streptococcus pneumonia* polysaccharides are given in Harding et al. [9].

2. *Power law relations.* It is possible to take advantage of the particular feature of the SEC-MALS-DPV system in providing an online record of $[\eta](V_e)$ and molar mass $M(V_e)$ as a function of elution volume V_e. The relation is the Mark-Houwink-Kuhn-Sakurada (MHKS) relation $[\eta] \sim M^a$, where the MHKS

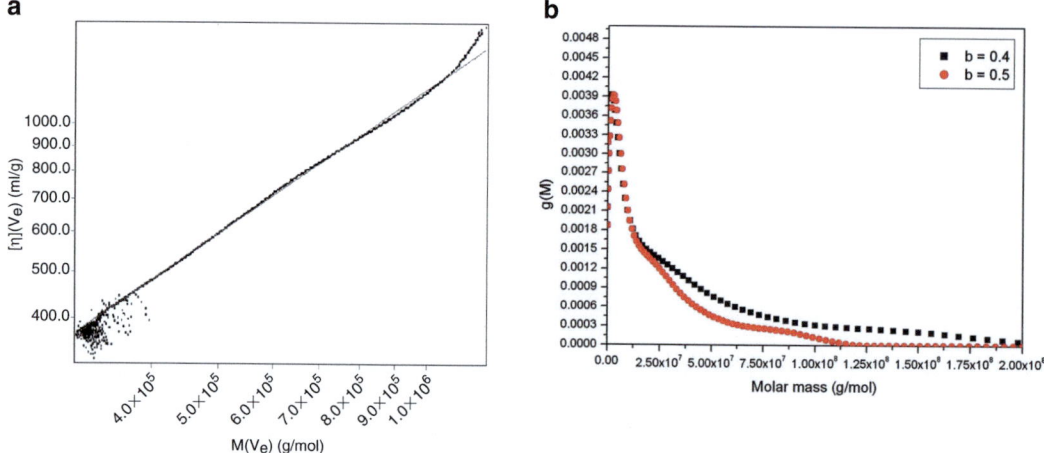

Fig. 3 Use of power law plots. (**a**) Example of a Mark-Houwink-Kuhn-Sakurada (MHKS) plot for capsular poly-saccharide SP (1) from the differential pressure viscometer coupled online to a size exclusion chromatography column and a multi-angle laser light scattering detector. The plot is of intrinsic viscosity [η] as a function of elution volume (ordinate) versus molar mass M as a function of elution volume. The *red line* represents the fit. Reproduced from Ref. 9 with permission from Elsevier. (**b**) Example of how the sedimentation power law equation can be used to generate a molecular weight distribution f(M) versus M for a large glycoconjugate from a sedimentation coefficient distribution using the Extended Fujita approach [24]. The plot shows the distribution for two different values of the power law coefficient b

exponent a has limits of 0 for a sphere, 1.8 for a rod, and 0.5–0.8 for a random coil. a can be found from a double logarithmic plot—and example for a *Streptococcal* capsular polysaccharide is given in Fig. 3a [9]. Power law relations also exist for other hydrodynamic coefficients: $s^o_{20,w} \sim M^b$, $D^o_{20,w} \sim M^{1-b}$. The sedimentation power law relation can also be put to good use for the evaluation of molar mass distributions, particularly those for large glycoconjugates whose sizes are beyond the resolution of SEC-MALS and too high for sedimentation equilibrium: this is the basis of the *Extended Fujita* method developed by Harding et al. [24] and has been applied to the determination of the molar mass distribution of a very large glycoconjugate (Fig. 3b).

3. *Persistence length L_p*. For a more quantitative estimate of chain flexibility, we can use the persistence length L_p, which has theoretical limits of 0 for a random coil and ∞ for a stiff rod. Practically the limits are ~1–2 nm for a random coil and ~200–300 nm for a very stiff rod shaped macromolecule. Persistence lengths L_p can be estimated using several different approaches using either intrinsic viscosity [5, 25, 26] or

sedimentation coefficient [27] measurements. For example the relation [25, 26]:

$$\left(\frac{M_w^2}{[\eta]}\right)^{1/3} = A_0 M_L \Phi^{-1/3} + B_0 \Phi^{-1/3} \left(\frac{2L_p}{M_L}\right)^{-1/2} M_w^{1/2} \qquad (2)$$

where Φ is the Flory-Fox coefficient (2.86×10^{23} mol^{-1}) and A_0 and B_0 are tabulated coefficients, and the Yamakawa-Fujii equation [27]:

$$s^0 = \frac{M_L\left(1-\bar{v}\rho_0\right)}{3\pi\eta_0 N_A} \times \left[1.843\left(\frac{M_w}{2M_L L_p}\right)^{1/2} + A_2 + A_3\left(\frac{M_w}{2M_L L_p}\right)^{-1/2} + \ldots\right] \qquad (3)$$

Yamakawa and Fujii showed that A_2 can be considered as $-\ln(d/2L_p)$ and $A_3 = 0.1382$ if the L_p is much larger than the chain diameter, d. Difficulties arise if the mass per unit length is not known, although both relations have now been built into an algorithm Multi-HYDFIT [28] which estimates the best range of values of L_p and M_L based on minimization of a target function Δ. An estimate for the chain diameter d is also required but extensive simulations have shown that the results returned for L_p are relatively insensitive to the value chosen for d (see, e.g., [29–32]). An example for a *Streptococcal* capsular polysaccharide is given in Fig. 4 yielding a value for L_p of (6.2 ± 0.6) nm—a semi-flexible structure [9].

4. *Sedimentation Conformation Zoning*
   A check for consistency of the above results can be obtained from a sedimentation conformation zone plot of $k_s M_L$ versus $[s]/M_L$ [33], where

$$[s] = \frac{s^0_{20,w}\eta_{20,w}}{\left(1-\bar{v}\rho_{20,w}\right)} \qquad (4)$$

The "zones" are based on a series of polysaccharides of known $[s]$. Figure 5 shows the results for capsular polysaccharides from *Streptococcus*—all those studied are clearly semi-flexible chains.

5. *Simple shape modeling of conjugation proteins.* The intrinsic viscosity and sedimentation coefficient can also give valuable shape information on the proteins used for the conjugation processes in terms of the axial ratio of the hydrodynamically equivalent ellipsoid. The software for this simple modeling—the "ELLIPS" suite of programs—is now very simple to use [34] and Fig. 6 shows such a representation for the tetanus toxoid protein. The representation looks remarkably like the "guessed" cartoon representation given earlier by Astronomo and Burton [35].

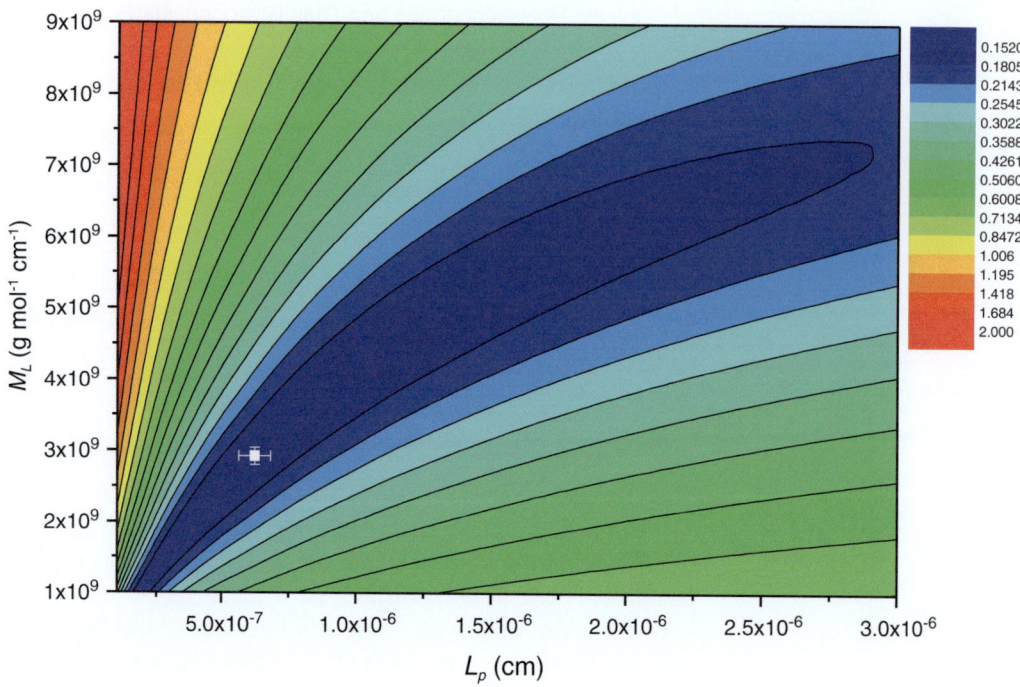

Fig. 4 Plot of persistence length L_p versus mass per unit length M_L for capsular polysaccharide SP (4). The plot yields $L_p \sim 6.2$ (nm) and $M_L \sim 2.92 \times 10^9$ (g.mol^{-1}.cm^{-1}) at the minimum target (error) function value of 0.15. Reproduced from Ref. 9 with permission from Elsevier

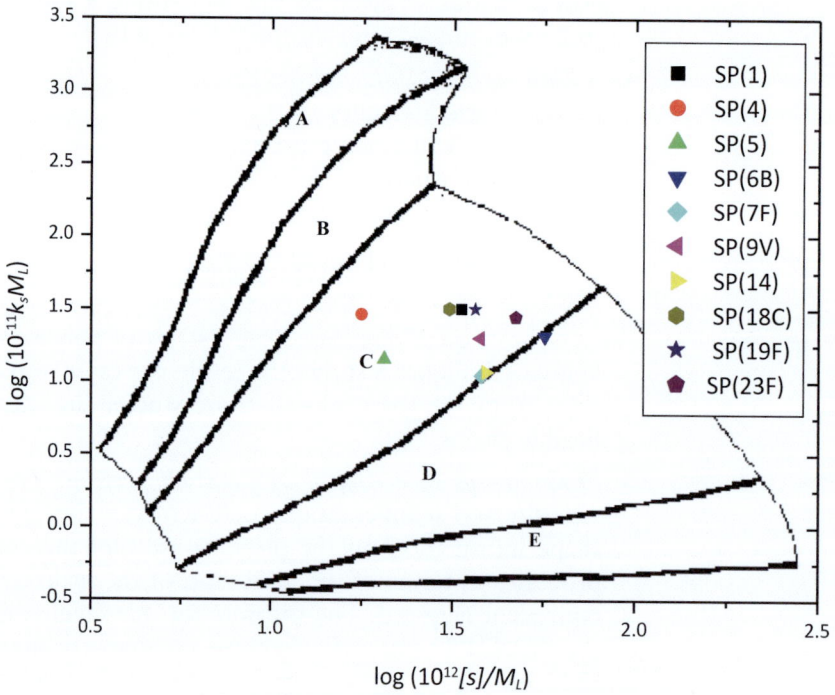

Fig. 5 Conformation zoning plot. k_s is the concentration dependence sedimentation coefficient (ml/g), M_L is the mass per unit length, and [s] is the intrinsic sedimentation coefficient. All ten *Streptococcal* capsular polysaccharides have conformations in Zone C (semi-flexible coil) or near the boundary with Zone D (highly flexible). Reproduced from Ref. 9 with permission from Elsevier

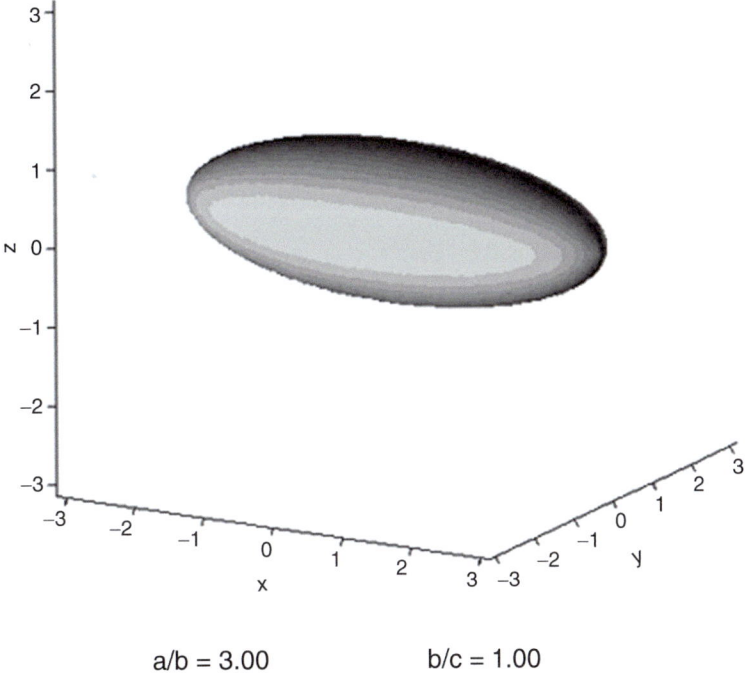

a/b = 3.00   b/c = 1.00

Fig. 6 Prolate ellipsoid representation for monomeric tetanus toxoid protein showing an asymmetric structure of axial ratio ~3. Reproduced from Ref. 10 with permission from Elsevier

8   Concluding Remarks

For a clear example or "case study" of how these methods can be applied to capsular polysaccharides and glycoconjugates, the reader is referred to a recent paper by Harding et al. [9] on the solution properties of capsular polysaccharides from *Streptococcus pneumonia*. For an example/case study of how the methods can be used to characterize the protein to which the polysaccharides are conjugated (such as the tetanus toxoid protein), the readers are referred to a companion paper by Abdelhameed et al. [10].

References

1. Harding SE (1994) Determination of macromolecular heterogeneity, shape and interactions using sedimentation velocity analytical ultracentrifugation. Methods Mol Biol 22:61–73, Jones, C, Mulloy B (eds)

2. Harding SE (1994) Determination of absolute molecular weights using sedimentation equilibrium analytical ultracentrifugation. Methods Mol Biol 22:75–84, Jones, C, Mulloy B (eds)

3. Harding SE (1994) Classical light scattering for the determination of absolute molecular weights and gross conformation of biological macromolecules. Methods Mol Biol 22:85–95, Jones, C, Mulloy B (eds)

4. Harding SE (1994) Determination of diffusion coefficients of macromolecules by Dynamic Light Scattering (DLS). Methods Mol Biol 22:97–108, Jones, C, Mulloy B (eds)

5. Harding SE (1997) The intrinsic viscosity of biological macromolecules. Progress in measurement, interpretation and application to structure in dilute solution. Prog Biophys Mol Biol 68:207–262

6. Dam J, Schuck P (2004) Calculating sedimentation coefficient distributions by direct modeling of sedimentation velocity concentration profiles. Methods Enzymol 384:185–212, Michael LJ, Ludwig B (eds); Academic Press

7. Harding SE (2005) Analysis of polysaccharide size, shape and interactions. In: Scott DJ, Harding SE, Rowe AJ (eds) Analytical ultracentrifugation: techniques and methods. The Royal Society of Chemistry, Cambridge, UK, pp 231–252

8. Schuck PJ, Gillis RB, Besong D et al (2014) SEDFIT-MSTAR: molecular weight and molecular weight distribution analysis of polymers by sedimentation equilibrium in the ultracentrifuge. Analyst 139:79–92

9. Harding SE, Abdelhameed AS, Morris GA et al (2012) Solution properties of capsular polysaccharides from Streptococcus pneumoniae. Carbohydr Polym 90:237–242

10. Abdelhameed AS, Morris GA, Adams GG et al (2012) An asymmetric and slightly dimerized structure for the tetanus toxoid protein used in glycoconjugate vaccines. Carbohydr Polym 90:1831–1835

11. Laue TM, Shah BD, Ridgeway TM et al (1992) Computer aided interpretation of analytical sedimentation data for proteins. In: Harding SE, Rowe AJ, Horton JC (eds) Analytical ultracentrifugation in biochemistry and polymer science. The Royal Society of Chemistry, Cambridge UK, pp 90–125

12. Creeth JM, Harding SE (1982) Some observations on a new type of point average molecular weight. J Biochem Biophys Methods 7:25–34

13. Harding SE, Vårum KM, Stokke BT et al (1991) Molecular weight determination of polysaccharides. In: White C (ed) Advances in carbohydrate analysis. JAI Press, Greenwich CT, USA, pp 63–114

14. Provencher SW (1992) Low-bias macroscopic analysis of polydispersity. In: Harding SE, Sattelle DB, Bloomfield VA (eds) Laser light scattering in biochemistry. Royal Society of Chemistry, Cambridge, UK, pp 92–111

15. Nobbmann U, Connah M, Fish B et al (2007) Dynamic light scattering as a relative tool for assessing the molecular integrity and stability of monoclonal antibodies. Biotechnol Genet Eng Rev 24:117–128

16. Pusey PN (1974) Macromolecular diffusion. In: Cummins HZ, Pike ER (eds) Photon correlation and light beating spectroscopy. Plenum, New York, NY, pp 387–429

17. Burchard W (1992) Static and dynamic light scattering approaches to structure determination of biopolymers. In: Harding SE, Sattelle DB Bloomfield VA (eds) Laser light scattering in biochemistry. Royal Society of Chemistry, Cambridge, pp 3–22

18. Horton JC, Harding SE, Mitchell JR (1991) Gel permeation chromatography – multi angle laser light scattering characterization of the molecular mass distribution of "Pronova" sodium alginate. Biochem Soc Trans 19: 510–511

19. Jumel K, Fiebrig I, Harding SE (1996) Rapid size distribution and purity analysis of gastric mucus glycoproteins by size exclusion chromatography/multi angle laser light scattering. Int J Biol Macromol 18:133–139

20. Wyatt PJ (2012) Multiangle light scattering from separated samples (MALs with SEC or FFF). In: Roberts G (ed) Encyclopedia of biophysics. Springer, Berlin

21. Wales M, van Holde KE (1954) The concentration dependence of the sedimentation constants of flexible macromolecules. J Polymer Sci 14:81–86

22. Creeth JM, Knight CG (1965) On the estimation of the shape of macromolecules from sedimentation and viscosity measurements. Biochim Biophys Acta 102:549–558

23. Creeth JM, Knight CG (1967) The macromolecular properties of blood-group substances. Sedimentation-velocity and viscosity measurements. Biochem J 105:1135–1145

24. Harding SE, Schuck P, Abdelhammed AS et al (2011) Extended Fujita approach to the molecular weight distribution of polysaccharides and other polymer systems. Methods 54:136–144

25. Bushin S, Tsvetkov V, Lysenko E et al (1981) The sedimentation-diffusion and viscometric analysis of the conformation properties and molecular rigidity of ladder-like polyphenyl siloxane in solution. Vysokomol Soedin 23A: 2494–2503

26. Bohdanecky M (1983) New method for estimating the parameters of the Wormlike Chain Model from the intrinsic viscosity of stiff-chain polymers. Macromolecules 16: 1483–1492

27. Yamakawa H, Fujii M (1973) Translational friction coefficient of wormlike chains. Macromolecules 6:407–415

28. Ortega A, Garcia de la Torre J (2007) Equivalent radii and ratios of radii from solution properties as indicators of macro-

molecular conformation, shape, and flexibility. Biomacromolecules 8:2464–2475

29. Morris GA, Patel TR, Picout DR et al (2008) Global hydrodynamic analysis of the molecular flexibility of galactomannans. Carbohydr Polym 72:356–360

30. Morris GA, García de la Torre J, Ortega A et al (2008) Molecular flexibility of citrus pectins by combined sedimentation and viscosity analysis. Food Hydrocoll 22:1435–1442

31. Kök MS, Abdelhameed AS, Ang S et al (2009) A novel global hydrodynamic analysis of the molecular flexibility of the dietary fibre polysaccharide konjac glucomannan. Food Hydrocoll 23:1910–1917

32. Patel TR, Morris GA, Ebringerová A et al (2008) Global conformation analysis of irradiated xyloglucans. Carbohydr Polym 74: 845–851

33. Pavlov GM, Rowe AJ, Harding SE (1997) Conformation zoning of large molecules using the analytical ultracentrifuge. Trends Anal Chem 16:401–405

34. Harding SE, Cölfen H, Aziz Z (2005) The ELLIPS suite of whole-body protein conformation algorithms for Microsoft Windows. In: Scott DJ, Harding SE, Rowe AJ (eds) Analytical ultracentrifugation. Techniques and methods. Royal Society of Chemistry, Cambridge, UK, pp 460–483

35. Astronomo RD, Burton DR (2010) Carbohydrate vaccines: developing sweet solutions to sticky situations? Nat Rev Drug Discov 9:308–324

Chapter 14

Glycoconjugate Vaccines: The Regulatory Framework

Christopher Jones

Abstract

Most vaccines, including the currently available glycoconjugate vaccines, are administered to healthy infants, to prevent future disease. The safety of a prospective vaccine is a key prerequisite for approval. Undesired side effects would not only have the potential to damage the individual infant but also lead to a loss of confidence in the respective vaccine—or vaccines in general—on a population level. Thus, regulatory requirements, particularly with regard to safety, are extremely rigorous. This chapter highlights regulatory aspects on carbohydrate-based vaccines with an emphasis on analytical approaches to ensure the consistent quality of successive manufacturing lots.

Key words Regulatory requirements, Quality control, World Health Organization, Pharmacopeias, Batch release

Abbreviations

(c)GMP	(current) Good manufacturing practice
CPMP	Committee for Proprietary Medicinal Products
CPS	Capsular polysaccharide
EMA	European Medicines Agency
EP	European Pharmacopoeia
EPI	Expanded Programme on Immunization
FDA	Food and Drug Administration
Hib	*Haemophilus influenzae* type b
HPAEC	High performance anion exchange chromatography
HPLC	High performance liquid chromatography
HPSEC	High performance size exclusion chromatography
ICH	International Conference on Harmonisation of Technical Requirements for Registration of Pharmaceuticals for Human Use
IS	International Standards

The content of this chapter represents the author's best understanding of current regulatory requirements. It is a personal opinion and is not a statement of NIBSC or MHRA policy.

Bernd Lepenies (ed.), *Carbohydrate-Based Vaccines: Methods and Protocols*, Methods in Molecular Biology, vol. 1331, DOI 10.1007/978-1-4939-2874-3_14, © Springer Science+Business Media New York 2015

LAL   *Limulus* amoebocyte lysate
MALLS   Multiple angle laser light scattering
Men   Meningococcal
NMR   Nuclear magnetic resonance
NRA   National Regulatory Authority
NRL   National Regulatory Laboratory
OCABR   Official Control Authority Batch Release
Pn   Pneumococcal
s.h.d.   Single human dose
SI   Système International d'Unités
TRS   WHO Technical report series
USP   United States Pharmacopeia
WHO   World Health Organization

1 Introduction

Most vaccines, including the currently available glycoconjugate vaccines, are administered to healthy infants, with the intention to prevent future disease. For that reason regulatory requirements, particularly with regard to safety, are extremely rigorous. Beyond the potential to damage the individual infant, the consequences at a population level of a loss of confidence in the safety of a vaccine, or vaccines in general, will lead to a reduction in uptake and to increased disease incidence. In addition, a reduction in the number of individuals immunized will lead to declining "herd protection" which protects those not able to be immunized, with increases in infection rates.

The regulatory requirements for licensed vaccine products can be considered in three parts. Firstly, there is the need to ensure a safe, reliable, and reproducible manufacturing process delivering a consistent and sterile product. This includes the use of characterized cell banks, qualification of starting materials, such as the media upon which bacteria are grown, equipment integrity, equipment and column matrix cleaning and sterilization, and ensuring final product sterility. There will be a wide range of monitoring equipment and in-process controls in place to ensure consistent performance. These requirements are general across a wide range of products and are codified in, for example, ICH Guidelines, WHO documents, pharmacopeial chapters, and Good Manufacturing Practice (GMP) requirements.

Secondly, there is the need to prove to national regulatory authorities (NRAs) through fully documented characterization and nonclinical (previously called preclinical) analyses, in vivo animal studies, and clinical trials that the proposed product is both safe and effective. There are a range of requirements which control the manner in which these tests are carried out—from the need to fully validate critical analytical methods though ethical considerations in the performance of clinical trials in infants to statistical evaluation of results—which are increasingly being harmonized between different

NRAs. This will include establishing the stability of the vaccine, storage and distribution requirements, and shelf life. There is also a need to define an immunization schedule for infants which is both appropriate for the age-related incidence of infection and compatible with the delivery of complementary vaccines. Only then will permission to market a product be granted (the license or marketing authorization). This process also allows implementation of improvements in the manufacturing process or vaccine presentation shown not to be deleterious to safety or efficacy. Many of these requirements are codified in documents from NRAs (the FDA in the USA and the EMA in Europe), in ICH documents [1], and, at a product-specific level, in some of the WHO Guidelines. Most countries have mechanisms in place to identify rare unwelcome side effects or to recall and investigate potentially problematic batches of vaccine.

Finally, there is a requirement on the manufacturer to demonstrate that successive manufacturing batches meet the requirements laid down in pharmacopeias and the marketing authorization (called "lot release" or "batch release"). The proposed testing regime and specifications will form part of the documentation supplied to the NRA. The expectation is that successive batches will be consistent with those shown to be effective in clinical trials. This is demonstrated by a combination of validation of the manufacturing process and by analytical testing of critical quality parameters in the product and key intermediates. Some of these lot release tests will be specific to the vaccine (such as potency or identity), whilst other such as pyrogenicity (or endotoxin) testing or adjuvant analysis are common across many or most vaccines. These aspects are covered in (confidential) licensing documents, public pharmacopeial monographs where available, and are commonly mentioned in the peer-reviewed literature. Regulatory agencies have mechanisms, usually for a fee, to provide potential manufacturers with advice about the requirements for licensing a product, including the lot release testing and clinical trials requirements, prior to submission of the formal application.

Within Europe, the USA, and many other countries, the use of vaccines in publicly sponsored mass pediatric immunization campaigns, the need to maintain public confidence, and the intense focus on the safety of these products have resulted in a general requirement for independent testing of critical quality attributes of these materials in government laboratories [2]. Testing in these laboratories, the FDA Center for Biologics Evaluation and Research (CBER) and the national Official Medicines Control Laboratories (OMCLs) in Europe, is a requirement for every manufacturing batch of product to be used in these regions. That batch is only allowed on the market after successful analysis. European expectations for the tests to be carried out by OMCLs are codified in the Official Control Authority Batch Release OCABR documents [3]. The scientific personnel in these laboratories are also the critical sources of expertise informing the licensing process and contributing to the development of WHO, pharmacopeial, and other international guidance documents.

Within the context of this chapter, the emphasis will be on the third of these requirements: the analytical approaches to ensure the consistent quality of successive manufacturing lots. Glycoconjugate vaccines can be considered a "product class," a group of products where the critical quality attributes and key analytical methods are common for a number of different products. This approach, developed by the United States Pharmacopeia (USP), considerably simplifies discussion and allows one to concentrate on what is important. However, differences in manufacturing processes for individual products may make some tests critical and render others less important. The manufacturing process for a conjugate vaccine and key points for testing are shown diagrammatically in Fig. 1.

For the purposes of this chapter, "vaccine" refers to the final product as delivered to the patient, consisting of the active glycoconjugates, excipients and adjuvants, preservatives if present, and contaminants, all sterile-filled into a suitable container. Table 1 contains a list of currently available and "expected soon" glycoconjugate vaccines. *Haemophilus influenzae* type b (Hib) conjugates are now mostly used as components of complex combination vaccines based on diphtheria and tetanus toxoids, with acellular or whole-cell pertussis and, often, either hepatitis b and/or inactivated polio

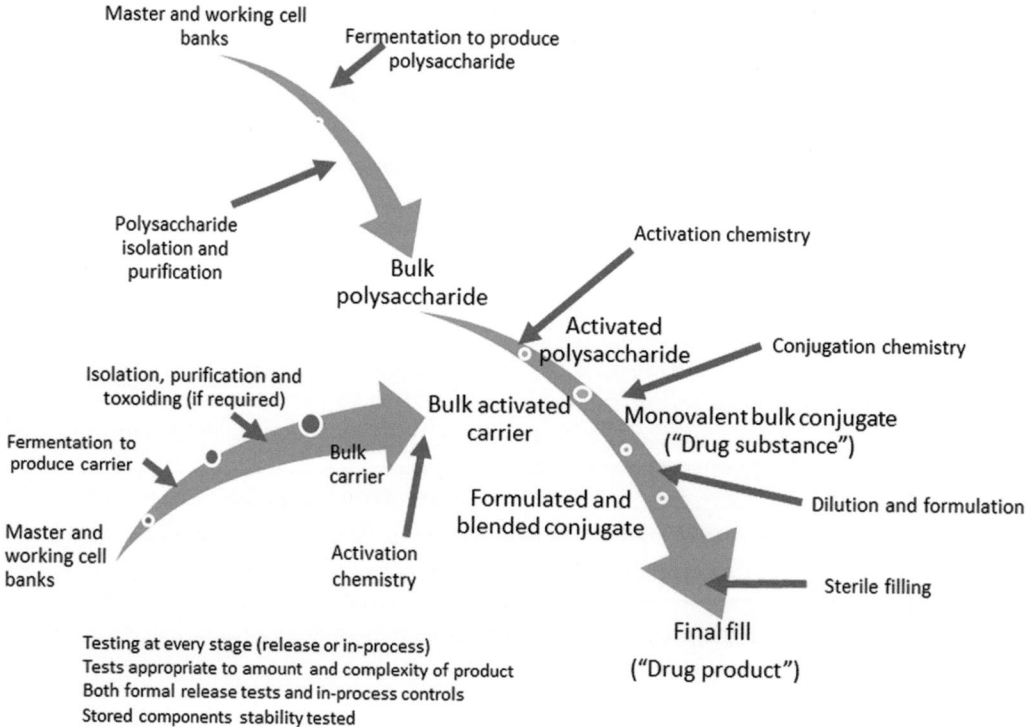

Fig. 1 Schematic showing the basic process for the production of a glycoconjugate vaccine, and the various stages at which validation or quality control procedures are appropriate. These include starting material, process materials, intermediates, and the final product

Table 1
Currently available and expected polysaccharide and glycoconjugate vaccines

Target organism	Polysaccharide or conjugate?	Components	Manufacturers
Haemophilus influenzae Type b	Conjugate	Frequently used a component of complex combination vaccines based on DTaP or DTwP, with IPV and/or Hep b components	Multiple
Neisseria meningitidis	Polysaccharide	Tetravalent polysaccharide (A, C, W135 and Y)	Multiple
	Conjugate	Monovalent Group A conjugates (for sub-Saharan Africa)	Serum Institute of India
		Monovalent Group C conjugates	Multiple manufacturers
		Tetravalent conjugate vaccines	Several, and others developing products
		Pentavalent conjugate vaccines (with X)	In development
Streptococcus pneumoniae	Polysaccharide	23-valent polysaccharide	One currently, but developing country manufacturers with product pipelines
	Conjugate	7-, 10 and 13-valent conjugates	Two manufacturers: others developing products
Salmonella enterica serovar Typhi	Polysaccharide	Monovalent Vi CPS Used with Hep A in a travelers vaccine	Many companies, especially in developing countries
	Conjugate	Several Vi conjugates currently being licensed	Bharat Biotech, but several other vaccines in development
Group B Streptococcus	Conjugate	Trivalent conjugate using CRM197 carrier	Phase III clinical trials planned to start Q1 2015. Maternal immunization required

DTaP diphtheria, tetanus, and acellular pertussis; *DTwP* diphtheria, tetanus, and whole-cell pertussis; *IPV* inactivated polio virus; *Hep A, Hep b* hepatitis a and b

virus immunogens. Discussion of the quality control of these combination vaccines is beyond the scope of this chapter. Pneumococcal glycoconjugate vaccines contain between 7 and (currently) 13 serotypes, meningococcal vaccines contain four serogroups,[1] and the current Vi conjugate is monovalent, although bivalent typhoid Vi/paratyphoid A conjugates are planned [4].

---

[1] Whilst pneumococcal strains expressing structurally distinct capsular polysaccharides are called serotypes, meningococcal strains expressing different capsules are called serogroups. Within the context of this chapter, "serotype" also refers to meningococcal serogroups.

2   Source Documents for Regulatory Requirements

These are of two types: those that are largely applicable to all products (whether pharmaceutical, biopharmaceutical, or vaccines) and which tend to cover manufacturing and licensing processes, and those that are specifically relevant to polysaccharide and glycoconjugate vaccines, and which are largely orientated to lot release procedures and specific clinical trial requirements.

2.1   Generally Applicable Across All Products, But with Focus on Vaccines

1. Good Manufacturing Practice (or current good manufacturing practice, cGMP). Whilst detailed regulations vary between countries, and are normally published on the websites of national regulatory authorities, the basic principles are consistent and set out concisely in the WHO Guidelines [5]. These cover buildings and equipment, personnel and training, record keeping, avoiding contamination, managing production changes and deviations, labeling, and quality control of product batches. There should be a system so that batches subsequently discovered to be unsatisfactory can be recalled. A similar series of requirements (collectively called GxP) refer to other aspects of the development, clinical trialing, licensing, and distribution also exist. All of these documents are subject to regular revision.

2. The *International Conference on Harmonisation of Technical Requirements for Registration of Pharmaceuticals for Human Use* (ICH) collates and formalizes input from regulators and industry in Europe, America, and Japan, to produce a series of high level requirements. These documents are freely available through the ICH website [1] and are divided into those dealing with Quality, Safety, Efficacy, and Multidisciplinary topics: the Q-, S-. E-, and M-series of documents. Not all are relevant for biological products or, more specifically, for vaccines. Within the Q series, Q5 and Q6B are specifically about biological products. ICH Q2 (R1) guideline provides high level requirements for the validation of critical analytical methods used in the pharmaceutical and biopharmaceutical industry. The E series of documents support harmonization in the design and analysis of clinical trials and in the reporting of adverse drug reactions. ICH act as custodians of the format for Modules 2–5 of the Common Technical Document (CTD) [6] which is a standardized format agreed between the major regulatory regions to assemble all the Quality, Safety, and Efficacy information required for drug registration.

3. The World Health Organization (WHO) publish a number of non-product-specific guidance documents, principally to

support NRAs in less well-developed countries relating to nonclinical and clinical evaluation of vaccines, stability assessment, and lot release. In doing so, they consolidate best practice in the design of these studies, explain to manufacturers the regulatory expectations, and support NRAs in the assessment of data.

4. The US Food and Drug Administration (FDA) requirements for all types of pharmaceutical products are published in Title 21 of the Code of Federal Regulations (21 CFR). The 600 series covers biological products, including vaccines. Specifically, 21 CFR 601 defines the licensing process for biologics and 21 CFR 606 describes cGMP in a US context. Some additional documents specifically relating to vaccine regulation are available through the FDA website [8].

5. In the European Union, biotechnology products, which would normally include glycoconjugate vaccines, are licensed at a European level, through the European Medicines Agency (EMA) through the Committee for Proprietary Medicinal Products (CPMP), rather than by individual nation states, although the procedures calls upon experts from individual countries to assess the file. The EMA website provides information on the process and access to scientific guidelines [9], although there is little information specifically relevant to glycoconjugate vaccines.

2.2 Glycoconjugate Vaccine-Specific Guidance

1. WHO develop product-specific Guidance documents for the different polysaccharide and glycoconjugate vaccines. These are published in the Technical Reports Series (TRS) and online, and cover all aspects of the production and quality control of these vaccines and, increasingly, guidance on the performance of clinical trials and analysis of the data [7]. These documents are drafted by committees of experts from a range of countries and manufacturers and are subject to a process of public comment before adoption. Table 2 contains a list of WHO Guidelines on the production and quality control of various glycoconjugate vaccines [10–19]. The process can take several years between the recognition of the need for a document and its final release. Compliance with these WHO Guidelines underpins the process of WHO prequalification, a combination of inspections and audits, which are a requirement for purchase of vaccines by WHO-associated bodies, such as the GAVI Alliance (formerly the "Global Alliance for Vaccines and Immunisation"), for use in mass vaccination campaigns in low income countries, the Expanded Programme on Immunization (EPI).

Table 2
WHO guidance for the production and quality control of polysaccharide and glycoconjugate vaccines

Vaccine	Date	Reference
Recommendations for the production and control of *Haemophilus influenzae type* b conjugate vaccines	2000	[10]
Recommendations to assure the quality, safety, and efficacy of group A meningococcal conjugate vaccines	2006	[11]
Recommendations for the production and control of meningococcal group C conjugate vaccines	2004	[12, 13]
Part C. Clinical evaluation of group C meningococcal conjugate vaccines (Revised 2007)	2007	[14]
Requirements for meningococcal polysaccharide vaccines (Revision 1999)	1999	[15]
Recommendations to assure the quality, safety, and efficacy of pneumococcal conjugate vaccines	2009	[16–18]
WHO/Health Canada Consultation on Serological Criteria for Evaluation and Licensing of New Pneumococcal Vaccines	2008	[19]
WHO Workshop on Standardization of Pneumococcal Opsonophagocytic Assay	2007	[20]
Requirements for Vi polysaccharide typhoid vaccine	1994	[21]
Guidelines on the quality, safety, and efficacy of typhoid conjugate vaccines	2013	[22]

Whilst these guidance documents are for existing products, the underlying principles are consistent and likely to be relevant for future polysaccharide and glycoconjugate vaccines, and hence they provide an excellent starting point when defining expectations for other products

2. Pharmacopeial monographs and chapters.[2] The European Pharmacopoeia (EP) has monographs for the currently licensed polysaccharide and glycoconjugate vaccines, e.g., [23], as well as combination vaccines which include, usually, Hib conjugates. Whilst these monographs have a common format and comparable requirements, they are functional and have little explanation. They largely cover only the analytical testing required for lot release of these vaccines. Specific analytical chapters also exist, for the analysis of the vaccines, components of the vaccine, excipients and adjuvants, and

---

[2] In the USP, pharmacopeial monographs are product-specific and contain specifications for key aspects of product quality which must be met. Chapters either provide more general information or describe analytical methods with system suitability and assay acceptance criteria which can be used for a range of products. Both monographs and analytical chapters are likely to be linked to reference standards. The European Pharmacopoeia makes a similar distinction.

impurities. The United States Pharmacopeia (USP) does not currently have monographs for vaccines, but two general chapters, <1235> *Vaccine for Human Use—General Consideration* [24] and <1238> *Vaccines for Human Use—Bacterial Vaccines* [25] and a "product class" chapter <1234> *Vaccines for Human Use—Polysaccharide and Glycoconjugate Vaccines—General considerations* [26], which became official in 2014. These are discursive and cover "background" requirements such as setting up and maintaining cell banks. USP is currently drafting detailed validated analytical methods to support <1234> *Vaccines for Human Use—Polysaccharide and Glycoconjugate Vaccines—General considerations.* Chapters under development include NMR identity testing of polysaccharides used in vaccine, polysaccharide quantification, and molecular size measurement. The proposed relationship between these chapters and relevant reference standards is shown in Fig. 2.

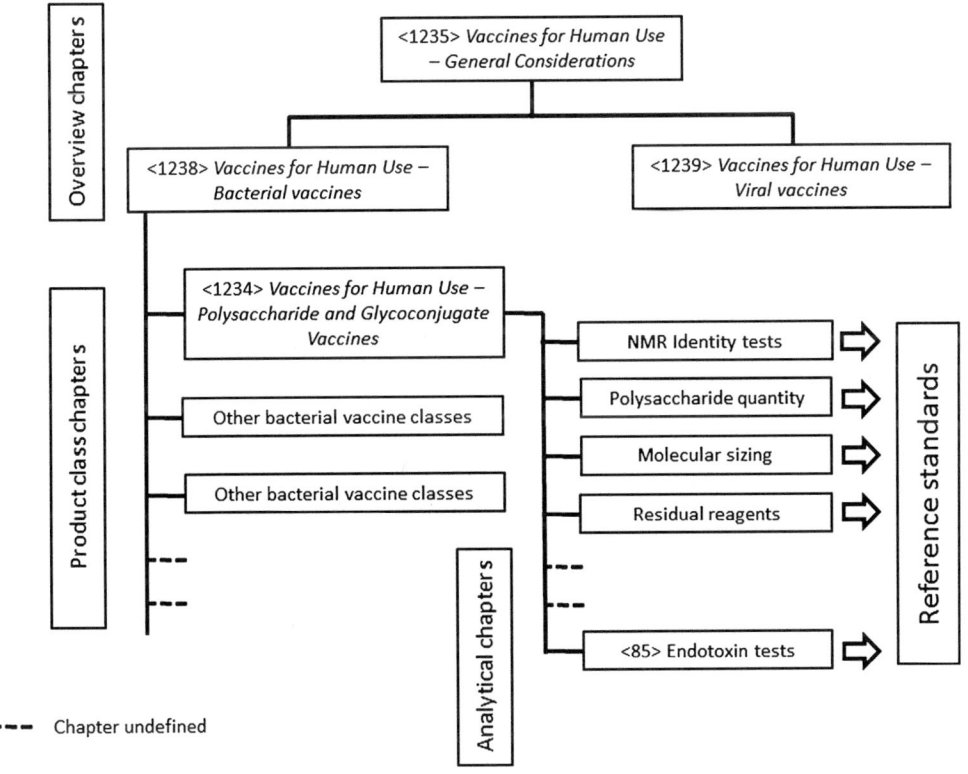

©2013 The United States Pharmacopeial Convention. Used with Permission.

Fig. 2 Overview of the expected development of chapters within the United States Pharmacopeia for guidance on the quality expectations for polysaccharide and glycoconjugate vaccines. This shows the relationship between overview chapters, a product class chapter, specific analytical chapters, and reference standards. This strategy is still in development and may be modified

3   Controlling the Manufacturing Process

Premises used for the manufacture of pharmaceuticals (including vaccines) must be licensed by an NRA and are subject to regular inspections by NRA of territories into which the vaccine is sold, and to audits to ensure compliance with general regulations and specific requirements. Often there is a mutual recognition system between NRAs for audits. There are a number of expectations for the control of the manufacturing process. Firstly, there is the qualification and testing of raw materials to be used, including well-characterized and validated cell banked materials. This is allied to a general desire to eliminate animal- (and, particularly, bovine-) derived materials from the manufacturing process. Changes in suppliers of critical raw materials should be associated with a validation process to demonstrate that there is no negative impact on the process or product. Secondly, there is validation of the process and assessment of its robustness, to establish process parameters that ensure a consistent product. This is linked to in-process monitoring, to highlight if the manufacturing process is moving outside of the conditions for which it is validated. Expected parameter values will have associated alert and action limits, triggering appropriate responses, and which are tracked to highlight a drift in the process. Thirdly, all processes should be carried out with the cGMP quality framework, which formalizes controls and defines the appropriate record keeping. Finally, all stored intermediates (including cell lines) and final products should be subject to a stability program to establish the timescales over which then can be kept before being used in the next stage of the manufacturing process.

Strict control of environmental conditions, such as filtered air handling and personal protective equipment to prevent contamination of the product, is required. Systems must be in place to document and respond to deviations from process requirements, and to investigate the causes and implications of such deviations. Staff must be appropriately trained and training records and competencies documented.

1. *Cell banks*: Cell lines used in the manufacture of both the polysaccharide and carrier protein should be fully characterized and stored in a validated master cell bank and working cell bank regime. Stability of the cell lines under these storage conditions should be validated, to ensure that cell growth and active component expression are consistent between successive manufacturing runs. The production scheme should be validated, with alert and action limits on factors such as cell growth, and dealt with within the expectations of GMP [27].

2. *Engineering and consistency lots*: Typically, a manufacturer will produce one or more "engineering lots" to confirm the correct

functioning of the equipment used. They will also produce several "consistency lots" to assess batch-to-batch variation prior to large-scale clinical use. These manufacturing scale batches provide the bulk material used in detailed characterization studies. If, after licensing, the manufacturer increasing the manufacturing scale, a similar operation should be considered. Once production is under way, monitoring schemes should be in place to ensure that product quality does not trend towards out-of-specification values.

3. *Third-party components*: In some cases a conjugate vaccine manufacturer may purchase active pharmaceutical ingredients (APIs) as intermediates (typically the bulk carrier protein or bulk polysaccharide) from another vaccine manufacturer or a contract manufacturing organization. In that case the purchaser is responsible for ensuring the quality and safety of that material. Similarly, quality provisions are required if a manufacturer moves material between sites (to a dedicated filling plant, for example), to ensure the integrity of the material.

4   Product Characterization, Clinical Trials, and Product Licensing

To be able to market a product, manufacturers must obtain a license (or marketing authorization) from the NRAs of the countries within which it will be used. To obtain this license manufacturers submit a dossier (typically the Common Technical Document—CTD [6]) which will be assessed by the NRA (or EMA in Europe) and which forms the basis for discussion and debate between the manufacturer and competent authority. Timelines for this process exist, with the opportunity to stop the clock whilst manufacturers respond to questions or obtain further data.

1. *Characterization of starting materials, intermediates and final products, and of their stability.* There is an expectation that cell lines, other starting materials, stored intermediates, product bulks, excipients and adjuvants, and final products will be well characterized in line with current technologies. This information forms a key part of the CTD. Required tests usually include demonstrating the identity of the various materials, identification and quantification of impurities, the molecular size of oligo- or polysaccharides conjugated to the carrier protein, quantification of oligomeric forms of the carrier protein, application of assays which determine, for example, the degree of activation of the polysaccharide and carrier proteins, integrity of the final conjugate, the polysaccharide-protein ratio of the conjugate, and proof of a covalent link between the polysaccharide and carrier protein. Residual reagents or uncapped activation sites in the final vaccine should be quantified.

Characterization studies should be carried out on at least three "consistency lots" to assess product variability: these data are important in ensuring that final product specifications are achievable at manufacturing scale. Full validation is not usually required for characterization assays, although they should be demonstrated to be fit-for-purpose.

2. *Induction of protective antibodies.* These vaccines protect at the individual level because they induce a repertoire of antibodies which support killing of the bacteria by host defence mechanisms [28–30]. A key factor therefore is to quantify the production of antibodies in a suitable animal model and to demonstrate that these antibodies, in the presence of other serum or cellular components, are bacteriocidal. Glycoconjugate vaccines have proven more effective that older polysaccharide vaccines because the immune response elicited is different from that induced by purified polysaccharide, especially in the induction of a T cell-dependent response, with isotype switching, immunological memory, and affinity maturation. An analysis of the pattern of antibodies and related cell-mediated responses is required. This will initially be in small animals, but as development continues larger animals, including primates, should be studied. Data on human responses to these vaccines will be collected from clinical trials.

For many of the glycoconjugate vaccines developed to date, including Hib, meningitis, and typhoid, humans are the only species which can be colonized by the organism and subsequently become infected. It is therefore difficult to undertake protection studies in nonhuman hosts.

3. *Mode of action and reduction of transmission*: The pathogens from which most current glycoconjugate vaccines offer protection are commensals, which harmlessly colonize the nasopharynx of a potential patient until cleared. Whilst colonization is very common, it is a rare event for the bacterium to cross into the bloodstream and establish an infection. In addition to the ability to elicit bacteriocidal anti-capsular polysaccharide antibodies alluded to above, glycoconjugate vaccines protect individuals and populations by two other mechanisms. Firstly, vaccinated individuals are much less likely to be colonized by the vaccine-targeted organism than the unvaccinated [31], which reduces the opportunity for an infection event to take place. Secondly, as a consequence of reduced colonization, transmission to another individual, vaccinated or not, is much reduced. Thus with high levels of immunization coverage even unvaccinated individuals have reduced incidence of infection [32], a phenomenon known as herd protection. Practically, the major sources of transmission for Hib and pneumococcal disease are the young, who are the first group

targeted for vaccination, but which results in much reduced disease incidence in the parent and grandparent generations [33]. Understanding the ability of a new vaccine to reduce colonization has become a key expectation to be assessed during clinical trials.

4. *Phase III clinical trials and surrogate markers of protection.* The usual expectation for vaccine licensure is a randomized, double-blinded Phase III clinical trial is that it will demonstrate protection—i.e., that the vaccinated group contracts less disease than the unvaccinated control group. For less frequent diseases, such as meningitis, this requires extremely large trial groups, and may not even be feasible. Once an effective vaccine is available, further reducing disease burdens and creating ethical tensions (as it would be unacceptable to have an unvaccinated control group), planning clinical trials with an efficacy endpoint becomes yet more difficult. If herd protection is likely to be a significant contributor to overall protection of the population, then the mechanism of randomization—of individuals or by geographic area—may become important. When an accepted surrogate of protection, such as the antibody levels associated with a reduction in disease, is known then protection-based trials have often been replaced by immunogenicity trials. This was the case with the introduction of the meningococcal Group C conjugate vaccine in the UK [34]. Protective antibody concentration data derived from natural infections and immunization with polysaccharide vaccines was used as a benchmark [35]. Once mass pediatric vaccination had been introduced it was possible to retrospectively analyze data on disease levels in UK infants to estimate vaccine efficiency [36]. After the successful protection trials of one pneumococcal conjugate vaccine, the abundant data on antibody levels was analyzed to create a correlate between protection and antibody levels [37]. These values, approximately 0.35 µg/mL of serotype-specific antibody, can be used to support licensing of new pneumococcal conjugate vaccines based on noninferiority of immunological response.

5   Lot Release Requirements

Glycoconjugates are unusual amongst vaccines in that, in the absence of an accessible routine animal model of protection or immunogenicity, lot release is heavily dependent on physicochemical methods. Dosage is specified in terms of the amount of saccharide present per single human dose (s.h.d.), in the SI-derived unit the microgram, in contrast to most vaccines where potency is in arbitrary units defined by, the related to the functional "International

Unit" defined by the relevant WHO International Standard and assayed in animal challenge or immunogenicity tests.

Lot release exists both as an internal process—certifying that key intermediates are fit for use in subsequent manufacturing steps—and as a final process required before vaccine is released to market. With the agreement of the NRA, it may be possible to omit some specific (and less critical) tests if the manufacturer can show through a validation study that the process is sufficiently well controlled that the intermediate or product can always be expected to meet the specification.

As a general principle, quality critical tests for the vaccine should be carried out at the latest possible stage in the manufacturing process—and on the final filled vaccine whenever possible. However, it is often easier to assay at an earlier bulk stage and demonstrate that the value is still valid for the final vaccine. Examples of this would be some identity tests or quantification of residual reagents from previous stages. Some of these tests, such as molecular sizing or quantification of unconjugated saccharide, are stability indicating. Routine lot release assays are typically a subset of the characterization assays which are either directly related to product quality or product consistency, although "lower-tech" methods may be used. All quality-related assays should be validated.

Aspects of the QC of the polysaccharide components of glycoconjugates have evolved from previous requirements for CPSs as components of purified polysaccharide vaccines. The following section is based around the guidance developed by the United States Pharmacopeia [26] which became official from December 2014. These documents are subject to periodic revision and update.

5.1 Polysaccharide and Carrier Protein Identity

The critical factors for the bulk polysaccharide are to confirm its identity and establish its purity. In-process controls to determine molecular weight, lack of endotoxin content, and low bio-burden ("nearly sterile")[3] are useful for the manufacturer. Identity can be established through one-dimensional ^{1}H NMR spectroscopy, by immunochemical methods, or by compositional analyses. NMR and compositional analyses also provide insights into purity. For many polysaccharides, the degree of O-acetylation is considered as aspect of identity, and should be shown to be consistent between manufacturing batches.

1. *Polysaccharide identity testing and purity.* The simplest method available, widely adopted by major manufacturers over the past 20 years, is to use one-dimensional ^{1}H NMR and to compare the

---

[3] A requirement for sterility necessitates extensive testing to confirm this, and sterility may not be essential at early production stages, as the final product may be sterile filtered at a later stage. The concept of low bio-burden implies that the manufacturing process is designed to exclude contaminating materials without the overhead of carrying out formal sterility testing.

test spectrum with a reference spectrum of an authentic sample obtained under identical experimental conditions. Spectrometers of at least 400 MHz are normally used. The major difference between groups using this approach is the temperature at which the spectrum is obtained: and different products may benefit from using different temperature [38–41]. The choice depends on polysaccharide stability, spectral line width, and positioning the resonance from residual water away from key polysaccharide spectral features. Comparison of spectra may be visual—the spectra should match in "resonance position, relative intensity and multiplicity" [42], or mathematical approaches, as developed by Merck [40]. In, notably, the meningococcal polysaccharides the spectra vary considerably due to variations in the degree of O-acetylation and its position: spontaneous migration of the O-acetyl group from Neu5Ac O-8 to O-7 occurs in the Men C CPS. In these cases, in-tube de-O-acetylation by addition of sodium hydroxide results in consistent spectra and comparison of the intensity of the resulting acetate anion resonance with a resonance from the polysaccharide provides quantitative information on the degree of O-acetylation [which are specifications for the meningococcal, S. Typhi Vi and some pneumococcal CPSs]. As the NMR spectrum is very information rich, it is possible to measure a wide range of other parameters from the same spectrum, including quantifying process-related impurities. The Hestrin assay [43] is still widely used to quantify O-acetyl content, returning a result in terms of µmol of O-acetyl per mg dry weight of polysaccharide (rather than number of O-acetyl groups per repeat unit from NMR methods).

For pneumococcal polysaccharides NMR spectroscopy has largely replaced a classical approach using a combination of semi-quantitative colorimetric tests with specificity for different classes of sugar (uronic acids, aminohexose) or substituents (the Hestrin assay for O-acetyl, total nitrogen or total phosphorus) and a suitable immunochemical assay, and to deduce identity and purity from these values. Increasingly, colorimetric assays have been replaced with approaches based on specific degradation [acid or base hydrolysis typically] and quantification of sugars and other components by HPLC or HPAEC. Pneumococcal capsular polysaccharides are almost always contaminated by a teichoic acid polysaccharide, called pneumococcal C-polysaccharide, and this can be quantified by NMR spectroscopy [44].

An alternative approach, preferably in manufacturing stages after conjugation and for blends containing multiple serotypes, is to use an immunochemical assay, such as rate nephelometry [45]. This requires access to appropriate specific poly- or monoclonal antibodies generated in-house or obtained from, for example, the Statens Serum Institut in Copenhagen. This approach can be linked to soft gel size exclusion chromatography, where individual fractions are probed with multiple mAbs to allow sizing of blends of multiple serotypes. Rate nephelometry provides identity information, but not purity information.

None of the above methods are good for quantification of host protein or nucleic acid process contaminants, for which limits of 1–2 % w/w are typically applied. These are typically quantified by a colorimetric assay (protein) and UV absorption spectroscopy (nucleic acid).

2. *Molecular sizing*: For unconjugated polysaccharides, high molecular weight is a requirement for them to be immunogenic, and molecular sizing was a critical quality assay. For historical reasons this requirement has tended to carry over for conjugate vaccines, on the basis that a polysaccharide component of a conjugate should meet the regulatory requirements for the related polysaccharide vaccine. However, as conjugates stimulate an immune response by a different mechanism than a pure polysaccharide, this requirement is questionable—many manufacturers deliberately depolymerize the CPS prior to conjugation. It remains an important measure of the consistency of manufacture of the CPS.

In cases where the saccharide is deliberately depolymerized, either chemically or physically, then sizing of the activated polysaccharide is important. Periodate oxidation of Hib PRP or the Men C CPS, for example, creates terminal aldehydic groups which allow conjugation. Similarly, controlled acid hydrolysis of some polysaccharides makes the nonreducing terminus available. The optimal molecular size for the oligosaccharides will have been determined in preclinical testing and, in such cases, molecular sizing is important and is, effectively, quantification of the degree of activation. In some other cases, manufacturers use physical methods to reduce molecular size of the high polysaccharide to reduce viscosity and improve its properties in the conjugation procedure: again sizing to ensure process consistency is advisable.

There is a trend to replace molecular sizing measurements with molecular mass measurements, typically determined by HPSEC-MALLS [46]. These have the advantage of being independent of the column matrix, but require more complex instrumentation and knowledge of the refractive index increment, dn/dc, for that polysaccharide in the specific elution solution. Existing molecular size specifications need to be correlated with a corresponding molecular weight specification.

3. *Carrier protein identity and purity*: Toxoid carrier proteins are typically already licensed products, and when used as components of conjugates should meet existing specifications, except that a higher degree of purity (expressed as antigenic purity) is

required. Purified protein or recombinant carriers should be of high purity and of demonstrated identity, and typically will have been characterized by the methods appropriate (peptide mapping, mass spectrometry, secondary structure determination, etc.) for such a recombinant of highly purified material.

4. *Activated polysaccharide and activated carrier protein*: Manufacturers should confirm the identity and ensure a consistent degree of activation of conjugation intermediates—either through oligosaccharide molecular sizing (where activation leads to depolymerization) [47, 48], quantitative spectroscopic approaches [49], or colorimetric assays to quantify the activation site/presence of a linker. Not all conjugation chemistries require activation of the carrier protein, but, if so, this should be quantified and shown to be consistent. If the manufacturing process mandates immediate use of activated intermediates without isolation and storage, the consistency of activation should be inferred from other downstream analyses (e.g., molecular sizing of the bulk monovalent conjugates).

5.2 Polysaccharide and Carrier Protein Quantity

Polysaccharide and carrier protein quantification is carried out as in-process controls (e.g., prior to conjugation to ensure a consistent product, or prior to blending monovalent conjugates for filling), and more formally to ensure that final lots meet specifications. Traditional colorimetric methods are valuable for polysaccharide bulks and monovalent conjugates (e.g., orcinol for ribose [50], resorcinol for sialic acid [51], phosphorus assays), whereas approaches based on degradation and HPLC/HPAEC have greater specificity and may also be used on simple blends [52, 53]. Care should be exercised in validation of these colorimetric approaches as frequently the standard curve is created with a monosaccharide and used to quantify a complex polysaccharide. Complex blends of glycoconjugate components in multivalent combination vaccines usually require quantitative immunochemical analyses, such as rate nephelometry, with access to qualified reference mono- or polyclonal antibody preparations. Full validation of these methods and access to appropriate in-house or pharmacopeial reference standards (*see* below) are needed.

5.3 Bulk Monovalent Conjugate

This material is the product of the conjugation reaction, after purification. At this stage, identity of the polysaccharide and carrier proteins moieties should be confirmed. The key tests at this stage are to determine the ratio of polysaccharide-to-protein (PS:protein ratio) and to confirm the integrity of the conjugate. At this stage assays for residual conjugation reagents and unconjugated ("free") polysaccharide and carrier protein become relevant. When the polysaccharide and/or the carrier protein has been activated, capping of the activation sites may be required. If uncapped activation sites are potentially clinically significant, they should be quantified.

1. *Polysaccharide-protein ratio*: The PS:protein ratio should have been optimized in nonclinical development. Conjugates with a low ratio contain excess carrier protein, and those with a high ratio may not be processed well by the antigen-presenting cell. This ratio may also affect the physical properties of the conjugate [54]. Consistency of this ratio is a sensitive indicator of the reproducibility of the conjugation step. The traditional approach is through independent determination of the quantities of the two moieties, using immunochemical, colorimetric, or chromatographic approaches. However, this ratio can also be estimated from a range of characterization tests. Less well explored are methods which give a direct ratio: research reports have described quantitative NMR approaches using denatured conjugates [55]. There is also the potential for HPSEC with dual refractive index and UV monitoring for total material and protein, and, in limited cases from circular dichroism measurements when the polysaccharide hapten makes a significant contribution to the spectrum of the conjugate (e.g., Typhoid Vi conjugates). Conjugates prepared from pure recombinant proteins and oligosaccharide haptens may also be accessible to mass spectrometric approaches to quantify PS:protein ratio.

   Some conjugation chemistries, such as reductive amination of lysine residues, result in markers which are stable to protein hydrolysis and can be quantified chromatographically. For example after hydrolysis of a CRM197-Hib conjugate, where the Hib has been activated and depolymerized by periodate oxidation, N_ε-(2-hydroxyethyl) lysine is produced and can be quantified by amino acid analysis [56]. This data confirms that the carrier protein has a consistent number of polysaccharide attachments sites, batch-to-batch.

2. *Conjugate integrity*: Since the immunological response to a conjugate is different from, and more protective than, that of a purified polysaccharide, there is a general requirement that manufacturers demonstrate that almost all of the saccharide is covalently attached to the carrier protein. Several ways to do this are possible. The simplest is that the molecular weight of the conjugate is higher than that of either component and can be separated and quantified by size exclusion chromatography: this may be difficult if one, or both, component is polydisperse. Alternatively, the conjugate can be separated from unconjugated polysaccharide (or protein) by exploiting differences in hydrophilicity [57] or by immunological precipitation of carrier protein-containing materials and the unconjugated (or "free") polysaccharide (or carrier protein) quantified.

During development it would be expected that the manufacturer shows through in vivo experiments that the immune response is that expected for a conjugate, with development of memory response, isotype switching, and affinity maturation. Both molecular sizing and free polysaccharide are stability-indicating assays and should be performed on final products to support setting of a shelf life. Free polysaccharide on the final product is a specification for all glycoconjugate vaccines.

The presence of unconjugated polysaccharide in the final product may impact negatively on the immunogenicity of the conjugate [58]: this can be probed in animal models during nonclinical development using samples spiked with additional unconjugated polysaccharide, but the relevance to immunization of a human infant may remain unclear.

3. *Capping of unreacted activation sites.* When polysaccharide aldehydes are coupled (e.g., the product of periodate oxidation of Hib PRP or Men C CPS), unreacted aldehydes can be capped using sodium borohydride treatment. Many other conjugation chemistries result in intermediates which are too unstable to present a problem at a later stage. Activation of the polysaccharide or carrier protein by addition of an ADH linker is the most problematic.

5.4 Quality Requirements for the Final Fill

The final fill material will contain defined quantities of the immunogens, adjuvants, and excipients to optimize the immune response of the vaccine whilst minimizing patient reactions, and to ensure vaccine stability. In some cases, usually multidose vials, antibacterial agents and a preservative may be present. The material needs to be sterile. To enhance the stability of the immunogen, the material may be lyophilized for re-dissolution at time of use, in which case a sterile diluent will also be required. Lyophilization has proven most important for meningococcal Group A vaccines. In this case the residual moisture must be controlled. Trace amounts of process residuals may be present and controlled. The manufacturer should have real time data for the stability of the immunogen and other components in the final fill, as this will be critical in assigning a suitable shelf life to the product. The influence of freezing and thawing (which can damage alum-based adjuvants or temporarily modify the pH of some buffers) should be studied and appropriate recommendations made.

1. *The active ingredient.* The critical assay is to quantify the saccharide content. Typically this is approximately 10 µg per s.h.d. in monovalent vaccines, but only 2–4 µg per serotype per s.h.d. in complex multivalent vaccines. In some of these, the conjugate is absorbed on the surface of an alum adjuvant. The approaches used are typically a colorimetric assay for monovalent vaccines, a combination of degradation and chromato-

graphic separation for low valency products (e.g. tetravalent meningococcal vaccines), and rate nephelometry for higher valency vaccines. These assays require quantitative polysaccharide reference standards and, in the case of nephelometry, standardized solution of appropriate monoclonal antibodies.

2. *Excipients and adjuvants, pH, isotonicity.* The manufacturer should confirm the identity and purity of excipients, and ensure that the pH and isotonicity of solution lies within specified ranges (for patient comfort and vaccine stability). In materials where aluminum-based adjuvants are used, there is a limit (up to a maximum of 1.25 mg per s.h.d) that is applicable to all such vaccines [59]. Where adjuvants are used, the degree of absorption to the adjuvant should be determined (or the process validated when measurements are made at an earlier stage).

3. *Residual process-related materials.* Residual formaldehyde, from carrier protein detoxification or reversion of the toxoid, is potentially present in the final conjugate and there are general limits for this. Similarly, clearance of process residuals from the conjugation chemistry (such as the urea residue 1-(3-dimethylaminopropyl)-3-ethylurea derived from hydrolysis of EDAC) should be confirmed either by direct analysis [60] or by process validation.

4. US Regulatory authorities typically require that each batch of a product is tested in vivo to ensure that there are no unexpected reactions. This is known as the General Safety Test.

5. *The final containers.* Final fills can be in vials or directly into single-use syringes. Pharmacopeial requirements exist for caps on vials, lubricants used to facilitate filling or for smooth movement of syringe plungers, and leachates derived from them. The quality of glass in vials and syringes is controlled, as these are typically heat sterilized prior to use. The vaccine should be supplied with a package insert that contains all the information that should be made available to the patient and to the administering person. The content is defined in regulatory documents and is controlled. Batch identification should be etched onto the vial so that problem batches can be readily identified and, if necessary, withdrawn from use.

Reference standards. Assays used in vaccine quality control require access to reference standards. Typically these include standards to demonstrate identity and quantity of the saccharide component, standards for the carrier protein, and standards relating to safety concerns (such as endotoxin levels) and residual reagents. Non-product-specific reference standards may be used to support molecular sizing assays or system suitability standards used to assure that an assay is functioning appropriately. Initially these standards are likely to be manufacturer specific, but they will be later supplied as WHO International Standards (intended to allow manufacturers to calibrate in-house standards against a common, internationally accepted reference), or pharmacopeial standards intended for routine use in batch release assays. Currently, relatively few reference standards are available specifically to support glycoconjugate vaccine regulation. The most important current need is for pharmacopeial standards to support quantification of the saccharide content of these vaccines: these standards will also support, for example, immunochemical identity tests. A summary showing expectations for routine testing and the characterization of polysaccharide-based vaccines is given in Fig. 3.

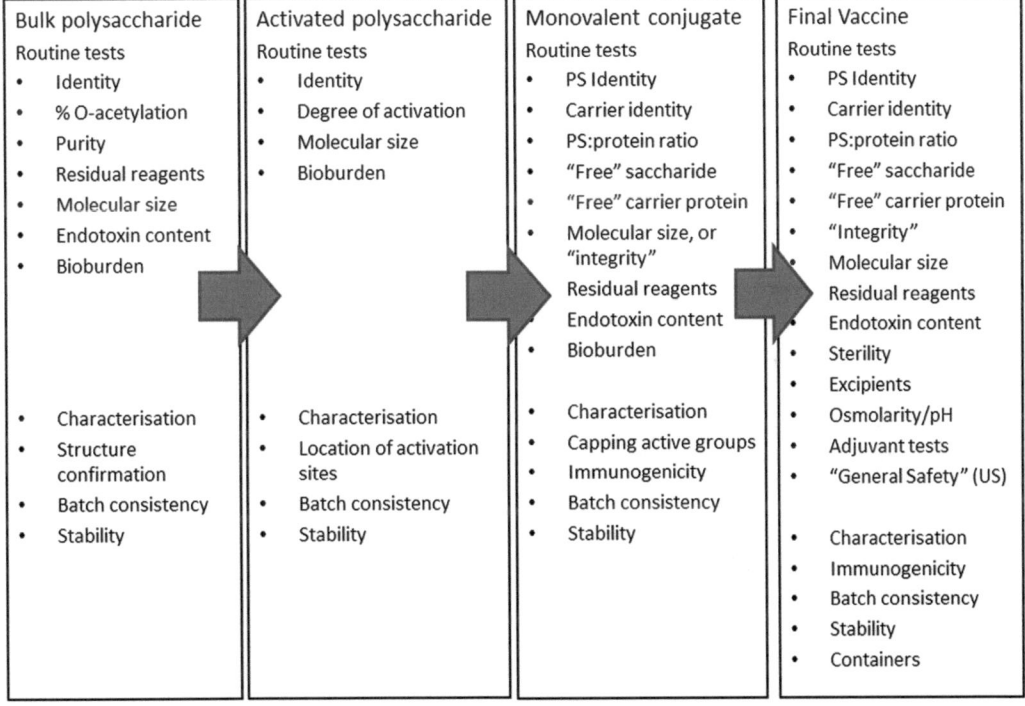

Fig. 3 Summary chart showing expectations for routine testing and characterization of the polysaccharide components of a monovalent glycoconjugate vaccine. This highlights the routine testing to be carried out, either for internal purposes or for formal regulatory compliance. Some aspects, such as estimating the degree of O-acetylation, are not relevant for all vaccines. For multivalent vaccines, the initial stages are required for each polysaccharide component. Characterization and routine testing of the blended and formulated bulk prior to filling, omitted here, may then become critical. Cell line characterization is omitted, and a similar chart can be constructed for the carrier protein moiety of the final vaccine. In some processes the activated polysaccharide may not be isolated, and consistency in its properties must be inferred from data on subsequent production materials

References

1. http://www.ich.org/products/guidelines/quality/article/quality-guidelines.html. Accessed 27 Aug 2014

2. http://www.who.int/biologicals/Guidelines_for_Lot_Release_AFTER_ECBS_27.1.2011.pdf?ua=1. Accessed 27 Aug 2014

3. http://www.edqm.eu/en/human-biologicals-611.html. Accessed 27 Aug 2014

4. Jones C, Lee CK, Ahn C et al (2013) WHO meeting report: Working group on quality, safety and efficacy of typhoid Vi capsular polysaccharide conjugate vaccines, Jeju, Republic of Korea, 5–7 September 2012. Vaccine 31:4466–4469

5. http://www.who.int/biologicals/publications/trs/areas/vaccines/gmp/WHO_TRS_822_A1.pdf and http://www.who.int/biologicals/vaccines/good_manufacturing_practice/en. Accessed 27 Aug 2014

6. http://www.ich.org/products/ctd.html. Accessed 27 Aug 2014

7. http://www.who.int/biologicals/technical_report_series/en/. Accessed 27 Aug 2014

8. http://www.fda.gov/BiologicsBloodVaccines/GuidanceComplianceRegulatoryInformation/Guidances/Vaccines/default.htm. Accessed 27 Aug 2014

9. http://www.ema.europa.eu/ema/index.jsp?curl=pages/regulation/landing/scientific_guideline_search.jsp&mid=WC0b01ac05804698db. Accessed 27 Aug 2014

10. WHO TRS (2000). 897:27–60

11. WHO TRS (2006). 962:115–172

12. WHO TRS (2001). 924:102–128

13. WHO TRS (2004). 926:90–94

14. WHO TRS (2007). 963:225–238

15. WHO TRS (2002). 904:94–155

16. WHO TRS (1980). 658:174–184

17. WHO TRS (1974). 594:50–75

18. WHO TRS (2009). 977:91–151

19. http://who.int/biologicals/publications/meetings/areas/vaccines/pneumococcal/Pneumo%20Meeting%20Report%20FINAL%20IK%2024_Dec_08.pdf?ua=1. Accessed 27 Aug 2014

20. http://who.int/biologicals/publications/meetings/areas/vaccines/pneumococcal/OPA%20meeting%20report-%20FINAL_June07.pdf?ua=1. Accessed 27 Aug 2014

21. WHO TRS (1994). 840:14–33

22. http://www.who.int/biologicals/areas/vaccines/TYPHOID_BS2215_doc_v1.14_WEB_VERSION.pdf?ua=1. Accessed 27 Aug 2014

23. *Haemophilus* Type b conjugate vaccine. European Pharmacopeia, Edition 8.2, European Pharmaocopiea, Strasbourg (2014)

24. <1235> Vaccines for Human Use—General Considerations: USP 37. United States Pharmacopeial Commission, Rockville, MD, 2014

25. <1238> Vaccines for human use—bacterial vaccines: USP 37. United States Pharmacopeial Commission, Rockville, MD, 2014

26. <1234> Vaccines for human use—polysaccharide and glycoconjugate vaccines: USP 37. United States Pharmacopeial Commission, Rockville, MD, 2014

27. http://www.ich.org/fileadmin/Public_Web_Site/ICH_Products/Guidelines/Quality/Q5D/Step4/Q5D_Guideline.pdf. Accessed 27 Aug 2014

28. Schlesinger Y, Granoff DM (1992) Avidity and bactericidal activity of antibody elicited by different *Haemophilus influenzae* type b conjugate vaccines. The Vaccine Study Group. JAMA 267:1489–1494

29. Frasch CE, Borrow R, Donnelly J (2009) Bactericidal antibody is the immunologic surrogate of protection against meningococcal disease. Vaccine 27(Suppl 2):B112–B116

30. Romero-Steiner S, Frasch CE, Carlone G et al (2006) Use of opsonophagocytosis for serological evaluation of pneumococcal vaccines. Clin Vaccine Immunol 13:165–169

31. Maiden MC, Ibarz-Pavón AB, Urwin R et al (2008) Impact of meningococcal serogroup C conjugate vaccines on carriage and herd immunity. J Infect Dis 197:737–743

32. Nurhonen M, Cheng AC, Auranen K (2013) Pneumococcal transmission and disease *in silico*: a microsimulation model of the indirect effects of vaccination. PLoS One 8:e56079

33. Pilishvili T, Lexau C, Farley MM (2010) Sustained reductions in invasive pneumococcal disease in the era of conjugate vaccine. J Infect Dis 201:32–41

34. Miller E, Salisbury D, Ramsay M (2001) Planning, registration, and implementation of an immunisation campaign against meningococcal serogroup C disease in the UK: a success story. Vaccine 20(Suppl 1):S58–S67

35. Balmer P, Borrow R (2004) Serologic correlates of protection for evaluating the response to meningococcal vaccines. Expert Rev Vaccines 3:77–87

36. Campbell H, Borrow R, Salisbury D et al (2009) Meningococcal C conjugate vaccine:

the experience in England and Wales. Vaccine 27(Suppl 2):B20–B29

37. Siber GR, Chang I, Baker S et al (2005) Estimating the protective concentration of anti-pneumococcal capsular polysaccharide antibodies. Vaccine 25:3816–3826

38. Lemercinier X, Jones C (2000) An NMR spectroscopic identity test for the control of the capsular polysaccharide from *Haemophilus influenzae* type b. Biologicals 28:75–83

39. Jones C, Lemercinier X (2002) Use and validation of an NMR test for the identity and O-acetyl content of capsular polysaccharides from *Neisseria meningitidis* used in vaccine manufacture. J Pharm Biomed Anal 30: 1233–1247

40. Abeygunawardana C, Williams TC, Sumner JS et al (2000) Development and validation of an NMR-based identity assay for bacterial polysaccharides. Anal Biochem 279:226–240

41. Lemercinier X, Martinez-Cabrera I, Jones C (2000) Use and validation of an NMR test for the identity and O-acetyl content the *Salmonella typhi* Vi capsular polysaccharide vaccine. Biologicals 28:17–24

42. 2.2.33 Nuclear Magnetic Resonance Spectrometry. European Pharmacopoeia, Edition 8.2, EDQM, Strasbourg, France, 2014

43. 2.5.19 O-acetyl groups in Polysaccharide Vaccines. European Pharmacopoeia, Edition 8.2, EDQM, Strasbourg, France, 2014

44. Xu Q, Abeygunawardana C, Ng AS et al (2005) Characterization and quantification of C-polysaccharide in *Streptococcus pneumoniae* capsular polysaccharide preparations. Anal Biochem 336:262–272

45. Lee CJ (1983) The quantitative immunochemical determination of pneumococcal and meningococcal capsular polysaccharides by light scattering rate nephelometry. J Biol Stand 11:55–64

46. MacNair JE, Desai T, Teyral J et al (2005) Alignment of absolute and relative molecular size specifications for a polyvalent pneumococcal polysaccharide vaccine (PNEUMOVAX 23). Biologicals 33:49–58

47. Ravenscroft N, Averani G, Bartoloni A et al (1999) Size determination of bacterial capsular oligosaccharides used to prepare conjugate vaccines. Vaccine 17:2802–2816

48. D'Ambra AJ, Baugher JE, Concannon PE et al (1997) Direct and indirect methods for molar-mass analysis of fragments of the capsular polysaccharide of *Haemophilus influenzae* type b. Anal Biochem 250:228–236

49. Xu Q, Klees J, Teyral J et al (2005) Quantitative nuclear magnetic resonance analysis and characterization of the derivatized *Haemophilus influenzae* type b polysaccharide intermediate for PedvaxHIB. Anal Biochem 337:235–245

50. 2.5.31 Ribose in Polysaccharide Vaccines. European Pharmacopoeia, Edition 8.2, EDQM, Strasbourg, France, 2014

51. 2.5.23 Sialic Acid in Polysaccharide Vaccines. European Pharmacopoeia, Edition 8.2, EDQM, Strasbourg, France, 2014

52. Tsai CM, Gu XX, Byrd RA (1994) Quantification of polysaccharide in *Haemophilus influenzae* type b conjugate and polysaccharide vaccines by high-performance anion-exchange chromatography with pulsed amperometric detection. Vaccine 12:700–706

53. Gudlavalleti SK, Crawford EN, Harder JD et al (2014) Quantification of each serogroup polysaccharide of *Neisseria meningitidis* in A/C/Y/W-135-DT conjugate vaccine by high-performance anion-exchange chromatography-pulsed amperometric detection analysis. Anal Chem 86:5383–5390

54. Cui C, Carbis R, An SJ et al (2010) Physical and chemical characterization and immunologic properties of *Salmonella enterica* serovar Typhi capsular polysaccharide-diphtheria toxoid conjugates. Clin Vaccine Immunol 17:73–79

55. Ravenscroft N (2000) The application of NMR spectroscopy to track the industrial preparation of polysaccharide and derives glycoconjugate vaccines. In: International Conference Biological Beyond 2000, European Pharmacopeia, Strasbourg, 2000. pp 131–144

56. Seid RC Jr, Boykins RA, Liu DF et al (1989) Chemical evidence for covalent linkages of a semi-synthetic glycoconjugate vaccine for *Haemophilus influenzae* type B disease. Glycoconj J 6:489–498

57. Lei QP, Shannon AG, Heller RK et al (2000) Quantification of free polysaccharide in meningococcal polysaccharide-diphtheria toxoid conjugate vaccines. Dev Biol (Basel) 103:259–264

58. Peeters CC, Tenbergen-Meekes AM, Poolman JT et al (1992) Immunogenicity of a Streptococcus pneumoniae type 4 polysaccharide–protein conjugate vaccine is decreased by admixture of high doses of free saccharide. Vaccine 10:833–840

59. 21 CFR 610.15. http://www.accessdata.fda.gov/scripts/cdrh/cfdocs/cfcfr/CFRSearch.cfm?fr=610.15. Accessed 27 Aug 2014

60. Lei QP, Lamb DH, Shannon AG et al (2004) Quantification of residual EDU (N-ethyl-N′-(dimethylaminopropyl) carbodiimide (EDC) hydrolyzed urea derivative) and other residual by LC-MS/MS. J Chromatogr B Analyt Technol Biomed Life Sci 813:103–112

INDEX

A

Adjuvant.. 11–23, 58, 61, 69, 91,
    159, 160, 173–186, 231, 232, 236, 239, 247, 248
Alum 19, 22, 174, 185, 247
8-Aminopyrene-1,3,6-trisulfonic acid trisodium salt
    (APTS)................ 125, 127, 128, 132–133, 137, 141
Antibody................... 2–6, 12, 14–16, 27–30, 32–38, 41–54,
    57–79, 81–91, 95, 135, 147, 162, 167, 168, 170, 175,
    177, 179–181, 184, 185, 240, 241, 243, 245, 248
Antigen...................................... 1, 3–5, 7, 11–23, 27–39, 41,
    42, 46, 50, 51, 57–79, 124, 145, 147, 159–163,
    173–176, 182–184, 189, 196, 199, 205, 244, 246
Antigen-presenting cell (APC) 3–5, 13, 19–21, 174, 246
AutoMap... 41–54
Azidination.. 148–152
Azo transfer... 146, 151

B

Bacterial culture... 84–85, 89, 90
B cell.. 4, 5, 14–16, 22, 69
B factor.. 50, 54
Blood collection.. 70, 83–85, 90
Bovine serum albumin (BSA).............................. 29, 30, 32,
    33, 38, 62, 68, 77, 161, 168, 170, 177

C

Candida albicans..................................... 6, 145, 147, 174, 195
Capillary gel electrophoresis (CGE)............................... 125
Capillary viscometry....................................... 211, 220–221
Capsular polysaccharide (CPS) 2–6, 8, 11–12, 14–18,
    82, 160, 162, 203, 211–225, 240, 242–244, 247
Carbohydrate............................. 1–8, 12–14, 16–23, 27–39,
    41–54, 57, 58, 82, 96, 123, 124, 146, 147, 156,
    159–163, 167, 170, 173–186, 189–207, 212
Carbohydrate-based vaccine 1–8, 41, 195
Carbohydrate recognition domain
    (CRD) 20, 174, 178
Carrier 2, 5, 15–17, 38, 58, 59, 63–69, 74, 77, 78,
    146, 159–170, 175, 196, 233, 238, 239, 242–249
Carrier induced epitopic suppression (CIES)........... 16, 159
Click chemistry 146, 147, 153
Clostridium difficile.. 6
Collision-induced dissociation (CID) 98–100, 110–112
Colony-forming unit (CFU) 85, 87, 91
Conformation assessment... 221–225

Conjugation

Conjugation... 5, 16, 58, 59, 63–64,
    66–67, 69, 77, 78, 146–149, 151–153, 161, 177, 182,
    189, 200, 223, 243–248
Cross-reacting mutant (CRM)............................. 5, 15, 17,
    18, 58, 66, 77, 78, 233, 246
Cryo-microscopy... 106–107
Cryptococcus neoformans 6, 195
C-type lectin receptor (CLR)............... 19–21, 23, 174–186
Cytokine... 4, 5, 13–16, 20,
    21, 23, 147, 174, 175, 183–185

D

Damage-associated molecular pattern (DAMP) 13, 174
Dectin-1 .... 21, 145–147
Deglycosylation 131–132, 135–137
Dendritic cell (DC) 5, 13, 14, 20,
    21, 145–147, 174, 175, 183–185
Dendritic cell-specific intercellular adhesion molecule-3-
    grabbing non-integrin (DC-SIGN) 20, 94,
    168, 169, 174, 179
Desialylation 109
Di(*N*-succinimidyl) adipate (DSAP)................... 58, 59, 63,
    64, 77, 177, 182
Diphtheria toxoid (DT)....................... 5, 15, 17, 58, 77, 232
Dynamic light scattering (DLS)...................... 211, 216–218

E

Electron microscopy 94, 95, 101, 103, 106, 107
Electrospray ionization (ESI) 96, 99, 108–111, 117
ELISA 81, 161, 167–170, 184, 185
Epitope.......... 8, 15, 17, 36, 37, 46, 57, 76, 95, 146, 160–163
European Pharmacopoeia (EP) 236

F

Fc fusion protein 176, 178–181, 186
Fingerprinting 123–142, 206
Fragmentation 16, 17, 96, 97,
    99–101, 108, 111–113, 117, 118, 145, 163, 185
Freund's adjuvant.. 58, 61, 69

G

Glycan 5–8, 11–23, 27–39, 52, 59, 67–72,
    74–76, 94–100, 102, 103, 107–109, 111–114,
    116–118, 124–128, 137, 160, 167, 174–182, 184
Glycan analysis 95, 140

Bernd Lepenies (ed.), *Carbohydrate-Based Vaccines: Methods and Protocols*, Methods in Molecular Biology, vol. 1331,
DOI 10.1007/978-1-4939-2874-3, © Springer Science+Business Media New York 2015

Glycan array 27–39, 175–178, 180–181
Glycochemistry.. 124
Glycoconjugate 5–6, 8, 11, 12, 15–18, 20, 21, 29, 58–61,
    63–68, 70, 77, 78, 82, 123, 145–156, 159–170, 175,
    183, 189, 196, 198, 199, 203, 207, 211–225, 229–249
Glyconanoparticle (GNP) 160–170
Glycoprotein.. 6, 11, 20, 29,
    38, 93–118, 123–142, 213, 218
Glycosylation...................... 6, 7, 93–118, 123–142, 206, 218
Good manufacturing practice (GMP)............. 230, 234, 238

H

Haemophilus influenzae type b (Hib)...................... 3, 5, 6, 8,
    15, 17, 123, 195–197, 205–207, 232, 233, 236, 240,
    244, 246, 247
Hemagglutinin 41, 127, 129, 130
High-performance liquid chromatography
    (HPLC).................................. 99, 110, 164, 243, 245
Hybridoma 59, 63, 67–69, 72–76
Hydrophilic interaction chromatography
    (HILIC) 124, 125, 127, 133, 137–139, 141

I

Immune response 1, 2, 4, 5, 7,
    13–22, 27, 28, 36, 58, 59, 67, 69–72, 81, 94, 145, 147,
    159, 160, 162, 163, 173–175, 240, 244, 247
Immunization.. 4, 15, 17, 18, 58–61,
    63–67, 69–72, 147, 160, 167, 175, 176, 178,
    184–185, 231, 233, 235, 240, 241
Immunogenicity 5, 57, 58, 124,
    145, 173, 189, 197, 241, 242, 247
Immunology 11–23, 95, 160, 174, 240, 241, 246
Immunoreceptor tyrosine-based activation motifs
    (ITAM)... 174
Immunoreceptor tyrosine-based inhibitory motifs
    (ITIMs) 174–175
Inflammasome..................................... 19, 22, 23
Influenza........ 19, 41, 124–127, 129–131, 134–136, 139, 140
Intrinsic viscosity measurement...................... 211, 220–221
Ion exchange chromatography (IEC)............................ 124
Ionization 58, 66, 96, 110, 111, 125
Ion mobility.. 93–118

K

Keyhole limpet haemocyanin (KLH) 16, 58, 77

L

Laminarin................................... 146–153, 156
Laser-induced fluorescence detection (LIF)........... 125, 128,
    130, 133, 137, 139
Leukocyte.............................. 81, 82, 85, 86, 89, 90
Lewis Y 42, 45–47, 49–51, 160
Ligand docking 44, 45

Lipoarabinomannan 174
Lipopolysaccharide (LPS) 4, 11, 15, 21, 23
Liquid chromatography (LC)................................. 124, 128

M

Maestro 44–47, 52
Mass spectrometry (MS)................................ 58, 93–118,
    124, 125, 128, 182, 186, 245, 246
Matrix-assisted laser desorption/ionization
    (MALDI)..................................... 58, 61, 64, 66–67,
    96, 104, 108–111, 117, 125
MDCK cell 129, 130, 134
Microarray................................. 28–39, 59, 61, 62,
    67–72, 74–77, 79, 177, 180, 181
Monosaccharide 12, 37, 67,
    68, 78, 96, 112–114, 118, 245
Multiplexed capillary gel electrophoresis with laser-induced
    fluorescence detection (xCGE-LIF)... 125, 128, 130,
    133, 137, 139
Multiplicity of infection (MOI) 85, 88, 89
Mycobacterium tuberculosis.................................... 174

N

Nanoparticle.................................... 159–170, 175
Neisseria meningitidis 3, 4, 6, 15–18,
    82, 160, 195, 198–199, 233
Neuraminidase (NA)........ 104, 109, 126, 127, 129, 136, 140
N-glycan................................. 96, 98, 99, 102, 103,
    107, 124–128, 132–133, 136, 137, 140, 141
NMR spectrometer 190–193
Nuclear magnetic resonance (NMR)
    spectroscopy.............................. 42, 59, 76,
    124, 161, 163–166, 170, 189–207, 237, 242, 243, 246

O

Oligosaccharide 2, 7, 8, 41, 57–79,
    146, 196, 198, 200, 244–246
Opsonophagocytosis assay (OPA).............................. 81–91
Ovalbumin (OVA)................................. 58, 77, 160,
    162, 163, 175–178, 182–185

P

Pathogen.. 1–3, 6, 11–14,
    18–21, 27, 28, 30, 33, 36, 41, 59, 82, 84, 86, 87, 90,
    124, 145, 173, 174, 195, 240
Pathogen-associated molecular pattern
    (PAMP)............................. 13, 20, 21, 174
Pattern recognition receptor
    (PRR) 13, 19–21, 23, 146
Polysaccharide 2–8, 11–12, 15–17, 21,
    23, 82, 123, 150, 160, 162, 189–191, 194–207,
    211–225, 233–237, 239–247, 249
Protein-ligand interaction 42, 47–49, 53

Protein-*N*-glycosidase F (PNGase F) 104, 107, 108, 116, 118, 124, 125, 131–132, 136–137, 141
PyMOL.. 44, 50

R

Reduction 17, 146, 147, 149–150, 152–154, 156, 162–164, 169, 230, 240–241

S

Salmonella typhi... 3, 243
Saponin 22
Sedimentation equilibrium analytical ultracentrifugation (SE AUC).. 211, 215–216
Sedimentation velocity analytical ultracentrifugation (SV AUC) 211–215
Serum 28–30, 32–35, 37, 38, 59, 62, 63, 68, 71, 74, 76, 77, 82, 84, 86–91, 101, 105, 107, 134, 148, 168–170, 177, 178, 185, 186, 233, 240, 243
Shimming... 193, 205
Sialic acid......................... 41, 96, 97, 99, 100, 108, 112, 245
Silico toolkit 44, 52
Site mapping 42, 46, 47, 49–53
Size exclusion chromatography (SEC)........... 124, 169, 211, 218–222, 244, 246
Sodium dodecyl sulfate-polyacrylamide gel electrophoresis (SDS-PAGE) 58, 60, 65–66, 95, 107, 116, 124–126, 131, 135, 136, 139, 140, 182, 186
Spectrophotometer... 84, 148, 161
Splenocyte 59, 70, 72, 73, 183–186
Streptococcus pneumoniae................................... 2, 12, 15, 17, 81, 82, 195, 199–203, 233
Subcloning... 74–76
Synthesis... 7, 8, 57, 58, 94, 95, 147–148, 150, 154, 155, 160, 161, 173

T

T cell 1, 4–6, 8, 13–17, 19–23, 147, 160–163, 174–178, 183–185, 240

Technical Reports Series (TRS) 235
Tetanus toxoid (TT)................................... 5, 15, 18, 58, 77, 145–153, 156, 213, 214, 223, 225, 232
Thomsen-Friedenreich (TF) antigen 159–160
Toll-like receptor (TLR) 15, 19–23
Transfection.. 176, 178, 179
Trehalose-6,6′-dimycolate (TDM) 21, 23, 174
Tumor-associated carbohydrate antigen (TACA) 7, 160

U

Ultracentrifugation 101, 105, 107, 115, 125, 129, 130, 134, 211, 212, 215, 218
United States Pharmacopeia (USP) 83, 232, 236, 237, 242
UV/Vis 65, 161

V

Vaccination 1–3, 5, 12, 15, 17, 27, 28, 30, 36, 37, 173, 235, 241
Vaccine 1–8, 11–23, 27, 28, 36, 41, 42, 50, 81, 82, 86, 89, 90, 95, 123, 124, 145–156, 159–170, 173, 174, 189–207, 229–249
Virus.. 2, 6, 8, 13, 20, 20, 57, 94, 95, 101–103, 105–107, 115, 116, 118, 123–127, 129–131, 134–137, 139, 140, 233
Virus purification 130–131, 134–135

W

Wales van-Holde ratio....................................... 221
Whole-blood... 70, 81–91

X

xCGE-LIF. *See* Multiplexed capillary gel electrophoresis with laser-induced fluorescence detection (xCGE-LIF)